Perspectives on Big Data Analysis:

Methodologies and Applications

CONTEMPORARY MATHEMATICS

622

Centre de Recherches Mathématiques Proceedings

C R M

Perspectives on Big Data Analysis:
Methodologies and Applications

International Workshop on Perspectives on High-dimensional Data Analysis II
May 30–June 1, 2012
Centre de Recherches Mathématiques,
Université de Montréal, Montréal

S. Ejaz Ahmed
Editor

American Mathematical Society
Providence, Rhode Island

Centre de Recherches Mathématiques
Montréal, Québec, Canada

2010 *Mathematics Subject Classification.* Primary 68T99, 62H25, 62J07, 62F05, 62G05, 62M05, 62G08, 60-XX, 62G99, 62H15.

Library of Congress Cataloging-in-Publication Data

International Workshop on Perspectives on High-Dimensional Data Analysis (2nd : 2012 : Montréal, Québec)

Perspectives on big data analysis : methodologies and applications : International Workshop on Perspectives on High-Dimensional Data Analysis II, May 30–June 1, 2012, Centre de recherches mathématiques, Université de Montréal, Montréal, Québec, Canada / S. Ejaz Ahmed, Editor.

pages cm.– (Contemporary mathematics / Centre de recherches mathématiques proceedings; volume 622)

Includes bibliographical references.

ISBN 978-1-4704-1042-1 (alk. paper)

1. Multivariate analysis–Congresses. 2. Artificial intelligence–Congresses. 3. Big data–Congresses. I. Ahmed, S. E. (Syed Ejaz), 1957-editor of compilation. II. Title.

QA278.I585 2012

519.5′35–dc23 2014000814

Contemporary Mathematics ISSN: 0271-4132 (print); ISSN: 1098-3627 (online)

DOI: http://dx.doi.org/10.1090/conm/622

∞ The paper used in this book is acid-free and falls within the guidelines
established to ensure permanence and durability.
Visit the AMS home page at http://www.ams.org/

10 9 8 7 6 5 4 3 2 1 19 18 17 16 15 14

Contents

Preface

This book comprises a collection of research contributions that were presented, with one exception, at the International Workshop on the Perspectives on High-Dimensional Data Analysis II, 2012 at the Centre de recherches mathématiques (CRM), Université de Montréal, Canada. One goal of the workshop was to improve the understanding of high-dimensional modeling from an integrative perspective and to bridge the gap among statisticians, computer scientists and applied mathematicians in understanding each other's tools. It provided a venue for participants to meet leading researchers of this field in a small group in order to maximize the chance of interaction and discussion. The objectives included: (1) to highlight and expand the breadth of existing methods in high-dimensional data analysis and their potential to advance both mathematical and statistical sciences, (2) to identify important directions for future research in the theory of regularization methods, in algorithmic development, and in methodology for different application areas, (3) to facilitate collaboration between theoretical and subject-area researchers, and (4) to provide opportunities for highly qualified personnel to meet and interact with leading researchers from countries around the world.

The discipline of statistical science is ever changing and evolving from investigation of classical finite-dimensional data to high-dimensional data analysis. Indeed, we are commonly experiencing data sets containing huge numbers of variables where in some cases the number of variables exceeds the number of sample observations. Many modern scientific investigations require the analysis of enormous, complex high-dimensional data that is beyond the classical statistical methodologies developed decades ago. For example, genomic and proteomic data, spatial-temporal data, social network data, and many others fall into this category. Modeling and making statistical decisions of high-dimensional data is a challenging problem. A range of different models with increasing complexity can be considered, and a model that is optimal in some sense needs to be selected from a set of candidate models. Simultaneous variable selection and model parameters estimation plays a central role in such investigations. There is a massive literature on variable selection, and penalized regression methods are becoming increasingly popular. Many interesting and useful developments have been published in recent years in scientific and statistical journals.

The application of regression models for high-dimensional data analysis is a challenging task. Regularization/penalization techniques have attracted much attention in this arena. Penalized regression is a technique for mitigating difficulties arising from collinearity and high dimensionality. This approach necessarily incurs an estimation bias, while reducing the variance of the estimator. A tuning parameter is needed to adjust the effect of the penalization so that a balance between model parsimony and goodness-of-fit can be achieved. Different forms of penalty functions have been studied intensively over the last two decades. More recently, some of the regularization techniques have been extended to deal with the estimation of large covariance matrices and the analysis of complex dependence structures such as networks and graphs.

However, the development in this area is still in its infancy. For example, methods may require the assumption of sparsity in the model where most coefficients are exactly 0 and nonzero coefficients are big enough to be separated from the zero ones. There are situations where noise cannot easily be separated from the signal. Furthermore, penalty estimators are not efficient when the number of variables is extremely large compared with the sample size. There are still many theoretical and computationally challenging problems when the number of variables increases at a nonpolynomial rate with the sample size.

This book collates applications and methodological developments in high-dimensional statistics dealing with interesting and challenging problems concerning the analysis of complex, high-dimensional data with a focus on model selection and data reduction. The chapters contained in this book deal with submodel selection and parameter estimation for an array of interesting models. We anticipate that the chapters published in this book will represent a meaningful contribution to the development of new ideas in big data analysis and will provide interesting applications. All the papers were thoroughly refereed. A brief description of the contents of each of the thirteen papers in this book follows.

Chapter 1 presents frameworks where principal component analysis (PCA) is an effective tool for the analysis of very high-dimensional data. In contrast to the many papers giving negative results on the lack of consistency of the covariance estimates that are the basis of PCA, this chapter shows that certain methods based on PCA are successful in detecting signals while controlling the probability of false discoveries. This chapter presents three such methods and justifies them by asymptotic arguments and by simulation results.

In particle physics, a system of equations known as the Lattice QCD equations plays a Nobel Prize-winning role. Such equations help physicists determine the weight of sub-atomic particles, like the famous Higgs Boson. These equations are archetypal because they also arise in biology, chemistry and engineering, to name a few. One characteristic of these equations is the feature that they form a large p (parameters) small n (number of equations) system. Chapter 2 introduces an unknown hidden parameter which makes the system of equations telescopic, and in doing so mitigates the effects of the large p, small n problem. A prior distribution is then imposed on the hidden parameter, and by using a Bayesian approach coupled with a hierarchical Markov Chain Monte Carlo algorithm, a solution to the system of equations is achieved.

A main goal of a Bayesian analysis of a spatial point pattern is to estimate the underlying intensity function. The estimation can proceed by either specifying a parametric form for the function or by a Bayesian nonparametric model. A common nonparametric model is the Poisson/gamma random field model introduced in 1998 by Wolpert and Ickstadt; they also provided a Levy construction algorithm to estimate the intensity function. In this chapter the authors propose a slice sampling algorithm for a hierarchical Poisson/gamma random field model for multi-type point patterns under the condition that multiple, independent, point patterns of each type are observed. In Chapter 3 it is demonstrated that the slice sampling algorithm is computationally more efficient than a Levy construction algorithm. The authors then demonstrate their model and algorithm on a functional neuroimaging meta-analysis of emotions.

Hidden Markov models (HMMs) have been used in a wide range of applications, including speech recognition and DNA sequence analysis. In some applications, the background of the problem readily indicates the number of states (or order) of the HMM to be fitted. Often, however, the knowledge of the researcher or practitioner is limited, and the order of the model needs to be estimated from the data. In Chapter 4, a new penalized quasi-likelihood method for order estimation in HMMs is proposed. Starting with an HMM with a large number of states, the method clusters and combines similar states through two penalty functions, yielding a model of a suitable lower order. The performance of the new method is assessed theoretically and via extensive simulation studies. Two well-known data sets are also examined to illustrate the use of the new method.

Chapter 5 considers estimation of the linear model regression coefficients as in a partially linear model (PLM) with a diverging number of predictors in a high-dimensional data analysis. In this chapter, a high-dimensional shrinkage estimation strategy is proposed to improve the prediction performance of a PLM based only on a predefined subset. The asymptotic property of the high-dimensional shrinkage estimator is developed, and its relative performances with respect to the full model and submodel estimators using the quadratic loss function are critically assessed. Furthermore, it is shown both analytically and numerically that the proposed high-dimensional shrinkage estimator performs better than the full model estimator, and in many instances, it performs better than penalty estimators.

Different kinds of shrinkage estimators have been proposed in situations where the number of variables dominates the sample size. Interestingly, ridge regression, which is one of the earliest shrinkage methods developed, has been studied minimally under such a paradigm. It would seem natural to do so since correlation between predictors abounds in such situations. Chapter 6 develops a novel geometric interpretation for a generalized ridge regression estimator which lends insights into its properties. Interestingly, it is shown that useful properties seem to exist in truly sparse settings but are not guaranteed in nonsparse settings. This chapter also develops a computationally efficient numerical algorithm for estimation and studies the performance of the procedure in a real data setting, looking at predictors of disease progression in diabetes.

Chapter 7 is concerned with hypothesis testing for a very general class of regression models where the number of covariates p is allowed to exceed the sample size n substantially, leading to ultra-high-dimensional problems. Specifically, motivated by large sample considerations, a new test statistic (which, in applications, can also be based on efficient scores) controlling for the family-wise error rate is developed. A novel multiplier bootstrap methodology with Rademacher weights is introduced for computing the critical values of the test.

Chapter 8 develops a new single-step multiple comparison procedure and derives the related simultaneous confidence intervals for arbitrary contrasts of means in high-dimensional repeated measures designs that are, by construction, consonant and coherent. The key approach is to combine parametric bootstrap with the modern matrix regularization techniques such as banding and thresholding for estimation of covariance matrices. The numerical studies indicate that the new regularized multiple contrast procedures deliver an accurate estimate of Type I error without sacrificing the power of the test, even for a small number of subjects and a substantial number of variables. The proposed testing procedure is also illustrated by an application to a sleep disorder study.

The asymptotic properties of cubic smoothing splines are well known. For example, when utilized as a nonparametric function estimation tool, the rate of convergence of the cubic smoothing splines can be proven to be nearly optimal. The derivation of the cubic smoothing splines depends on a special property of the penalty that is the integral of the squared second-order derivative. Such a definition requires that the second derivative of the underlying function is square integrable, which could sound arbitrary and restrictive to some researchers. Chapter 9 shows that an alternative approach, whose derivation is completely based on data, hence circumventing the aforementioned restriction, can preserve nearly all the nice asymptotic properties that the cubic smoothing splines enjoy, such as conditioning on the predictor variable following an equally spaced design. The key idea in this chapter is to derive the decaying rate of eigenvalues in a newly derived smoothing matrix. The derivation uses some results from functional analysis and operator theory. Numerical experiments demonstrate better performance of the proposed method, as compared with existing counterparts.

Chapter 10 concerns the question of whether the SCAD procedure is applicable to logistic regression models where the dimension of covariates diverges in an exponential rate of the sample size, and theoretically justifies this applicability. In addition, an optimization algorithm is developed by combining the concave convex procedure and the coordinate descent algorithm in solving regularization problems. The numerical study shows the promise of the proposed procedure in various high-dimensional logistic regression settings.

Chapter 11 considers the variable selection and estimation problems for a logistic regression model using the shrinkage and penalty methods. The shrinkage method relies on prior information of inactive predictors when estimating the coefficients of active predictors. On the other hand, the penalty methods identifies inactive variables by producing zero solutions for their associated regression coefficients. Both methods are shown to have higher efficiency than the classical methods for a wide class of models. A large sample theory for the shrinkage estimators, including asymptotic distributional biases and risks, is developed. A Monte Carlo simulation study is conducted for different combinations of inactive predictors, and the performance of each method is evaluated in terms of a simulated mean squared error. This study indicates that if the number of inactive predictors is correctly specified, the shrinkage method would be expected to do better than the penalty method, but if the number of inactive predictors is incorrectly specified, the penalty methods would be expected to do better than the shrinkage methods. A real data example is presented to illustrate the proposed methodologies.

In Chapter 12 a dimensionality reduction method is proposed which has a low computational cost. This method is inspired by the observation that reasonably large chunks of a high-dimensional dataset can be approximated by low-dimensional patches over its underlying manifold.

In summary, several directions for statistical inference in high-dimensional statistics were highlighted by the talks, papers, and the discussion. This volume conveys some of the surprises, puzzles and success stories in big data analysis.

As an ending thought, I would like to thank all the authors who submitted their papers for possible publication in this special issue as well as all the reviewers for their valuable input and constructive comments on all submitted manuscripts. I will take this opportunity to thank all the participants for their amazing contributions and support for the workshop. I would like to express my special thanks to the superb staff at CRM for the encouragement and support in the organization of this workshop. The hospitality was unparalleled and equally appreciated by participants and organizers. Louis Pelletier provided superb local arrangements for the workshop, and André Montpetit outstanding technical support for the production of this volume. Last but not least, I am thankful to Galia Dafni, Christine Thivierge and the staff of the AMS for their support in the completion of this volume.

S. Ejaz Ahmed
St. Catharines, Canada
November 2013

Contemporary Mathematics
Volume **622**, 2014
http://dx.doi.org/10.1090/conm/622/12430

Principal Component Analysis (PCA) for high-dimensional data. PCA is dead. Long live PCA

Fan Yang, Kjell Doksum, and Kam-Wah Tsui

ABSTRACT. Sample covariances and eigenvalues are famously inconsistent when the number d of variables is at least as large as the sample size n. However, when $d \gg n$, genomewide association studies (GWAS) that apparently are based on principal component analysis (PCA) and use sample covariances and eigenvalues are famously successful in detecting genetic signals while controlling the probability of false discoveries. To reiterate: "PCA is dead, long live PCA," or "PCA is the worst of methods, PCA is the best of methods." We outline recent work **(Yang, 2013)** that reconciles the worst/best dichotomy by acknowledging that PCA is indeed inconsistent for many classical statistical settings, but for settings that are natural in genomic studies, PCA produces effective methods. The dichotomy can in part be explained by how models are viewed and the goal of the study being carried out. We compare the effectiveness of three PCA methods for testing the association between covariates and a response in a framework with continuous variables. These methods are based on adjusting the data using PCA, then applying Pearson, Spearman and normal scores correlation tests.

1. Introduction

Because of the importance of the covariance matrix Σ and its eigenvalues to statistical analysis their accurate estimation is an important goal in statistics. With high-dimensional data where the dimension d of the random vector $\mathbf{x}$ is at least as large as the sample size n, the sample covariance matrix S may fail to be consistent. Because large data sets are becoming common, this is an important problem. A number of recent articles that address the problem of constructing consistent estimates of Σ in the $d \geq n$ case start by referring to the inconsistency of the sample covariance S. A typical example is "It is now well understood that in such a setting the standard sample covariance matrix does not provide satisfactory performance and regularization is needed." (**Cai and Zhou, 2012**). Other articles that start by referring to S as unsatisfactory and address the large d problem using regularization methods such as banding, tapering, thresholding, shrinking and penalization are by Wu and Pourahmadi (**2003**), Zou, Hastie, and Tibshirani (**2006**), Bickel and Levina (**2008a**, **2008b**), El Karoui (**2008**), Amini and Wainwright (**2009**), Cai,

2010 *Mathematics Subject Classification.* Primary 62H25.
Key words and phrases. Eigenstrat, Eigensoft, GWAS, rank tests, stratification, dual principal components.

Zhang, and Zhou (**2010**), Lam and Fan (**2009**), Johnstone and Lu (**2009**), Ahmed and Raheem (**2012**), Ma (**2012**), and Deng and Tsui (**2013**), among others.

On the other hand, in genomics, PCA based on sample covariance matrices and their eigenvalues have been used to construct effective tests of association between genetic marker scores and disease indicators. One collection of genomewide association studies (GWAS) based on the methodology "Eigenstrat," or its updated and expanded version "Eigensoft," started with the papers by Price et al. (**2006**) and Patterson et al. (**2006**). For a statistical examination of GWAS for case-control studies, see **Lin and Zeng (2011)**.

The discrepancy between PCA in High Dimensional Data Analysis (HDDA) being "unsatisfactory" in statistics and "effective" in genomics can be explained by the phase "in such a setting" in the **Cai and Zhou** above quote. Here we examine settings where PCA is effective. In particular, we show that PCA is effective in HDDA when (i) the response vector is split into a low-dimensional vector containing the responses of initial interest and a high-dimensional vector of potentially confounding covariates, and (ii) the sample is drawn from a population made up of unknown subpopulations or strata and this population stratification has the potential to create confounding variables that lead to spurious correlation between a response and predictors.

Sections 2, 3, and 4 provide a summary of our framework taken from **Yang (2013)**. Section 5 uses simulations to show and compare the effectiveness of these PCA methods.

2. Association regression models based on PCA

2.1. Principal components. Population PCA for the random vector $\mathbf{x} = (X_1, \ldots, X_d)^T$ first produces a measure of the variability of $\mathbf{x}$ by finding the linear combination $\mathbf{e}^T\mathbf{x}$ that has maximal normalized variance $\mathrm{Var}(\mathbf{e}^T\mathbf{x})/\|\mathbf{e}\|^2$. Let Σ denote the covariance matrix of $\mathbf{x}$, then $\mathbf{e}_1$, the first eigenvector, is

$$\mathbf{e}_1 = \arg\max_{\mathbf{e}:\|\mathbf{e}\|=1}\{\mathbf{e}^T\Sigma\mathbf{e}\} \tag{2.1}$$

and the first eigenvalue and the first principal component (PC_1) are

$$\lambda_1 = \mathbf{e}_1^T\Sigma\mathbf{e}_1, \quad \mathrm{PC}_1 = \mathbf{e}_1^T\mathbf{x}.$$

The second eigenvector $\mathbf{e}_2$, second eigenvalue λ_2, and second PC are obtained in the same way except $\mathbf{e}_2$ is found by maximizing (2.1) over $\mathbf{e}$ orthogonal to $\mathbf{e}_1$. To obtain $\mathbf{e}_k$, λ_k and PC_k, (2.1) is maximized over $\mathbf{e}$ orthogonal to $\mathbf{e}_1, \ldots, \mathbf{e}_{k-1}$. This process produces the principal components $\mathrm{PC}_1, \ldots, \mathrm{PC}_d$ that capture much of the variability of $\mathbf{x}$ in the sense that $\mathrm{Var}(\mathrm{PC}_j) = \lambda_j$ and $\sum_{j=1}^d \lambda_j = \sum_{j=1}^d \mathrm{Var}(X_j)$.

2.2. Regression and association studies. Suppose Y is a response variable and that $\mathbf{x} \in \mathbb{R}^d$ is a random predictor. In association studies, the null hypothesis H_{0k} that Y and X_k are independent is tested for one X_k at a time. Thus what is needed is a test statistic T_k whose null distribution is known; at least asymptotically, when the null hypothesis H_{0k} holds, $k = 1, \ldots, d$. In this framework, $\mathbf{x}_{-k} = \{X_j : 1 \le j \le d, j \ne k\}$ are confounding variables that could lead to spurious correlation between X_k and Y. Linear analysis based on the linear model

$$Y = \alpha_k + \beta_k X_k + \sum_{j\neq k}\beta_j X_j + \epsilon \tag{2.2}$$

does not provide stable estimates unless a sample of size $n \gg d$ is available. This has led to the introduction of shrinkage methods, penalty methods and methods based on models with sparse covariance matrices. Some of the references can be found in Section 1.

In this paper we consider the case where confounding is due to population stratification and use PCA applied to $\mathbf{x}_{-k}$ to correct for such stratification. In particular, the $\sum_{j \neq k} \beta_j X_j$ term in (2.2) will be replaced by a sum $\sum_{j=1}^{q} \eta_j Z_j$ where the Z_j represents principal components based on $\mathbf{x}_{-k}$ and $q \leq 10$. To find the Z_j, we use dual PCA, which we introduce in the next section. Under certain assumptions, these Z_j's are effective indicators of which stratum an individual belongs to.

3. Dual eigenanalysis and models for stratified populations

3.1. Stratified populations. A stratified population with K strata or subpopulations $S_1, \ldots, S_K$ is such that when one member of the population is selected, the probability that the member is from subpopulation S_k is π_k, where $\sum_{k=1}^{K} \pi_k = 1$, $\pi_k > 0$, $1 \leq k \leq K$. Consider n independent draws and let N_k be the number of draws from S_k, then $\mathbf{N} = (N_1, \ldots, N_K)^T$ follows the multinomial distribution $\mathrm{MN}(n, \pi_1, \ldots, \pi_K)$ where $\sum_{k=1}^{K} N_k = n$. This strata information is not available. Instead we have n independent draws from a population that contain K unknown strata. That is, K and $\mathbf{N}$ are unobservable.

Consider a random vector $(X_1, \ldots, X_d)^T$ whose covariance matrix Σ is assumed to exist. We have available a $n \times d$ random data matrix $\mathbf{X} = (X_{ij})_{n \times d}$ where the random vectors $\mathbf{x}_i = (X_{i1}, \ldots, X_{id})^T$, $1 \leq i \leq n$, are independent and identically distributed. When the $\mathbf{x}_i$ are drawn from a stratified population the major variability of $\mathbf{X} = (X_{ij})$ as we change i is due to this stratification, and this variability can be examined by considering the $n \times n$ dual covariance matrix defined by

$$\widehat{\Sigma}_D = d^{-1}(\mathbf{X} - \overline{\mathbf{X}})(\mathbf{X} - \overline{\mathbf{X}})^T, \tag{3.1}$$

where $\mathbf{X} - \overline{\mathbf{X}}$ is the $n \times d$ matrix with entries $(X_{ij} - \overline{X}_j)$, and $\overline{X}_j = n^{-1} \sum_{i=1}^{n} X_{ij}$.

To interpret $\widehat{\Sigma}_D$, let $\mathbf{W}^{(d)} = (W_1^{(d)}, \ldots, W_n^{(d)})^T$ be the result of one random draw from the collection of n-vectors

$$\{(X_{1j}, \ldots, X_{nj})^T : 1 \leq j \leq d\}$$

Then

$$\widehat{\Sigma}_D = \mathrm{Cov}(\mathbf{W}^{(d)} - \overline{\mathbf{W}}^{(d)}) \equiv E[(\mathbf{W}^{(d)} - \overline{\mathbf{W}}^{(d)})(\mathbf{W}^{(d)} - \overline{\mathbf{W}}^{(d)})^T],$$

where $\overline{\mathbf{W}}^{(d)} = (n^{-1} \sum_{i=1}^{n} W_i^{(d)})\mathbf{1}$ and $\mathbf{1}$ is a n-vector of 1's.

It is known that $\widehat{\Sigma}_D$ has the same nonzero eigenvalues, up to a constant d/n, as the usual covariance matrix

$$\widehat{\Sigma} = n^{-1}(\mathbf{X} - \overline{\mathbf{X}})^T(\mathbf{X} - \overline{\mathbf{X}}).$$

There is another simple relationship between PCA of $\widehat{\Sigma}$ and $\widehat{\Sigma}_D$: let $\hat{\lambda}_q > 0$ be the qth largest eigenvalue of $\widehat{\Sigma}$, then the qth principal component of $\widehat{\Sigma}$ evaluated at $\mathbf{x}_i$ equals the ith entry of the qth eigenvector of $\widehat{\Sigma}_D$, up to a constant.

One advantage of $\widehat{\Sigma}_D$ is that in HDDA its dimension $n \times n$ is much smaller than the dimension $d \times d$ of $\widehat{\Sigma}$. Another advantage is that if we explicitly model stratification, then we find that even though $\widehat{\Sigma}_D$ is computed without using strata information, a conditional eigenanalysis of $\widehat{\Sigma}_D$ reveals the unknown population stratification and provides methods for adjusting for stratification. This is because for models that include stratification, the variability measured by $\widehat{\Sigma}_D$ is mainly due to the $\mathbf{x}_i$ coming from different strata. To show this we have to change our model framework and evaluate strata-blind methods from the point of view of an evaluator who knows the strata information. We do this in the next subsection.

3.2. The conditional model framework. Because $X_{ij} - \overline{X}_j = X_{ij} - \mu_j + O_p(n^{-1})$, $W_1^{(d)} - \overline{W}^{(d)}, \dots, W_n^{(d)} - \overline{W}^{(d)}$ are nearly independent, and the covariance matrix $\widehat{\Sigma}_D$ would appear to be nearly diagonal. Thus eigenanalysis of $\widehat{\Sigma}_D$ would appear to lack the ability to measure stratification effects. Using **Jung and Marron (2009)**, we can prove that unconditional eigenanalysis of $\widehat{\Sigma}_D$ leads to inconsistent methods. But here is where we change "the setting" of the usual statistical eigenanalysis: our methods do not use the strata information, but we evaluate the performance of the methods assuming that N_k of the $W_i^{(d)}$, $1 \le i \le n$, come from stratum k, $k = 1, \dots, K$. Without loss of generality, we order $W_i^{(d)}$ so that $W_1^{(d)}, \dots, W_{N_1}^{(d)}$ are from stratum 1, and so on. In this setting where we condition on the unobservable $\mathbf{N} = (N_1, \dots, N_K)^T$,

$$E_{\mathbf{N}}(W_i^{(d)}) = \mu_k, \quad N_k + 1 \le i \le N_{k+1},\ 0 \le k \le K-1,$$

where μ_k is the kth stratum mean and $N_0 = 0$. It follows that in this setting, for $\mu = \sum_{k=1}^K \hat{\pi}_k \mu_k$, where $\hat{\pi}_k = N_k/n$,

$$E_{\mathbf{N}}(W_i^{(d)} - \overline{W}^{(d)}) = \mu_k - \mu, \quad N_k + 1 \le i \le N_{k+1}, 0 \le k \le K-1. \tag{3.2}$$

This implies that for stratified populations, the dual covariance matrix measures the deviation of the individual strata means from the overall mean, and in this setting both the diagonal and off-diagonal elements of

$$\Sigma_D^{(\mathbf{N})} = E_{\mathbf{N}}[(\mathbf{W}^{(d)} - \overline{\mathbf{W}}^{(d)})(\mathbf{W}^{(d)} - \overline{\mathbf{W}}^{(d)})^T]$$

measure the extent of the stratification. For the ith individual in a sample of n individuals, the variables $X_{i1}, \dots, X_{id}$ provides information about which strata the individuals is likely to belong to, and this information can be obtained from the eigenanalysis of $\widehat{\Sigma}_D$.

REMARK 3.1. In our conditional framework, $\Sigma_D^{(\mathbf{N})}$ is not a covariance matrix because $\mathbf{W}^{(d)} - \overline{\mathbf{W}}^{(d)}$ does not have mean zero (see (3.2)). This makes our analysis different from the statistical literature that reports on the failure of PCA.

3.3. Dual eigenanalysis for $d \gg n$. To examine the $d \gg n$ case, we let $d \to \infty$ and let n be finite or we let $n \to \infty$ in the $d = \infty$ universe. The elements of $\widehat{\Sigma}_D$ are averages of d terms. A basic assumption is that these averages satisfy the weak law of large numbers. That is, as $d \to \infty$, we assume that there exists $\mathbf{W} = (W_1, \dots, W_n)^T$ such that with $\overline{\mathbf{W}} = (n^{-1}\sum_{i=1}^n W_i)\mathbf{1}$,

$$\widehat{\Sigma}_D \xrightarrow{P} \Sigma_D^{(\mathbf{N})} \equiv E_{\mathbf{N}}[(\mathbf{W} - \overline{\mathbf{W}})(\mathbf{W} - \overline{\mathbf{W}})^T], \tag{3.3}$$

where W_i is interpreted to be the probability limit of $W_i^{(d)}$ as $d \to \infty$ for the conditional model in Section 3.2.

From (3.2), we see that because $E_{\mathbf{N}}(W_i^{(d)} - \overline{W}^{(d)})$ is the same for all d,

$$E_{\mathbf{N}}(W_i - \overline{W}) = \mu_k - \mu, \quad N_k + 1 \le i \le N_{k+1},\ 0 \le k \le K-1.$$

For the ith individual in a sample of size n, we have available an infinite sequence $X_{i1}, X_{i2}, \dots$ that provides information about which strata the individual belongs to. This information is provided by W_i. It is known (**Yang, 2013**) that when sampling from a stratified population with K strata, $\Sigma_D^{(\mathbf{N})}$ has $(K-1)$ eigenvalues of order n, one that equals zero and the rest that are constant. Moreover, the $(K-1)$ eigenvectors of $\Sigma_D^{(\mathbf{N})}$ corresponding to the $(K-1)$ largest eigenvalues $\lambda_1 > \cdots > \lambda_{K-1}$ capture the variability of $W_1, \dots, W_n$ due to stratification. In genomics, Patterson et al. (**2006**) and Price et al. (**2006**) argue convincingly using real data that the number of strata is at most eleven. In the next section we will use the eigenvectors corresponding to the q largest eigenvalues of $\Sigma_D^{(\mathbf{N})}$, where $q \leq 10$.

REMARK 3.2. Dual eigenanalysis can also be accomplished using singular value decomposition. The approach taken here is more in line with the intuition provided by Section 3.3.

3.4. Classical eigenanalysis for $d \gg n$. The above is a brief summary of the PCA of $\widehat{\Sigma}_D$ and $\Sigma_D^{(\mathbf{N})}$ for the $d \gg n$ case from **Yang (2013)**. Next we turn to a summary of HDDA PCA for $\widehat{\Sigma}$ and Σ. It is known that the m largest eigenvalues $\hat{\lambda}_1 \geq \hat{\lambda}_2 \geq \cdots \geq \hat{\lambda}_m > 0$ of $\widehat{\Sigma}$ and $\widehat{\Sigma}_D$ are the same. Let $\widehat{\mathrm{PC}}_q$ be the n-vector obtained by evaluating the qth PC of $\widehat{\Sigma}$ at $\mathbf{x}_1, \dots, \mathbf{x}_n$, and let $\hat{\mathbf{e}}_{Dq}$ be the eigenvector of $\widehat{\Sigma}_D$ corresponding to $\hat{\lambda}_q$, then (**Yang, 2013**)

$$\hat{\lambda}_q^{-1/2}\widehat{\mathrm{PC}}_q = \hat{\mathbf{e}}_{Dq}, \quad q = 1, \dots, m. \tag{3.4}$$

That is, the PC's of $\widehat{\Sigma}$ are equivalent to the eigenvectors of $\widehat{\Sigma}_D$. Because (3.4) holds for all d, we represent the $d \gg n$ PC's of $\widehat{\Sigma}$ for the conditional model as the equivalence class $\{c\mathbf{e}_{Dq}^{(\mathbf{N})} : c > 0\}$, where $\mathbf{e}_{Dq}^{(\mathbf{N})}$ is the qth eigenvector of $\Sigma_D^{(\mathbf{N})}$. In practice any c will work. We set $c = 1$ and represent the $d \gg n$ PC of $\widehat{\Sigma}$ as

$$\mathbf{PC}_q^{(\mathbf{N})} = \mathbf{e}_{Dq}^{(\mathbf{N})}.$$

Simulation results (**Price et al., 2006**; **Yang, 2013**) have shown that for stratified populations, when $d \gg n$, creating a sparse model where $\mathbf{x}_{-k}$ is replaced by $\mathbf{e}_{Dj}^{(\mathbf{N})}$, $1 \leq j \leq 10$, controls for the confounding effects of $\mathbf{x}_{-k}$ on the correlation between Y and X_k.

3.5. Regression models for stratified populations. Based on the previous discussion, we consider sparse models based on the first q PC's based on the usual sample covariance matrix $\widehat{\Sigma}$, where $q \leq 10$. To simplify notation, in this subsection, let X be the X_k whose association with Y is being examined and let $\mathbf{X} \in \mathbb{R}^{d-1}$ be the vector of confounding predictors. The discussion in Section 3.3 shows that a reasonable sparse model is

$$Y_i = \alpha + \beta X_i + \sum_{j=1}^{q} \eta_j Z_{ij} + \epsilon_i, \quad 1 \leq i \leq n \tag{3.5}$$

where Z_{ij} is the $d = \infty$ version of the ith entry of the jth dual population eigenvector based on $\mathbf{X}$. Extending results of **Patterson et al. (2006)**, Yang (**2013**) has shown that when $K < q$, Z_{ij} is an indicator of which strata j the ith observation comes from for $j = 1, \ldots, K$; while Z_{ij} is "noise" for $j = K+1, \ldots, q$. Assume a trivariate normal distribution, or more generally assume that,

$$Y_i = a + \mathbf{b}^T \mathbf{Z}_i + \epsilon_{1i}, \quad X_i = c + \mathbf{d}^T \mathbf{Z}_i + \epsilon_{2i},$$

where $\mathbf{Z}_i = (Z_{i1}, \ldots, Z_{iq})^T$. Then we can rewrite this model as (see **Bickel and Doksum (1976**, Example 6.2.1))

$$Y_i - Y_{i0} = \alpha + \beta(X_i - X_{i0}) + \delta_i, \tag{3.6}$$

where the δ_i have mean zero and the same variance τ^2, and X_{i0} and Y_{i0} are the linear predictors of X_i and Y_i based on $\mathbf{Z}$ that minimize the mean squared prediction error.

4. Association test statistics

When the variables X_i and δ_i in model (3.6) are Gaussian, the test statistic based on the MLE of β is a 1-1 function of the Pearson correlation coefficient of the strata adjusted variables

$$X^A = X - X_0, \quad Y^A = Y - Y_0.$$

Here X_0 and Y_0 are unknown and will be replaced by their predictors

$$\widehat{X}_0 = \overline{X} + \left(\widehat{\mathrm{Cov}}^{-1}(\widehat{\mathbf{Z}})\widehat{\mathrm{Cov}}(\widehat{\mathbf{Z}}, X)\right)^T \left(\widehat{\mathbf{Z}} - \overline{\widehat{\mathbf{Z}}}\right), \tag{4.1}$$

$$\widehat{Y}_0 = \overline{Y} + \left(\widehat{\mathrm{Cov}}^{-1}(\widehat{\mathbf{Z}})\widehat{\mathrm{Cov}}(\widehat{\mathbf{Z}}, Y)\right)^T \left(\widehat{\mathbf{Z}} - \overline{\widehat{\mathbf{Z}}}\right), \tag{4.2}$$

where $\widehat{\mathrm{Cov}}$ refers to sample covariance matrix and $\widehat{\mathbf{Z}}$ is the $q \times 1$ vector of sample PC's. Our test statistics will be based on the stratum adjusted variables

$$\widehat{X}^A = X - \widehat{X}_0, \quad \widehat{Y}^A = Y - \widehat{Y}_0. \tag{4.3}$$

For a sample $\{(\mathbf{x}_i, Y_i), 1 \leq i \leq n\}$, we compute $\{(\widehat{X}_{ij}^A, \widehat{Y}_i^A), 1 \leq i \leq n\}$ according to (4.3) and define

$$R_{\mathrm{P}} = \text{Sample corr}\{(\widehat{X}_{ij}^A, \widehat{Y}_i^A), 1 \leq i \leq n\}.$$

This is the Pearson's type correlation coefficient. Because the adjusted X and Y are not Gaussian, it makes sense to consider robust statistics such as the Spearman and the normal scores rank statistics

$$R_{\mathrm{S}} = \text{Sample corr}\{(Q_i, R_i) : 1 \leq i \leq n\},$$

$$R_{\mathrm{NS}} = \text{Sample corr}\left\{\left(\Phi^{-1}\left(\frac{Q_i}{n+1}\right), \Phi^{-1}\left(\frac{R_i}{n+1}\right)\right) : 1 \leq i \leq n\right\},$$

where Q_i and R_i are the ranks of $\widehat{X}_{ij}^A$ and $\widehat{Y}_i^A$ among $(\widehat{X}_{1j}^A, \ldots, \widehat{X}_{nj}^A)$ and $(\widehat{Y}_1^A, \ldots, \widehat{Y}_n^A)$, respectively, and Φ is the cumulative distribution function of $\mathrm{N}(0,1)$. All these statistics have the asymptotic standard normal distribution $\mathrm{N}(0,1)$ under the null hypothesis H_0: $\beta = 0$. Thus we use standard normal critical values for our hypothesis testing.

5. Simulations

In this section we compare the Type I error probabilities and power of the three methods of Section 4 for testing the association between a predictor X_k and a response Y when sampling from a population with $K = 2$ strata. The strata information is unobservable, but researchers typically use methods based on $q = K - 1 = 10$. Thus we also compare properties of the methods based on $q = 1$ and $q = 10$ to examine whether using q too large leads to a loss in performance. Another question is what are the properties of the tests if stratification is ignored, that is, the statistician uses $q = 0$. It is found that in this case the Type I error probabilities are far off target, so a power comparison when $q = 0$ is not of interest. Section 4 type tests with $q = 0$ should not be used for stratified populations.

We generate a $n \times d$ data matrix $\mathbf{X}_{n\times d}$ for a population with two strata column by column. Let n_1 and n_2 be positive integers such that $n_1 + n_2 = n$. For $j = 1, \ldots, d$, we generate the jth column as follows. We first fix the value of ϕ and then generate μ_1 and μ_2 i.i.d. from $\mathrm{N}(0, \phi^2)$. Then n_1 independent values for sample members are generated from the first stratum as $\mathrm{N}(\mu_1, 1)$, and independently, n_2 independent values for sample members are generated from stratum 2 as $\mathrm{N}(\mu_2, 1)$. The jth column consists of these n_1 and n_2 generated values. The d columns are generated independently. For our simulation experiment, we set $n_1 = 600$, $n_2 = 400$, (hence $n = n_1 + n_2 = 1000$) and ϕ is set as 1, 2 or 3.

Next we generate $\mathbf{Y} = (Y_1, \ldots, Y_n)^T$ using the model

$$Y_i = \alpha_i + \beta X_{ik} + \eta Z_{i1} + \epsilon_i, \quad 1 \le i \le n \tag{5.1}$$

where $\epsilon \sim \mathrm{N}(0, \sigma_\epsilon^2)$ with $\sigma_\epsilon^2 = 0.75, 1$ or 1.25, X_{ik} is the ith observation for the variable X_k that is generated as in the previous paragraph, and Z_{i1} is the ith entry of $\mathbf{Z}_1$.

In order to compute the strata adjusted values (4.3), which the test statistics are based on, we compute the normalized sample PC vectors using simulated data, denoted by $\widehat{\mathbf{Z}}_1 = \widehat{\mathbf{PC}}_1/\|\widehat{\mathbf{PC}}_1\|, \ldots, \widehat{\mathbf{Z}}_q = \widehat{\mathbf{PC}}_q/\|\widehat{\mathbf{PC}}_q\|$. The model is fixed with the true number of strata $K = 2$, but the statistician does not know K and checks the performance of his/her methods using $q = 0$ (ignoring the stratification), $q = 1$ (using the correct number of PC's) or $q = 10$ (being overly conservative).

To investigate Type I error, we generate data using model (5.1) with $\alpha = 1$, $\beta = 0$, $\eta = 1$, $n = 1{,}000$ and $d = 10{,}000$. Repeat the above experiment for 50 times and compute the proportion

$$a = \frac{\#\,\text{Rejections}}{10{,}000 \times 50}$$

as the estimate of the probability of Type I error.

To examine Power $= 1 - P(\text{Type II error})$, we generate data sets with $n = 1{,}000$, 10,000 irrelevant variables ($\beta = 0$) and 200 relevant variables using model (5.1) with $\alpha = 1$, $\beta = 0.1$ and $\eta = 1$. We then have $d = 10{,}200$ variables. Repeat the above experiment for 50 times. Then the estimated power is

$$b = \frac{\#\,\text{Rejections among the 200 relevant variables}}{200 \times 50}.$$

Table 1 shows that without the adjustment for stratification, the three tests fail to achieve the correct significance level, while all three succeed when using one or ten PC's to correct for stratification.

TABLE 1. Monte Carlo Type I error probabilities for Gaussian N(0, 1) errors. The nominal significance level is 10^{-3}.

	$\phi = 1$	$\phi = 2$	$\phi = 3$
Pearson 0 PC	2.92×10^{-3}	5.04×10^{-3}	6.39×10^{-3}
Spearman 0 PC	2.81×10^{-3}	4.49×10^{-3}	5.10×10^{-3}
Normal Scores 0 PC	2.86×10^{-3}	4.22×10^{-3}	4.85×10^{-3}
Pearson 1 PC	1.08×10^{-3}	1.04×10^{-3}	1.04×10^{-3}
Spearman 1 PC	1.05×10^{-3}	1.06×10^{-3}	1.05×10^{-3}
Normal Scores 1 PC	1.04×10^{-3}	1.04×10^{-3}	1.04×10^{-3}
Pearson 10 PC	1.14×10^{-3}	1.14×10^{-3}	1.15×10^{-3}
Spearman 10 PC	1.07×10^{-3}	1.10×10^{-3}	1.10×10^{-3}
Normal Scores 10 PC	1.14×10^{-3}	1.14×10^{-3}	1.14×10^{-3}

TABLE 2. Power for Gaussian $N(0, \sigma_\epsilon^2)$ errors. The significance level is 10^{-3}.

	$\sigma_\epsilon = 0.75$	$\sigma_\epsilon = 1$	$\sigma_\epsilon = 1.25$
Pearson 1 PC	0.8175	0.4390	0.2171
Spearman 1 PC	0.7597	0.3792	0.1851
Normal Scores 1 PC	0.8154	0.4366	0.2164
Pearson 10 PC	0.8156	0.4391	0.2175
Spearman 10 PC	0.7576	0.3821	0.1847
Normal Scores 10 PC	0.8130	0.4358	0.2150

TABLE 3. Power for $\gamma\,N(0,1) + (1-\gamma)\,\mathrm{Unif}(-2,2)$ errors. The significance level is 10^{-3}.

	$\gamma = 0$	$\gamma = 0.5$	$\gamma = 0.7$	$\gamma = 0.9$
Pearson 1 PC	0.2833	0.3618	0.3968	0.4281
Spearman 1 PC	0.2420	0.3153	0.3466	0.3769
Normal Scores 1 PC	0.6530	0.5519	0.5018	0.4642
Pearson 10 PC	0.2850	0.3637	0.3967	0.4273
Spearman 10 PC	0.2338	0.3085	0.3419	0.3714
Normal Scores 10 PC	0.5825	0.5141	0.4780	0.4561

Table 2 shows that the Pearson and normal scores tests perform equally well and better than the Spearman test for Gaussian errors. Moreover, very little power is lost by using 10 PC's instead of one, which is the "true" number of PC's.

We also tried contaminated distributions for the error ϵ and found that for $0.9\,N(0,1) + 0.1\,N(0,4)$ and $0.9\,N(0,1) + 0.1\,N(10,9)$, the results are very similar to Tables 1 and 2. However, for the model where the distribution of ϵ has the form $\gamma\,N(0,1) + (1-\gamma)\,\mathrm{Unif}(-2,2)$ with $\gamma \leq 0.9$, the normal scores test is best, Pearson is second best, and Spearman has the lowest power. See Table 3. (The Type I errors are still controlled at the significance level for the non-Gaussian errors (not shown).) All the tables show that when the true number of strata is 2, using methods that assume 11 strata ($q = 10$) do not lose much in performance when compared to methods using the correct $q = 1$.

References

Ahmed, S. E. and S. M. E. Raheem. 2012. *Shrinkage and absolute penalty estimation in linear models*, WIREs Computational Statistics **4**, no. 6, 541–553, DOI 10.1002/wics.1232.

Amini, A. A. and M. J. Wainwright. 2009. *High-dimensional analysis of semidefinite relaxations for sparse principal components*, Ann. Statist. **37**, no. 5B, 2877–2921, DOI 10.1214/08-AOS664. MR2541450 (2010j:62163)

Bickel, P. J. and K. A. Doksum. 1976. *Mathematical statistics*, Holden-Day Inc., San Francisco, Calif. Basic ideas and selected topics; Holden-Day Series in Probability and Statistics. MR0443141 (56 #1513)

Bickel, P. J. and E. Levina. 2008a. *Regularized estimation of large covariance matrices*, Ann. Statist. **36**, no. 1, 199–227, DOI 10.1214/009053607000000758. MR2387969 (2009a:62255)

Bickel, P. J. and E. Levina. 2008b. *Covariance regularization by thresholding*, Ann. Statist. **36**, no. 6, 2577–2604, DOI 10.1214/08-AOS600. MR2485008 (2010b:62197)

Cai, T. T. and H. H. Zhou. 2012. *Minimax estimation of large covariance matrices under ℓ_1-norm*, Statist. Sinica **22**, no. 4, 1319–1349. MR3027084

Cai, T. T., C.-H. Zhang, and H. H. Zhou. 2010. *Optimal rates of convergence for covariance matrix estimation*, Ann. Statist. **38**, no. 4, 2118–2144, DOI 10.1214/09-AOS752. MR2676885 (2011h:62215)

Deng, X. and K.-W. Tsui. 2013. *Penalized covariance matrix estimation using a matrix-logarithm transformation*, J. Comput. Graph. Statist. **22**, no. 2, 494–512, DOI 10.1080/10618600.2012.715556.

d'Aspremont, A., L. El Ghaoui, M. I. Jordan, and G. R. G. Lanckriet. 2007. *A direct formulation for sparse PCA using semidefinite programming*, SIAM Rev. **49**, no. 3, 434–448 (electronic), DOI 10.1137/050645506. MR2353806 (2008f:90079)

El Karoui, N. 2008. *Operator norm consistent estimation of large-dimensional sparse covariance matrices*, Ann. Statist. **36**, no. 6, 2717–2756, DOI 10.1214/07-AOS559. MR2485011 (2010d:62132)

Johnstone, I. M. and A. Y. Lu. 2009. *On consistency and sparsity for principal components analysis in high dimensions*, J. Amer. Statist. Assoc. **104**, no. 486, 682–693, DOI 10.1198/jasa.2009.0121. MR2751448

Jolliffe, I. T., N. T. Trendafilov, and M. Uddin. 2003. *A modified principal component technique based on the LASSO*, J. Comput. Graph. Statist. **12**, no. 3, 531–547, DOI 10.1198/1061860032148. MR2002634

Jung, S. and J. S. Marron. 2009. *PCA consistency in high dimension, low sample size context*, Ann. Statist. **37**, no. 6B, 4104–4130, DOI 10.1214/09-AOS709. MR2572454 (2010m:62173)

Lam, C. and J. Fan. 2009. *Sparsistency and rates of convergence in large covariance matrix estimation*, Ann. Statist. **37**, no. 6B, 4254–4278, DOI 10.1214/09-AOS720. MR2572459 (2011d:62064)

Lin, D. Y. and D. Zeng. 2011. *Correcting for population stratification in genomewide association studies*, J. Amer. Statist. Assoc. **106**, no. 495, 997–1008, DOI 10.1198/jasa.2011.tm10294. MR2894759 (2012k:92063)

Ma, Z. 2012. *Accuracy of the Tracy-Widom limits for the extreme eigenvalues in white Wishart matrices*, Bernoulli **18**, no. 1, 322–359, DOI 10.3150/10-BEJ334. MR2888709

Patterson, N., A. L. Price, and D. Reich. 2006. *Population structure and eigenanalysis*, PLoS Genetics **2**, no. 12, e190, DOI 10.1371/journal.pgen.0020190.

Price, A. L., N. J. Patterson, R. M. Plenge, M. E. Weinblatt, N. A. Shadick, and D. Reich. 2006. *Principal components analysis corrects for stratification in genome-wide association studies*, Nature Genetics **38**, 904–909, DOI 10.1038/ng1847.

Shen, H. and J. Z. Huang. 2008. *Sparse principal component analysis via regularized low rank matrix approximation*, J. Multivariate Anal. **99**, no. 6, 1015–1034, DOI 10.1016/j.jmva.2007.06.007. MR2419336 (2009m:62184)

Ulfarsson, M. O. and V. Solo. 2008. *Sparse variable PCA using geodesic steepest descent*, IEEE Trans. Signal Process. **56**, no. 12, 5823–5832, DOI 10.1109/TSP.2008.2006587. MR2518261

Witten, D. M., R. Tibshirani, and T. Hastie. 2009. *A penalized matrix decomposition, with applications to sparse principal components and canonical correlation analysis*, Biostatistics **10**, no. 3, 515–534, DOI 10.1093/biostatistics/kxp008.

Wu, W. B. and M. Pourahmadi. 2003. *Nonparametric estimation of large covariance matrices of longitudinal data*, Biometrika **90**, no. 4, 831–844, DOI 10.1093/biomet/90.4.831. MR2024760 (2004j:62148)

Yang, F. 2013. *High dimensional data analysis and biomedical genomics*, Ph.D. Thesis, University of Wisconsin–Madison.

Zou, H., T. Hastie, and R. Tibshirani. 2006. *Sparse principal component analysis*, J. Comput. Graph. Statist. **15**, no. 2, 265–286, DOI 10.1198/106186006X113430. MR2252527

Department of Statistics, University of Wisconsin-Madison, Madison, Wisconsin 53706

E-mail address: fyang@stat.wisc.edu

Department of Statistics, University of Wisconsin-Madison, Madison, Wisconsin 53706

E-mail address: doksum@stat.wisc.edu

Department of Statistics, University of Wisconsin-Madison, Madison, Wisconsin 53706

E-mail address: kwtsui@stat.wisc.edu

Contemporary Mathematics
Volume **622**, 2014
http://dx.doi.org/10.1090/conm/622/12437

Solving a System of High-Dimensional Equations by MCMC

Nozer D. Singpurwalla and Joshua Landon

Abstract. In this paper we introduce an archetypal system of equations that are common to many branches of science and engineering, and propose a statistical approach for solving them via a Markov Chain Monte Carlo algorithm. The equations are high dimensional in a sense to be described in the paper. We invite the attention of other researchers to address the kind of problem given here via approaches alternate to ours.

1. Introduction: Background

This is an expository paper, the essence of which has been abstracted and imported from **Landon et al. (2011)**. Its purpose is to introduce mathematical scientists to a system of equations encountered by natural and physical scientists in a wide variety of applications, and to suggest a statistical approach for solving them. The number of equations is infinite; however, they are nonlinear and high dimensional, in the sense that they entail an infinite number of parameters. Statisticians refer to such a system of equations as small n, large p. The proposed statistical approach entails a Markov Chain Monte Carlo (MCMC) type exercise. There are two key ideas which help reduce the underlying dimensionality of the equations. One is the introduction of a new (unknown) parameter into the system, and the other capitalizing on the resulting telescopic pattern that the parameter infuses into the system. This is in contrast to the traditional approach in dimension reduction wherein one endeavors to discover and to focus on the key variables that influence a response. The unknown parameter is endowed with an uncertainty distribution—a prior distribution—which gets updated to a posterior distribution as the MCMC iteration progresses. Our hope is that this paper will motivate other applied mathematicians, probabilists, and statisticians to capitalize on the idea of this paper and to develop methods superior to ours for addressing a class of commonly occurring problems in science and engineering.

Our exposure to the equations arose out of a problem in particle physics, which goes under the name of lattice quantum chromodynamics—abbreviated lattice QCD. We will therefore introduce the equations of interest using the notions

2010 *Mathematics Subject Classification.* Primary 60-XX, 62-XX, 82-XX.

Key words and phrases. Exponential peeling, dimension reduction, Markov Chain Monte Carlo, lattice QCD.

This research was supported by the Army Research Office grant W911NF-09-1-0039 and by the National Science Foundation grant DMS-0915156 with the George Washington University.

All correspondence pertaining to this paper should be directed to the first author.

and notation of particle physics and then point out the other scenarios wherein similar equations also arise.

2. The underlying equations

In the lattice QCD system, the left-hand side of each equation is the result of a physics based Monte Carlo run, each run taking an exorbitantly long time to complete. For example, a *meson correlator* $G(t \mid \bullet)$ takes the form

$$G(t \mid \bullet) = \sum_{n=1}^{\infty} A_n \mathrm{e}^{-E_n t}, \quad \text{for } t = 0, 1, 2, \ldots, \tag{2.1}$$

where A_n and E_n are parameters and $0 < E_1 < E_2 < \cdots < E_n < \cdots$.

Interest centers around the estimation of the A_n's and the E_n's, $n = 1, 2, \ldots,$ based on $G(t \mid \bullet)$, estimated by the physics based Monte Carlo runs as $\widehat{G}(t \mid \bullet)$, for $t = 0, 1, \ldots, k$, for $k < \infty$. The finiteness of k is due to the time consuming nature of the Monte Carlo runs; usually k is quite small. The physics based codes which generate the $\widehat{G}(t \mid \bullet)$'s do not involve the A_n's and the E_n's. The $\widehat{G}(t \mid \bullet)$'s are autocorrelated—thus the label "correlator"—and the physics based codes also provide an estimate of the autocorrelation matrix.

2.1. Relevance to scenarios in science and engineering. The lattice QCD architecture of (2.1) is not unique to physics. They also occur in other scenarios such as biology, chemistry, and engineering. A few of these are highlighted below.

Mathematical biology: *Exponential peeling in compartment systems.* When considering radioactive tracers used for studying transfer rate of substances in living systems (cf. **Robertson (1957)** and **Rubinow (1975**, p. 125)), sums of exponentials are encountered. Here, the $G(t \mid \bullet)$ of (2.1) represents the concentration of a substance, the t's are integer values of time, the A_i's and the E_i's are constants that need to be estimated. Here interest generally centers around A_1, A_2, E_1 and E_2 of (2.1), whose coefficients are negative. An ad hoc graphical procedure called *the method of exponential peeling* is used to estimate the parameters (cf. **Smith and Morales (1944)**, **Perl (1960)**, and **Van Liew (1967)**).

Some other scenarios in biology where the lattice QCD type equations appear are in bone metabolism studies and cerebral blood flow (cf. **Glass and de Garreta (1967)**), and in biological decay (cf. **Foss (1969)**). In the latter context, Dyson and Isenberg (**1971**) consider, for fluorescence decay, an equation of the type

$$y(t) = \sum_{j=1}^{m} \alpha_j \exp(-t/\tau_j), \quad 0 \leq t \leq T,$$

where $y(t)$ represents "moments of the fluorescence", α_j's the amplitudes [the A_n's of (2.1)], and the τ_j's are time constants corresponding to the E_n's of (2.1). Here the α_j's are zero for $j \geq m + 1$.

Gene expression data. When considering a time series of gene expression data (cf. **Giurcăneanu et al. (2005)**), a system of equations paralleling that of (2.1) arises again. In this context $G(t \mid \bullet)$ represents "mRNA concentrations" as a function of time, and the parameters A_n and E_n describe interactions between the genes. In the gene expression context, as in the lattice QCD context, the parameters E_n are increasing in n.

Nuclear magnetic resonance (NMR). NMR experiments often generate data that are modeled as the sum of exponentials (cf. **Bretthorst et al. (2005)**). Experiments relying on NMR to probe reaction kineticis, diffusion, molecular dynamics, and xenobiotic metabolism are some of the applications where parameter estimates provide insight into chemical and biological processes. See, for example, **Paluszny et al. (2008/09)** who study brain tissue segmentation from NMR data.

Here one considers equations of the type:

$$d_i = C + \sum_{j=1}^{m} A_j \exp\{-\alpha_j t_i\} + n_i,$$

where m is the number of exponentials and d_i a data value sampled at t_i. The parameters of interest are the decay rate constants α_j, the amplitudes A_j, and the constant offset C. The n_i's are the error terms.

Electromechanical oscillations in power systems. Equations entailing the sum of exponentials are also encountered in the context of low frequency electromechanical oscillations of interconnected power systems, the impulse response of linear systems in networks, ultracentrifuge and a host of other relaxation and time-decay phenomena (cf. **Dyson and Isenberg (1971)**). For example, in the electromagnetic oscillations scenario, Sanchez-Gasca and Chow (**1999**) encounter an equation analogous to our equation (2.1) with $G(t \mid \bullet)$ denoting a signal and A_n connoting a signal residue associated with the "mode" E_n.

To summarize, the relationships of the type given by (2.1) arise in so many contexts of science and engineering that it seems to be quintessential, and almost some kind of a law of nature. The lattice QCD problem considered here can therefore be seen as a prototype and a convenient platform to exposit a problem of general applicability. In many of the application scenarios described above, statistical methods have been used, many ad hoc, some empirical Bayesian, and a few Bayesian (under the rubric of maximum entropy). Many of these methods have not exploited an underlying telescopic pattern in these equations which makes an appearance when a latent parameter is introduced into the system, and inference about the latent parameter made.

3. Reparameterization using a latent variable

An examination of (2.1) yields the following boundary conditions. $G(0 \mid \bullet) = \sum_{n=1}^{\infty} A_n$, implying that the A_n's are constrained. When $t \to \infty$, $G(t \mid \bullet) = 0$, which implies that for large values of t, A_n and the E_n cannot be individually estimated. Thus obtaining $G(t \mid \bullet)$, or estimating it, for large t does not have a payback.

Since the E_n's increase with n, we may, as a start, reparameterize the E_n's as $E_n - E_{n-1} = c$, for some *unknown* latent parameter c, $c > 0$, for $n = 2, 3, \ldots$. Thus

(3.1) $$E_n = E_1 + (n-1)c, \quad n = 2, 3, \ldots,$$

with E_1 and c unknown. With the above assumption in place, a parsimonious version of (2.1) takes the form:

(3.2) $$G(t \mid \bullet) = \mathrm{e}^{-E_1 t} \sum_{n=1}^{\infty} A_n \mathrm{e}^{-(n-1)ct}, \quad t = 0, 1, 2, \ldots.$$

With c fixed, the parsimonious model given above reveals the following features:

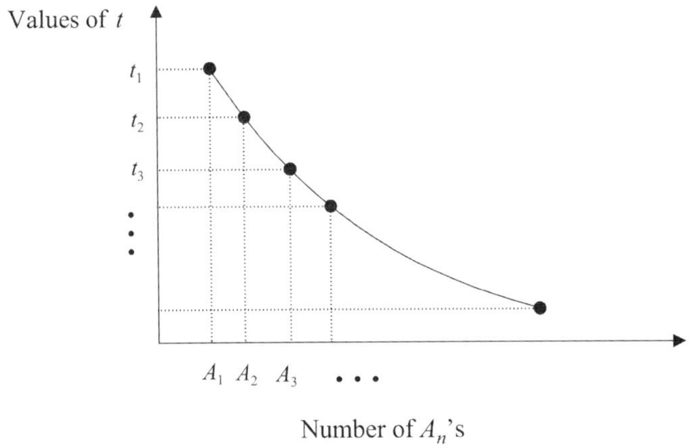

FIGURE 1. Number of A_n's as a function of t.

(a) When t is *small*, the number of A_n's entering (3.2) is large; indeed, infinite when $t = 0$.

(b) When t is *large*, the number of A_n's we need to consider is small, because the combination of a large t with any n, will make the term $A_n \exp(-(n-1)ct)$ get small enough to be ignored.

(c) Moderate values of t and n will also make the above term small, causing A_n to be irrelevant.

Figure 1 illustrates the feature that as t gets large, the number of A_n's one needs to consider gets small.

As a consequence of the above, for any fixed c, we can find a t_1 such that in the right-hand side of (3.2), i.e., in the expression

$$e^{-tE_1}[A_1 + A_2 e^{-ct} + A_3 e^{-2ct} + \cdots + A_n e^{-(n-1)ct} + \cdots],$$

all the terms, save for $A_1 e^{-tE_1}$, are essentially zero.

Similarly, we can find a t_2, $t_2 < t_1$, such that all the terms save for A_1 and $A_2 e^{-ct_2}$ get annihilated. Continuing in this vein, there exists a sequence $t_k < t_{k-1} < \cdots < t_2 < t_1$, such that all that matters are the terms associated with $A_1, A_2, \ldots, A_k$. In what follows, we suppose that k is *specified*.

Thus, for any fixed c and k, with $t_1 > t_2 > \cdots > t_k$ chosen in the manner described above, our parsimonious version of (2.1) *telescopes* as follows:

$$\begin{aligned}
G(t_1 \mid \bullet) &= e^{-E_1 t_1} A_1, \\
G(t_2 \mid \bullet) &= e^{-E_1 t_2}(A_1 + A_2 e^{-ct_2}), \\
G(t_3 \mid \bullet) &= e^{-E_1 t_3}(A_1 + A_2 e^{-ct_3} + A_3 e^{-2ct_3}) \\
&\vdots \\
G(t_k \mid \bullet) &= e^{-E_1 t_n}(A_1 + A_2 e^{-ct_k} + \cdots + A_k e^{-(k-1)ct_k}).
\end{aligned} \tag{3.3}$$

To summarize, by introducing the constant c, fixing a k, and identifying an underlying pattern in the lattice QCD equations, we have reduced the problem to the case of k equations and $(k+2)$ unknowns, $A_1, \ldots, A_k, E_1$ and c. The choice of what

k to choose is determined by the ease with which $G(t \mid \bullet)$ or its estimate $\widehat{G}(t \mid \bullet)$, $t = 0, 1, \ldots, k$, can be obtained.

4. The statistical model and its solution

In many applications, particularly in particle physics, much progress is achieved if trustworthy estimates of just A_1, A_2, E_1 and E_2 can be had. The other pairs (A_3, E_3), $(A_4, E_4), \ldots$, can be considered later; see Section 5. Thus we start by focusing attention on the first two equalities of (3.3); that is, the case $k = 2$ and some fixed c. Specifically, we consider:

$$\begin{aligned} G(t_1|A_1, E_1) &= e^{-E_1 t_1} A_1, \\ G(t_2|A_1, E_1, A_2, c) &= e^{-E_1 t_2}(A_1 + A_2 e^{-ct_2}). \end{aligned} \tag{4.1}$$

If $y_i = \widehat{G}(t_i \mid \bullet)$, $i = 1, 2$, then our aim is to estimate A_1, E_1, A_2 and c, in the light of y_1 and y_2. To set up our likelihoods, we assume that

$$\begin{aligned} Y_1 &= G(t_1 \mid \bullet) + \epsilon_1, \\ Y_2 &= G(t_2 \mid \bullet) + \epsilon_2, \end{aligned} \tag{4.2}$$

where $\epsilon_i \sim N(0, \sigma_i^2)$, $i = 1, 2$, and $\text{Corr}(\epsilon_1, \epsilon_2) = \rho_{12}$, where y_1 and y_2 are given and σ_1^2, σ_2^2 and ρ_{12} assumed known.

As a consequence, the statistical model boils down to the bivariate normal distribution:

$$\begin{bmatrix} Y_1 \\ Y_2 \end{bmatrix} \sim N\left(\begin{bmatrix} G(t_1 \mid \bullet) \\ G(t_2 \mid \bullet) \end{bmatrix}, \begin{bmatrix} \sigma_1^2 & \rho_{12}\sigma_1\sigma_2 \\ \rho_{12}\sigma_1\sigma_2 & \sigma_2^2 \end{bmatrix} \right), \tag{4.3}$$

Writing out a likelihood function for the unknowns A_1, E_1, A_2 and c, based on (4.3) is a straightforward matter. However, we need to bear in mind that since the parameters A_1 and E_1 appear in both $G(t_1 \mid \bullet)$ and $G(t_2 \mid \bullet)$, both y_1 and y_2 provide information about A_1 and E_1, with y_2 providing information about A_2 and c as well. To exploit this feature, we construct our likelihoods based on the marginal distribution of Y_1, and the conditional distribution of Y_2 given Y_1. That is, on:

$$Y_1 \sim N(A_1 e^{-E_1 t_1}, \sigma_1^2), \tag{4.4}$$

and

$$(Y_2 \mid Y_1 = y_1) \sim N\Big(G(t_2 \mid \bullet) + \rho_{12}\frac{\sigma_2}{\sigma_1}(y_1 - G(t_1 \mid \bullet)), \sigma_2^2(1 - \rho_{12}^2) \Big). \tag{4.5}$$

Specifically, the likelihood of A_1 and E_1, with y_1 fixed, is:

$$\mathcal{L}(A_1, E_1; y_1) \propto \exp\left[-\frac{(y_1 - A_1 e^{-E_1 t_1})^2}{2\sigma_1^2} \right], \tag{4.6}$$

and the likelihood of A_1, E_1, A_2 and c, with y_2 specified, and the effect of y_1 incorporated via the distribution of A_1 and E_1, posterior to a knowledge of y_2, is of the form:

$$\begin{aligned} &\mathcal{L}(A_1, E_1, A_2, c; y_1, y_2) \\ &\quad \propto \exp\left[-\frac{\{y_2 - (e^{-E_1 t_2}(A_1 + A_2 e^{-ct_2})) + \rho_{12}(\sigma_2/\sigma_1)(y_1 - A_1 e^{-E_1 t_1})\}^2}{2\sigma_2^2(1 - \rho_{12}^2)} \right]. \end{aligned} \tag{4.7}$$

4.1. Specification of the prior distributions. To implement a Bayesian approach we need to make assumptions about conditional independence, and assign prior distributions for the unknown parameters. The priors that we end up choosing should be proper priors based on an appreciation of the underlying science.

The A_i's are supposedly between 0 and 1, and no relationship between them has been claimed. Thus it is natural to assume that A_1 and A_2 are a priori independent, and have a beta distribution on $(0, 1)$ with parameters (α, β); we denote this as $\mathcal{B}(A_i; \alpha, \beta)$, $i = 1, 2$. The relationship between E_1 and c is less straightforward. We conjecture that the larger the E_1, the smaller the c, and that E_1 can take values over $(0, \infty)$. It is therefore reasonable to assume that the prior on E_1 is a gamma distribution with scale parameter η and shape parameter λ; we denote this by $\mathcal{G}(E_1; \eta, \lambda)$. Some other meaningful choices for a prior on E_1 could be a Weibull, or a Pareto, the latter being noteworthy as a fat-tailed distribution. To encapsulate the dependence between E_1 and c, we suppose that given E_1, c has a uniform distribution over $(0, \omega/E_1)$, for some $\omega > 0$. Finally, we also assume that E_1 and c are independent of all the A_i's.

The above choice of priors, with user specified hyperparameters $\alpha, \beta, \omega, \lambda$, and η, is illustrative. In principle, any collection of meaningful priors can be used, since the ensuing inference is done numerically via a Markov Chain Monte Carlo (MCMC) approach.

4.2. An outline of the MCMC excercise. The telescopic nature of the lattice QCD equations suggests that the MCMC will have to be conducted in the following three phases.

Phase I. Using $E_1^{(0)}$ as a starting value and y_1 as data, simulate $A_1^{(1)}$ via (4.6) as the likelihood using $E_1^{(0)}$, and then simulate $E_1^{(1)}$ using $A^{(1)}$, and repeat this for MCMC for 1000 iterations, thus obtaining the posterior distribution of A_1 and E_1.

Phase II. Using $c^{(0)}$ as a starting value, and y_2 as data, obtain a sample from the posterior distribution of A_1, E_1, A_2 and c via the likelihood of (4.7), and 1000 iterations of the MCMC run. Sample values of A_1 and E_1 from their posterior distributions obtained in Phase I, will serve as the priors of A_1 and E_1 in Phase II. Since the parameters A_1 and E_1 reappear in the likelihood of (4.7) as the mean of y_2, Phase II of the MCMC run captures the effect of y_2 on these parameters. The effect of y_1 was captured in Phase I.

Phase III. Repeat Phase I and Phase II m times using new starting values of E_1 and c to produce a sample of size m from the posterior distribution of A_1, E_1, A_2 and c, with y_1 and y_2 as the data.

The MCMC exercise described above is routine, but computer intensive and entails 12 steps, six in each phase (see **Landon et al. (2011)** for details on the 12-step method), and this too for a highly curtailed version of the equations of interest. The details of how this is done could be interesting, because they involve some discretization of the simulated posterior distributions, and working with individual sampled values reminiscent of that done in particle Kalman filtering (cf. **Gordon et al. (1993)**). Thus we label our approach as *Particle MCMC*. More details are given in **Landon (2007)**, and the required software can be downloaded at `http://www.gwu.edu/~stat/irra/Lattice_QCD.htm`.

4.3. A caveat of the proposed scheme. The caveat mentioned here stems from the features that c has been fixed, and that the MCMC runs are centered around fixed values of y_1 and y_2. To see why, recall that our parsimonious version of the equations—see (3.3)—is based on those t_i's for which the exponential terms vanish; however, the t_i's are determined by a fixed value of c. Thus any change in the value of c will bring about a change in the values of t_i, and as a consequence the equations will also have to be different. This would tantamount to obtaining new values of the y_i's. However, all the likelihoods in the MCMC runs are based on fixed values of the y_i's—see (4.6) and (4.7). But a change in the value of c is inevitable, because in Phase II of the MCMC run one iterates around sampled values from the posterior distribution of c, so that the initial $c^{(0)}$ systematically gets replaced by $c^{(1)}, c^{(2)}, \ldots, c^{(1000)}$.

A way to overcome this caveat is to recognize that for any $c^{(i)} > c^{(0)}$, $i = 1, 2, \ldots$, the exponential terms mentioned above will continue to vanish, so that any specified values of y_i will continue to satisfy the right-hand side of (4.2).

A strategy to ensure that the successively generated values of $c^{(i)}$, $i = 1, 2, \ldots$, will tend to be greater than $c^{(0)}$, is to pick small values of $c^{(0)}$ for each of the m iterations of Phase III of the MCMC algorithm. During the course of the MCMC runs, should one encounter a generated value of $c^{(i)}$ that is smaller than $c^{(0)}$, then one should *discard* the so generated value $c^{(i)}$, and generate another value of $c^{(i)}$. Whereas this does not necessarily imply that the prior distribution of c conditional on E_1 is in some sense invalid, an excessive number of discarded $c_i^{(i)}$'s would raise the question of the meaningfulness of the said prior.

5. Extending the approach

The approach outlined in Sections 3 and 4 has some limitations. The purpose of this section is to prescribe strategies for overcoming these. By far, the most noteworthy limitation is that the model of (4.1) restricts attention to a consideration of the parameters A_1, E_1, A_2 and c, whereas the underlying equations have an infinite number of A_i's and E_i's. The second concern pertains to the fact that in Section 4, data associated with the t's intermediate to t_1 and t_2 are not used in the MCMC algorithm. The proposed approach therefore does not exploit all the available data y_t. Finally, there is a question of assuming a constant spacing c of the E_i's. What is the effect of unequally spaced E_i's on inference? Recall that the role played by c is important. First, it imparts parsimony by eliminating all the E_i's save for E_1. Second, it gives birth to the telescopic series which is central to our approach. It turns out that the effect of c is transitionary (it is a nuisance parameter) and that inferences about $E_2, A_3, E_3, A_4, \ldots$, is possible if we exploit the result observed in Section 4.

5.1. Inferences for E_2, A_3 and beyond. Our ability to extend the approach of Sections 3 and 4 to the case of $E_2, A_3, E_3, A_4, \ldots$, is driven by the feature that t_2 (i.e., y_6) does not contribute much toward the inferences for A_1 and E_1, beyond that provided by t_1 (i.e., y_{12}). Thus the effect of t_3 (say y_4) will be less so, making it possible for us to do the following:

Rewrite (2.1) as

$$G(t \mid \bullet) - A_1 e^{-E_1 t} = \sum_{n=2}^{\infty} A_n e^{-E_n t}, \tag{5.1}$$

and let $G^*(t \mid \bullet) = G(t \mid \bullet) - \hat{A}_1 e^{-\widehat{E}_1 t}$, where $\hat{A}_1$ and $\widehat{E}_1$ are the modes (means) of the posterior distributions of A_1 and E_1. Thus

$$G^*(t \mid \bullet) \approx \sum_{n=2}^{\infty} A_n e^{-E_n t},$$

and setting $A_n^* = A_{n+1}$ and $E_n^* = E_{n+1}$, for $n = 2, 3, \ldots$, we have

$$G^*(t \mid \bullet) \approx \sum_{n=1}^{\infty} A_n^* e^{-E_n^* t}, \tag{5.2}$$

The right-hand side of (5.2) parallels the right-hand side of (2.1), save for the fact that A_n^* and E_n^* replace A_n and E_n. The material of the previous sections now applies, but with the caveat that since (5.2) is an approximation, whereas (2.1) is exact, the variance of the error terms associated with the former should be larger than those associated with the latter.

The posterior distributions of E_1^* and A_2^* will be the posterior distributions of E_2 and A_3. The role of c as a nuisance parameter is a now apparent. The posterior distribution of A_1^* will serve as a revised posterior distribution of A_2. Indeed for the MCMC runs associated with the treatment of (5.2), we may sample from the posterior distribution of A_2 to generate the posterior distribution of A_1^*.

We may continue in the above vein to estimate E_3 and A_4 by defining $G^{**}(t \mid \bullet) = G^*(t \mid \bullet) - \hat{A}_1^* e^{-\widehat{E}_1^* t}$, where $\hat{A}_1^*$ and $\widehat{E}_1^*$ are the modes of the posterior distributions of A_1^* and E_1^*, respectively. Similarly with $(E_4, A_5), (E_5, A_6)$, and so on.

5.2. Using additional Y_t's. For enhanced inferences about the parameters A_1 and E_1 we may want to use all values of Y_t intermediate to those associated with the labels t_1 and t_2. Similarly, with the Y_t's intermediate to the ones associated with the labels t_2 and t_3, and so on. What makes this possible is the fact that t_1 is the *largest* value of t for which $(A_2, E_2), (A_3, E_3), \ldots,$ gets annihilated, whereas t_2 is the largest value of t at which $(A_3, E_3), (A_4, E_4), \ldots,$ gets annihilated, and so on. Thus values of t intermediate to t_1 and t_2 will continue to annihilate $(A_2, E_2), (A_3, E_3), \ldots,$ and those intermediate to t_2 and t_3 will annihilate $(A_3, E_3), (A_4, E_4), \ldots,$ and so on.

Let $y_{11}, y_{12}, \ldots, y_{15}$ denote the Y_t's intermediate to those associated with the labels t_1 and t_2. Then, to incorporate the effect of $y_{11}, \ldots, y_{15}$ for enhanced inference about A_1 and E_1, the iterative scheme described in Phase I of Section 4.2 will have to be cycled five more times, each cycle involving a use of the y_{1j}, $j = 1, \ldots 5$, before proceeding to Phase II, wherein the effect of y_6 and y_7 comes into play. Similarly, with y_{21}, the single value intermediate to that associated with the labels t_2 and t_3.

6. Summary and conclusions

The Bayesian approach proposed by us has been facilitated by the fact that by introducing a latent parameter, the architecture of the nonlinear equations reveals an attractive pattern. This pattern boils down to our consideration of a truncated telescopic series of equations, each equation being the sum of a finite number of exponential functions. Similar sets of equations also arise in other arenas of science, as mentioned in Section 2.1. The nonlinear nature of the equations mandates that our proposed approach—which entails stylized proper priors—be implemented by a particle style Markov Chain Monte Carlo (MCMC) approach. Such a procedure turns out to be computationally very intensive—about one million iterations for making inference about three parameters.

The proposed procedure, when invoked on simulated data, is able to reproduce the input parameters. This is one way to claim the validity of our approach. The procedure when invoked on some real data pertaining to the quark masses of protons and pions, is also able to produce results that are in agreement with the results produced using alternate physics based methods. However, the physics based methods are able to obtain only partial results. By contrast, our approach can produce estimates of as many parameters as is desired—but there is no way to validate these against alternate approaches or actual numbers, because these are not available.

Future work in this arena will entail enhancements to gain computational efficiencies and the choice of proper priors that are motivated by a consideration of the underlying physics. This means that an undertaking such as this will call for insights and skills that go beyond mathematics, statistics, and computing.

Acknowledgements. The detailed comments of a referee are gratefully acknowledged.

References

Bretthorst, G. L., W. C. Hutton, J. R. Garbow, and J. J. H. Ackerman. 2005. *Exponential parameter estimation (in NMR) using Bayesian probability theory*, Concepts in Magnetic Resonance Part A **27A**, no. 2, 55–63, DOI 10.1002/cmr.a.20043.

Dyson, R. D. and I. Isenberg. 1971. *Analysis of exponential curves by a method of moments, with special attention to sedimentation equilibrium and fluorescence decay*, Biochemistry **10**, no. 17, 3233–3241, DOI 10.1021/bi00793a012.

Foss, S. D. 1969. *A method for obtaining initial estimates of the parameters in exponential curve fitting*, Biometrics **25**, 580–584.

Giurcăneanu, C. D., I. Tăbuş, and J. Astola. 2005. *Clustering time series gene expression data based on sum-of-exponentials fitting*, EURASIP J. Appl. Signal Process. **8**, 1159–1173. MR2168617

Glass, H. I. and A. C. de Garreta. 1967. *Quantitative analysis of exponential curve fitting for biological applications*, Physics in Medicine and Biology **12**, no. 3, 379–388.

Gordon, N. J., D. J. Salmond, and Smith A. F. M. 1993. *Novel approach to nonlinear/non-Gaussian Bayesian state estimation*, IEE Proceedings F **140**, no. 2, 107–113.

Landon, J. 2007. *A problem in particle physics and its Bayesian analysis*. Thesis (Ph.D.)–The George Washington University. MR2710570

Landon, J., F. X. Lee, and N. D. Singpurwalla. 2011. *A problem in particle physics and its Bayesian analysis*, Statist. Sci. **26**, no. 3, 352–368, DOI 10.1214/11-STS364. MR2917960 (2012m:62345)

Paluszny, M., M. Martín-Landrove, G. Figueroa, and W. Torres. 2008/09. *Boosting the inverse interpolation problem by a sum of decaying exponentials using an algebraic approach*, Electron. Trans. Numer. Anal. **34**, 163–169. MR2597808 (2011a:65032)

Perl, W. 1960. *A method for curve-fitting by exponential functions*, The International Journal of Applied Radiation and Isotopes **8**, no. 4, 211–222.

Robertson, J. S. 1957. *Theory and use of tracers in determining transfer rates in biological systems*, Physiological Reviews **37**, no. 2, 133–157.

Rubinow, S. I. 1975. *Introduction to mathematical biology*, Wiley, New York.

Sanchez-Gasca, J. J. and J. H. Chow. 1999. *Performance comparison of three identification methods for the analysis of electromagnetic oscillations*, IEEE Transactions on Power Systems **14**, no. 3, 995–1002, DOI 10.1109/59.780912.

Smith, R. E. and M. F. Morales. 1944. *On the theory of blood-tissue exchanges.* II: *Applications*, Bull. Math. Biology **6**, no. 4, 133–139, DOI 10.1007/BF02478430.

Van Liew, H. D. 1967. *Method of exponential peeling*, J. Theoret. Biol. **16**, no. 1, 43–53, DOI 10.1016/0022-5193(67)90052-5.

Department of Systems Engineering and Engineering Management, City University of Hong Kong, 83 Tat Chee Avenue, Kowloon Tong, Hong Kong

E-mail address: nsingpur@cityu.edu.hk

Department of Statistics, George Washington University, 801 22nd St. NW, Washington, DC 20052

E-mail address: jlandon@gwu.edu

Contemporary Mathematics
Volume **622**, 2014
http://dx.doi.org/10.1090/conm/622/12434

A slice sampler for the hierarchical Poisson/Gamma random field model

Jian Kang and Timothy D. Johnson

ABSTRACT. We propose a new algorithm for the posterior inference on the hierarchical Poisson/Gamma random field model for the analysis of multi-type spatial point patterns. We develop a stick-breaking construction of the model and introduce a set of auxiliary variables such that the intensity function can be represented by a finite number of components. We illustrate the proposed algorithm on a simulated data set and a functional neuroimaging meta-analysis data set.

1. Introduction

The Poisson/gamma random field (PGRF) model [**WI98**] is a Bayesian hierarchical, nonparametric spatial point process model (a Cox process). The underlying intensity function is modeled as a convolution of a spatial kernel and a gamma random field, resulting in a positively autocorrelated intensity measure. The PGRF model accounts for both spatial variation and uncertainty in the underlying intensity. Thus, this model has found widespread use due to its robustness and computational efficiency in the modeling of spatial point patterns that exhibit clustering or aggregation [**IW99, BIW00, BIW⁺02, SP00, NF10, WWO10**]. The PGRF model assumes that the data arise as a single realization of a (single-type) spatial point process. However, there is growing interest in analyzing multi-type spatial point patterns [**Bad10**] motivated by problems in the biological and medical sciences such as disease mapping [**LBB⁺00, DR94, BTMH05**]; studies in spatial epidemiology; the analyses of plant ecologies [**BM01, Dig81, Dig83, DM83, vLB99**]; and the meta-analysis of functional neuroimaging studies [**KBJ⁺08, ELG⁺09, WLN⁺09, KJNW11**].

[**Kan11**] proposed an approach for fitting multivariate spatial point patterns by extending the PGRF model to an hierarchical PGRF (HPGRF) model motivated by the meta-analysis of multi-type functional neuroimaging data. This generalization is analogous to the extension of the Dirichlet process to the hierarchical Dirichlet process [**TJBB06**]. Specifically, [**Kan11**] considered each type of spatial point pattern as a realization of a PGRF process where the gamma random field for each type is a realization from a population level gamma random field. The random intensity measures for the different types are related, thus allowing not only

2010 *Mathematics Subject Classification.* Primary 62G05.
Key words and phrases. Bayesian spatial point processes, slices sampler, hierarchical model.

aggregation of points within a type, but aggregation of points across types. For the posterior computation, [**Kan11**] proposed a Lévy construction of the model which leads to a truncation of an infinite sum in practice. The number of points, M, in the gamma random field usually needs to be very large to achieve a desired level of accuracy for the intensity estimates, thus the computation cost can be high. This motivates the needs for a more efficient posterior simulation algorithm. To this end, we propose a new computational algorithm for the HPGRF model by leveraging the relationship between the gamma process and the Dirichlet process [**Fer73**].

Many Markov chain Monte Carlo (MCMC) algorithms have been developed for the mixture of Dirichlet process models. The pioneering work was developed by [**EW95**], [**Esc95**] and [**BM96**] based on a Gibbs sampler, where [**BM96**] introduced the remixing step to resample the locations of the Dirichlet process. [**IJ01**] developed a Gibbs sampling scheme using the more general stick-breaking priors as a direct extension of [**EW95**]'s approach. [**Nea00**] and [**JN04**] also developed several efficient Metropolis-Hasting algorithms to obtain better convergence of the posterior simulations. [**Wal07**] proposed a slice sampler by introducing a set of latent variables which allow a finite random number of mixture components to be sampled within each iteration of the MCMC algorithm. In a similar fashion, we develop a stick-breaking construction for the HPGRF model, utilizing the connection between the gamma process and the Dirichlet process. We introduce auxiliary variables such that the intensity function of the point process is represented by a finite number of mixture components. This new algorithm vastly reduces the computational cost compared with the Lévy construction algorithm proposed by [**Kan11**].

The rest of the paper is organized as follows: In Section 2, we review the HPGRF model and describe the stick-breaking representation of the model. In Section 3, we introduce latent variables for the sampling strategy and propose our slice sampling algorithm. We illustrate our algorithm on a simulation study and a meta-analysis of functional neuroimaging data in Section 4. In Section 6, we conclude the paper with a short discussion.

2. Hierarchical Poisson/gamma random fields

2.1. The model. Suppose we have J distinct types of spatial point patterns on $\mathcal{B}$ (for our purpose $\mathcal{B} \subset \mathbb{R}^3$, and, in particular, will represent the brain) and n_j independent realizations of each type. Let $\mathbf{y}_{i,j}$, $i = 1, \ldots, n_j$, denote the ith, type j, observed point pattern and assume that each $\mathbf{y}_{i,j}$ is a realization from a Cox process [**Cox55**] $\mathbf{Y}_j$ driven by a common random intensity measure $\Lambda_j(dy)$. By the definition of a Cox process, we note that conditional on $\Lambda_j(dy)$, $\mathbf{Y}_j$ is a Poisson point process. We assume $\Lambda_j(dy)$ is a convolution of a gamma random field (a random measure) $G_j(d\mathrm{x})$ with a finite kernel measure $K_{\sigma_j^2}(dy, \mathrm{x})$, where σ_j^2 is the kernel variance: $\Lambda_j(dy) = \int_{\mathcal{B}} K_{\sigma_j^2}(dy, \mathrm{x}) G_j(d\mathrm{x})$. The $G_j(d\mathrm{x})$ are mutually independent with common base measure $G_0(d\mathrm{x})$ and scale parameter τ. To introduce dependence between types, let $G_0(d\mathrm{x})$ be a gamma random field with base measure $\alpha(d\mathrm{x})$ and scale parameter β. In summary, we have

$$\begin{aligned} [\mathbf{Y}_j \mid \Lambda_j(dy)] &\sim \mathcal{PP}\{\mathcal{B}, \Lambda_j(dy)\}, \\ [G_j(d\mathrm{x}) \mid G_0(d\mathrm{x}), \tau] &\overset{\text{iid}}{\sim} \mathcal{GRF}\{G_0(d\mathrm{x}), \tau\}, \\ [G_0(d\mathrm{x}) \mid \alpha(d\mathrm{x}), \beta] &\sim \mathcal{GRF}\{\alpha(d\mathrm{x}), \beta\}. \end{aligned} \tag{2.1}$$

Here, $\mathcal{PP}\{A, \Lambda(dx)\}$ denotes a Poisson point process on A, with intensity measure $\Lambda(dx)$, and $\mathcal{GRF}\{a(dx), b\}$ denotes a gamma random field with base measure $a(dx)$ and scale parameter b. The kernel variances, σ_j^2, base measure $\alpha(dx)$ and scale parameters τ and β, are all given hyperprior distributions that we define later. Suppose that $K_{\sigma_j^2}(dy, x)$ and $\Lambda_j(dy)$ are both dominated by Lebesgue measure, ℓ: that is, there exist Lebesgue measurable functions $k_{\sigma_j^2}(y, x)$ and $\lambda_j(y)$ such that $K_{\sigma_j^2}(B, x) = \int_B k_{\sigma_j^2}(y, x)\ell(dy)$ and $\Lambda_j(B) = \int_B \lambda_j(y)\ell(dy)$ for any Lebesgue measurable set $B \subseteq \mathcal{B}$. We choose $K_{\sigma_j^2}(dy, x)$ to be a probability measure on $\mathcal{B}$ for each $x \in \mathcal{B}$. This implies that $k_{\sigma_j^2}(y, x) \geq 0$. Thus, by Fubini's Theorem we have that the intensity function $\lambda_j(y) = \int_{\mathcal{B}} k_{\sigma_j^2}(y, x)G_j(dx)$.

2.2. The stick-breaking construction. In this section, we discuss an alternative representation of model (2.1). Note that if $G_0(dx) \sim \mathcal{GRF}(\alpha(dx), \beta)$, then $G_0(\mathcal{B}) \sim \text{Gamma}(\alpha(\mathcal{B}), \beta)$, where $\text{Gamma}(a, b)$ denotes a gamma distribution with shape a and scale b. Let $\overline{G}_0(dx) = G_0(dx)/G_0(\mathcal{B})$ and $\bar{\alpha}(dx) = \alpha(dx)/\alpha(\mathcal{B})$, then $\overline{G}_0(dx) \sim \mathcal{D}ir(\alpha(\mathcal{B}), \bar{\alpha}(dx))$, where $\mathcal{D}ir(a, b(dx))$ denotes a Dirichlet process with precision parameter a and probability base measure $b(dx)$. Using the stick-breaking construction [**IJ01**], we write $\overline{G}_0(dx) = \sum_{m=1}^{\infty} v_m^* \delta_{\theta_m}(dx)$, where $v_m^* = \phi_m \prod_{k=1}^{m-1}(1 - \phi_k)$, $\phi_m \sim \text{Beta}(1, \alpha(\mathcal{B}))$ and $\theta_m \sim \bar{\alpha}(dx)$, where $\text{Beta}(a, b)$ denotes a beta distribution with parameters a and b. The set $\{\theta_m\}_{m=1}^{\infty}$ are an infinite number of common jump points for the gamma random fields G_j for $j = 1, \ldots, J$. The set $\{v_m^*\}_{m=0}^{\infty}$ are weights for the jump points of $G_0(dx)$ where $\sum_{m=1}^{\infty} v_m^* = 1$. The set of parameters $\{\phi_m\}_{m=0}^{\infty}$ defines the prior model of $\{v_m^*\}$. This implies the following representation of $G_0(dx)$.

$$G_0(dx) = G_0(\mathcal{B}) \sum_{m=1}^{\infty} v_m^* \delta_{\theta_m}(dx). \tag{2.2}$$

Since $G_j(dx) \sim \mathcal{GRF}(G_0(dx), \tau)$, we have $G_j(dx) = \sum_{m=1}^{\infty} \mu_{j,m}\delta_{\theta_m}(dx)$, where the set $\{\mu_{j,m}\}$ defines the jump height for the jump points for $G_j(dx)$ and it has a gamma prior, i.e., $\mu_{j,m} \sim \text{Gamma}(G_0(\mathcal{B})v_m^*, \tau)$. Note that $G_j(\mathcal{B}) = \sum_{m=1}^{\infty} \mu_{j,m} \sim \text{Gamma}(G_0(\mathcal{B}), \tau)$. We define $\mu_{j,m}^* = \mu_{j,m}/G_j(\mathcal{B})$ which are the weights for the jump points of $G_j(\mathcal{B})$ and $\sum_{m=1}^{\infty} \mu_{j,m}^* = 1$. It is straightforward to show that

$$\mu_{j,m}^* = \lambda_{j,m} \prod_{k=1}^{m-1} (1 - \lambda_{j,k}), \quad \lambda_{j,m} \sim \text{Beta}\left(G_0(\mathcal{B})v_m^*, G_0(\mathcal{B})\left(1 - \sum_{k=1}^{m} v_k^*\right)\right),$$

where $\lambda_{j,m}$ defines the prior of $\mu_{j,m}^*$. Then a representation of $G_j(dx)$ is given by

$$G_j(dx) = G_j(\mathcal{B}) \sum_{m=1}^{\infty} \mu_{j,m}^* \delta_{\theta_m}(dx). \tag{2.3}$$

In summary, combining (2.2) and (2.3), a stick-breaking representation of (2.1) is given by

$$G_j(dx) = \zeta_j \sum_{m=1}^{\infty} \mu^*_{j,m} \delta_{\theta_m}(dx), \quad G_0(dx) = \zeta_0 \sum_{m=1}^{\infty} \nu^*_m \delta_{\theta_m}(dx),$$

$$\zeta_j \sim \text{Gamma}(\zeta_0, \tau), \quad \zeta_0 \sim \text{Gamma}(\alpha(\mathcal{B}), \beta),$$

$$\mu^*_{j,m} = \lambda_{j,m} \prod_{k=1}^{m-1} (1 - \lambda_{j,k}), \quad \lambda_{j,m} \sim \text{Beta}\left(\alpha(\mathcal{B}) v^*_m, \alpha(\mathcal{B})\left(1 - \sum_{k=1}^{m} v^*_k\right)\right), \tag{2.4}$$

$$\nu^*_m = \phi_m \prod_{k=1}^{m-1} (1 - \phi_k), \quad \phi_m \sim \text{Beta}(1, \alpha(\mathcal{B})),$$

$$\theta_m \sim \alpha(dx)/\alpha(\mathcal{B}), \quad \beta \sim \text{Gamma}(a_\beta, b_\beta), \quad \tau \sim \text{Gamma}(a_\tau, b_\tau),$$

where $\zeta_j = G_j(\mathcal{B})$ for $j = 1, 2, \ldots, J$. The parameter τ controls the deviations about the mean of ζ_js, and β controls the deviations about the mean of ζ_0. In this paper, the hyperparameters a_β, b_β and a_τ are fixed at 0.001.

3. The algorithm

The stick-breaking construction (2.4) still involves an infinite number of components which is not applicable for the posterior simulation without truncation. To introduce a more efficient sampling scheme, we consider the following data augmentation scheme that is inspired by the slice sampler for mixtures of Dirichlet process prior models introduced by [**Wal07**] (thus, this algorithm is also a slice sampler).

3.1. Data augmentation. The data distribution of the HPGRF model (2.1) is

$$[\mathbf{Y}_j \mid \Lambda_j(dy)] \sim \mathcal{PP}\left\{\mathcal{B}, \int_{\mathcal{B}} K_{\sigma_j^2}(dy, x) G_j(dx)\right\}.$$

From the stick-breaking construction (2.4), we have

$$[\mathbf{Y}_j \mid \zeta_j, \{(\mu^*_{j,m}, \theta_m)\}_{m=1}^{\infty}, \sigma_j^2] \sim \mathcal{PP}\left\{\mathcal{B}, \zeta_j \sum_{m=1}^{\infty} \mu^*_{j,m} K_{\sigma_j^2}(dy, \theta_m)\right\}.$$

To make the posterior simulation more efficient, we attach a mark $U_{j,l}$ to each $Y_{i,j} \in \mathbf{Y}_j$, which defines clusters of the $Y_{i,j}$. We write $(\mathbf{Y}_j, \mathbf{U}_j) = \{(Y_{j,l}, U_{j,l}), Y_{j,l} \in \mathbf{Y}_j\}$. Now,

$$[(\mathbf{Y}_j, \mathbf{U}_j) \mid \bullet] \sim \mathcal{PP}\left\{\mathcal{B}, \zeta_j \sum_{m=1}^{\infty} I[u < \mu^*_{j,m}] K_{\sigma_j^2}(dy, \theta_m) \ell(du)\right\}, \tag{3.1}$$

where $\ell(du)$ is Lebesgue measure. Let $A_j(u) = \{m : \mu^*_{j,m} > u\}$. The intensity function of $(\mathbf{Y}_j, \mathbf{U}_j)$ is equal to

$$\zeta_j \sum_{m \in A_j(u)} K_{\sigma_j^2}(dy, \theta_m) \ell(du),$$

Note that $A_j(u)$ is a finite set, since we can find $\sum_{m=1}^{M_j} \mu^*_{j,m} > 1 - u$, this implies that for any $m > M_j$, $\mu^*_{j,m} < 1 - \sum_{m'=1}^{M_j} \mu^*_{j,m'} < u$. To resolve the above finite mixture, we attach another mark $S_{j,l}$ to $(Y_{j,l}, U_{j,l}) \in (\mathbf{Y}_j, \mathbf{U}_j)$ to indicate the membership of $(Y_{j,l}, U_{j,l})$ to the jump points $\{\theta_m\}_{m=0}^{\infty}$. Write $(\mathbf{Y}_j, \mathbf{U}_j, \boldsymbol{S}_j) = \{(Y_{j,l}, U_{j,l}, S_{j,l}), (Y_{j,l}, U_{j,l}) \in (\mathbf{Y}_j, \mathbf{U}_j)\}$. Then,

$$[(\mathbf{Y}_j, \mathbf{U}_j, \boldsymbol{S}_j) \mid \bullet] \sim \mathcal{PP}\left\{\mathcal{B}, \zeta_j K_{\sigma_j^2}(dy, \theta_s)\ell(du) \sum_{k \in A_j(u)} \delta_k(ds)\right\}, \tag{3.2}$$

where $I[u < \mu_{j,s}] = I[s \in A_j(u)] = \sum_{k \in A_j(u)} \delta_k(ds)$. Note that integrating out $\mathbf{U}_j$ and $\boldsymbol{S}_j$ in (3.2) results in (2.1).

3.2. Posterior computation. We develop the posterior inference algorithm for model (3.2). Let $(\mathbf{y}_{i,j}, \mathbf{u}_{i,j}, \mathbf{s}_{i,j})$, $i = 1, \ldots, n_j$ be the multiple realizations of the marked point process $\{(\mathbf{Y}_j, \mathbf{U}_j, \boldsymbol{S}_j)\}$. The joint density of $\{\{\mathbf{u}_{i,j}, \mathbf{s}_{i,j}\}_{i=1}^{n_j}\}_{j=1}^{J}$, $\{\boldsymbol{\mu}_j\}_{j=1}^{J}$, $\{\sigma_j^2\}_{j=1}^{J}$, $\boldsymbol{\varphi} = \{\phi_m\}_{m=1}^{\infty}$, $\boldsymbol{\theta}$, τ and β given $\{\{\mathbf{y}_{i,j}\}_{i=1}^{n_j}\}_{j=1}^{J}$ is proportional to

$$\begin{aligned}
&\prod_{j=1}^{J}\prod_{i=1}^{n_j} \exp\left(-\zeta_j \sum_{m=1}^{\infty} \mu_{j,m}^* K_{\sigma_j^2}(\mathcal{B}, \theta_m)\right) \zeta_j^{m_{i,j}} \prod_{l=1}^{m_{i,j}} k_{\sigma_j^2}(y_{i,j,l}, \theta_{s_{i,j,l}}) I[u_{i,j,l} < \mu_{j,s_{i,j,l}}^*] \\
&\quad \times \prod_{j=1}^{J} \frac{\tau^{\zeta_0}}{\Gamma(\zeta_0)} \zeta_j^{\zeta_0 - 1} \exp(-\tau\zeta_j) \prod_{m=1}^{\infty} \frac{1}{B(\alpha(\mathcal{B})v_m^*, \alpha(\mathcal{B})(1 - \sum_{k=1}^{m} v_k^*))} \\
&\quad \times \lambda_{j,m}^{\alpha(\mathcal{B})v_m^* - 1}(1 - \lambda_{j,m})^{\alpha(\mathcal{B})(1 - \sum_{k=1}^{m} v_k^*) - 1} \\
&\quad \times \frac{\alpha(\mathcal{B})^{\beta}}{\Gamma(\alpha(\mathcal{B}))} \zeta_0^{\alpha(\mathcal{B})} \exp(-\beta\zeta_0) \prod_{m=1}^{\infty} \frac{1}{B(1, \alpha(\mathcal{B}))}(1 - \phi_m)^{\alpha(\mathcal{B}) - 1} \\
&\quad \times \frac{b_\beta^{a_\beta}}{\Gamma(a_\beta)} \beta^{a_\beta - 1} \exp(-b_\beta\beta) \frac{b_\tau^{a_\tau}}{\Gamma(a_\tau)} \tau^{a_\tau - 1} \exp(-b_\tau\tau) \\
&\quad \times \prod_{j=1}^{J} \frac{b_\sigma^{a_\sigma}}{\Gamma(a_\sigma)} \sigma_j^{-2(a_\sigma + 1)} \exp(-b_\sigma\sigma_j^{-2}).
\end{aligned}$$

This implies the following posterior simulation algorithm:

(1) **Sampling $u_{i,j,l}$:**

$$u_{i,j,l} \sim U(0, \mu_{j,s_{i,j,l}}^*). \tag{3.3}$$

(2) **Sampling $\lambda_{j,m}$:** Let $s^* = \max_{i,j,l}(s_{i,j,l})$. For $m = 1, \ldots, s^*$, the full conditional distribution of $\lambda_{j,m}$ is proportional to

(3.4)

$$\begin{aligned}
&\exp\left[-n_j\zeta_j\left(\lambda_{j,m}\prod_{k=1}^{m-1}(1 - \lambda_{j,k})K_{\sigma_j^2}(\mathcal{B}, \theta_m) + \sum_{l > m} \lambda_{j,l} \prod_{k=1}^{l-1}(1 - \lambda_{j,k})K_{\sigma_j^2}(\mathcal{B}, \theta_l)\right)\right] \\
&\quad \times \prod_{i=1}^{n_j} \prod_{l : s_{i,j,l} \geq m} I\left(\lambda_{j,m}\prod_{k=1}^{m-1}(1 - \lambda_{j,k}) > u_{i,j,l}\right) \lambda_{j,m}^{\alpha(\mathcal{B})v_m^* - 1}(1 - \lambda_{j,m})^{\alpha(\mathcal{B})(1 - \sum_{k=1}^{m} v_k^*) - 1} \\
&= C_{j,m}^{\lambda_{j,m}} D_{j,m}^{1 - \lambda_{j,m}} \lambda_{j,m}^{\alpha(\mathcal{B})v_m^* - 1}(1 - \lambda_{j,m})^{\alpha(\mathcal{B})(1 - \sum_{k=1}^{m} v_k^*) - 1} I(a_{j,m} < \lambda_{j,m} < b_{j,m}).
\end{aligned}$$

where

$$C_{j,m} = \exp\left(-n_j\zeta_j \prod_{k=1}^{m-1}(1-\lambda_{j,k})K_{\sigma_j^2}(\mathcal{B},\theta_m)\right),$$

$$D_{j,m} = \exp\left(-n_j\zeta_j \sum_{l>m}\lambda_{j,l} \prod_{\substack{k\neq m\\ k<l}}(1-\lambda_{j,k})K_{\sigma_j^2}(\mathcal{B},\theta_l)\right),$$

$$a_{j,m} = \max_i \max_{l:s_{i,j,l}=m} \frac{u_{i,j,l}}{\prod_{k=1}^{m-1}(1-\lambda_{j,k})},$$

$$b_{j,m} = 1 - \max_i \max_{m<s_{i,j,l}} \frac{u_{i,j,l}}{\lambda_{j,s_{i,j,l}}\prod_{\substack{k<s_{i,j,l}\\ k\neq m}}(1-\lambda_{j,k})}.$$

To sample from this full conditional we use a Metropolis–Hastings (M–H) algorithm: first draw $\lambda^*_{j,m}$ from

$$F(\lambda^*_{j,m}) = \frac{(1-a_{j,m})^{\alpha(\mathcal{B})(1-\sum_{k=1}^m v_k^*)} - (1-\lambda^*_{j,m})^{\alpha(\mathcal{B})(1-\sum_{k=1}^m v_k^*)}}{(1-a_{j,m})^{\alpha(\mathcal{B})(1-\sum_{k=1}^m v_k^*)} - (1-b_{j,m})^{\alpha(\mathcal{B})(1-\sum_{k=1}^m v_k^*)}}.$$

Update $\lambda_{j,m} = \lambda^*_{j,m}$ with probability $\min\{1, E_{j,m}^{\lambda^*_{j,m}}\lambda_{j,m}^{*\alpha(\mathcal{B})v_m^*-1}/E_{j,m}^{\lambda_{j,m}}\lambda_{j,m}^{\alpha(\mathcal{B})v_m^*-1}\}$, where $E_{j,m} = C_{j,m}/D_{j,m}$. For $m > s^*$, the full conditional of $\lambda_{j,m}$ is proportional to

$$E_{j,m}^{\lambda_{j,m}}\lambda_{j,m}^{\alpha(\mathcal{B})v_m^*-1}(1-\lambda_{j,m})^{\alpha(\mathcal{B})(1-\sum_{k=1}^m v_k^*)-1}, \tag{3.5}$$

for which draw $\lambda^*_{j,m} \sim \text{Beta}(\alpha(\mathcal{B})v_m^* - 1, \alpha(\mathcal{B})(1-\sum_{k=1}^m v_k^*))$, then update $\lambda_{j,m}$ with probability $\min\{1, E_{j,m}^{\lambda^*_{j,m}}/E_{j,m}^{\lambda_{j,m}}\}$.

(3) **Sampling $\boldsymbol{\phi_m}$:** For $m = 1, 2, \ldots$, the full conditional of ϕ_m is proportional to

$$g(\phi_m) = \frac{D_{j,m}^{\phi_m}(1-\phi_m)^{\alpha(\mathcal{B})-1}}{\Gamma(\alpha(\mathcal{B})\phi_m\prod_{k=1}^{m-1}(1-\phi_k))\Gamma\left(\alpha(\mathcal{B})(1-\sum_{k=1}^m \phi_k\prod_{l=1}^{k-1}(1-\phi_l))\right)},$$

where $D_{j,m} = \left(\lambda_{j,m}/(1-\lambda_{j,m})\right)^{\alpha(\mathcal{B})\prod_{k=1}^{m-1}(1-\phi_k)}$. We use a random walk proposal by sampling $\phi_m^* \sim N(\phi_m, \sigma_\phi^2)$. If $\phi_m \in (0,1)$, set $\phi_m = \phi_m^*$ with probability $\min\{1, g(\phi_m^*)/g(\phi_m)\}$.

(4) **Sampling $\boldsymbol{s_{i,j,l}}$:** for $l = 1, \ldots, m_{i,j}$, $i = 1, \ldots, n_j$ and $j = 1, \ldots, J$.

$$\Pr(s_{i,j,l} = m \mid \bullet) \propto k_{\sigma_j^2}(y_{i,j,l}, \theta_m)I[m \in A_j(u_{i,j,l})]. \tag{3.6}$$

The possible choices of $\{s_{i,j,l}\}$ is a finite set, which is $\{m : \mu^*_{j,m} > u_{i,j,l}\}$. For each i, j, l, we can find $M_{i,j,l}$ such that

$$\sum_{m=1}^{M_{i,j,l}} \mu^*_{j,m} > 1 - u_{i,j,l} \quad \text{and} \quad \sum_{m=1}^{M_{i,j,l}-1} \mu^*_{j,m} \leq 1 - u_{i,j,l}.$$

Then $\widetilde{M} = \max_{i,j,l} M_{i,j,l}$ and $\tilde{u} = \min_{i,j,l} u_{i,j,l}$. Furthermore, we have

$$\sum_{m=1}^{\widetilde{M}} \mu^*_{j,m} > 1 - \tilde{u}, \quad \text{for any } j.$$

This implies that for any i, j and l, if $m > \widetilde{M}$, then $\mu^*_{j,m} < u_{i,j,l}$. Thus,

$$\widetilde{M} = \max_{i,j,l} \sum_{m=1}^{\infty} I\{\mu^*_{j,m} > u_{i,j,l}\}.$$

Therefore, the possible choices of $s_{i,j,l}$ are $\{1, \ldots, \widetilde{M}\}$.

(5) **Sampling $\boldsymbol{\theta_m}$:** for $m = 1, 2, \ldots$. The full conditional of θ_m is proportional to

$$\prod_{j=1}^{J} \prod_{i=1}^{n_j} \exp(-\zeta_j \mu^*_{j,m} K_{\sigma_j^2}(\mathcal{B}, \theta_m)) \prod_{l: s_{i,j,l}=m} k_{\sigma_j^2}(y_{i,j,l}, \theta_m) I[u_{i,j,l} < \mu^*_{j,m}].$$

If we assume that

$$k_{\sigma_j^2}(y_{i,j,l}, \theta_m) = \frac{1}{(2\pi\sigma_j^2)^{3/2}} \exp\left(-\frac{1}{2\sigma_j^2} \sum_{k=1}^{3} (y_{i,j,l,k} - \theta_{m,k})^2 \right)$$

then

$$\exp\left(-\sum_{j=1}^{J} n_j \zeta_j \mu^*_{j,m} K_{\sigma_j^2}(\mathcal{B}, \theta_m) - \sum_{j=1}^{J} \sum_{i=1}^{n_j} \sum_{l: s_{i,j,l}=m} \sum_{k=1}^{3} \frac{(y_{i,j,l,k} - \theta_{m,k})^2}{2\sigma_j^2} \right) \tag{3.7}$$

$$\propto \exp\left(-\sum_{j=1}^{J} n_j \zeta_j \mu^*_{j,m} K_{\sigma_j^2}(\mathcal{B}, \theta_m) - \sum_{k=1}^{3} \frac{(\theta_{m,k} - \tilde{y}_{m,k})^2}{2/\tilde{d}_m} \right)$$

where

$$\tilde{y}_{m,k} = \frac{1}{\tilde{d}_m} \sum_{j=1}^{J} \frac{1}{\sigma_j^2} \sum_{i=1}^{n_j} \sum_{l: s_{i,j,l}=m} y_{i,j,l,k}$$

$$\tilde{d}_m = \sum_{j=1}^{J} \frac{1}{\sigma_j^2} \sum_{i=1}^{n_j} \sum_{l=1}^{m_{i,j}} I[s_{i,j,l} = m].$$

This implies that we can update θ_m by sampling $\theta^*_{m,k}$, for $k = 1, 2, 3$,

$$\theta^*_{m,k} \sim N(\tilde{y}_{m,k}, 2/\tilde{d}_m).$$

We update $\theta_m = \theta^*_m$ with probability

$$\min\left\{ 1, \frac{\exp\{\sum_{j=1}^{J} n_j \zeta_j \mu^*_{j,m} K_{\sigma_j^2}(\mathcal{B}, \theta_m)\}}{\exp\{\sum_{j=1}^{J} n_j \zeta_j \mu^*_{j,m} K_{\sigma_j^2}(\mathcal{B}, \theta^*_m)\}} \right\}.$$

(6) **Sampling $\boldsymbol{\zeta_j}$:** For $j = 1, 2, \ldots, J$, we have

$$\zeta_j \sim \text{Gamma}\left(\sum_{i=1}^{n_j} m_{i,j} + \zeta_0, \sum_{m=1}^{\widetilde{M}} \mu^*_{j,m} K_{\sigma_j^2}(\mathcal{B}, \theta_m) + \tau \right).$$

(7) **Sampling $\boldsymbol{\zeta_0}$:** Full conditional of ζ_0 is proportional to

$$h(\zeta_0) = \left(\tau^J \prod_{j=1}^{J} \zeta_j \right)^{\zeta_0} \zeta_0^{\alpha(\mathcal{B})} \exp(-\beta \zeta_0).$$

We use a random walk to update ζ_0 by drawing $\zeta_0^* \sim N(\zeta_0, \sigma^2_{\zeta_0})$. If $\zeta_0^* > 0$, then set $\zeta_0 = \zeta_0^*$ with probability $\min\{1, h(\zeta_0^*)/h(\zeta_0)\}$.

(8) **Sampling σ_j^2:** For $j = 1, 2, \ldots, J$, the full conditional of σ_j^2 is proportional to

$$\prod_{i=1}^{n_j} \exp\left(-\zeta_j \sum_{m=1}^{\infty} \mu_{j,m}^* K_{\sigma_j^2}(\mathcal{B}, \theta_m)\right) \prod_{l=1}^{m_{i,j}} k_{\sigma_j^2}(y_{i,j,l}, \theta_{s_{i,j,l}})$$

$$\propto \exp\left(-n_j\zeta_j \sum_{m=1}^{\infty} \mu_{j,m}^* K_{\sigma_j^2}(\mathcal{B}, \theta_m) - \frac{1}{2\sigma_j^2} \sum_{i=1}^{n_j} \sum_{l=1}^{m_{i,j}} \|y_{i,j,l} - \theta_{s_{i,j,l}}\|\right)$$

$$\propto \sigma_j^{-2(a_\sigma + 3/2 \sum_{i=1}^{n_j} m_{i,j})+1} \exp\left\{-\sigma_j^{-2}\left(b_\sigma + \frac{1}{2} \sum_{i=1}^{n_j} \sum_{l=1}^{m_{i,j}} \|y_{i,j,l} - \theta_{s_{i,j,l}}\|\right)\right\},$$

$$\times \exp\left(-n_j\zeta_j \sum_{m=1}^{\infty} \mu_{j,m}^* K_{\sigma_j^2}(\mathcal{B}, \theta_m)\right).$$

This implies that we can update σ_j^2 by sampling

$$\sigma_j^{2*} \sim \left(a_\sigma + \frac{3}{2} \sum_{i=1}^{n_j} m_{i,j}, b_\sigma + \frac{1}{2} \sum_{i=1}^{n_j} \sum_{l=1}^{m_{i,j}} \|y_{i,j,l} - \theta_{s_{i,j,l}}\|\right).$$

We accept it with probability

$$\min\left\{1, \exp\left(n_j\zeta_j \sum_{m=1}^{\widetilde{M}} \mu_{j,m}^* (K_{\sigma_j^2}(\mathcal{B}, \theta_m) - K_{\sigma_j^{2*}}(\mathcal{B}, \theta_m))\right)\right\}.$$

(9) **Sampling β:**

$$\beta \sim \text{Gamma}\Big(a_\beta, b_\beta + \zeta_0 - \log\big(\alpha(\mathcal{B})\big)\Big).$$

(10) **Sampling τ:**

$$\tau \sim \text{Gamma}\left(a_\tau + J\zeta_0, b_\tau + \sum_{j=1}^{J} \zeta_j\right).$$

4. Examples

In this section, we report results from simulation studies that demonstrate the ability of the proposed algorithm for the HPGRF in terms of accurate intensity estimates and computational cost.

4.1. Simulated data sets. We simulate three-dimensional spatial point patterns on a region $A = [1, 91] \times [1, 109] \times [1, 91]$ from three different types of points based on Poisson point processes with mixture intensity functions, i.e., for $i = 1, \ldots, N$,

$$[\mathbf{Y}_{i,j} \mid \boldsymbol{\mu}, \boldsymbol{\Sigma}] \sim \mathcal{PP}\{A, \lambda_j(\mathrm{x})d\mathrm{x}\}, \quad j = 1, 2, 3. \tag{4.1}$$

where $\lambda_j(\mathrm{x}) = \epsilon + \sum_{(\theta,\mu,\Sigma)\in(\boldsymbol{\theta},\boldsymbol{\mu},\boldsymbol{\Sigma})_j} \theta\phi_3(\mathrm{x}; \mu, \Sigma)$ and $\phi_3(\mathrm{x}; \mu, \Sigma)$ denotes the *three-dimensional* Gaussian density at x with mean μ and covariance Σ. We set intensity parameters such that point patterns from different types show clustering on four regions, where the intensity parameters are summarized in Table 1 and where $\epsilon =$

TABLE 1. Parameters for true intensity functions

Region j	1	2	3	4
θ_j	100	50	150	150
μ_j	$(60,30,20)^T$	$(10,50,70)^T$	$(20,20,20)^T$	$(45,60,20)^T$
Σ_j	$\begin{pmatrix} 65 & 15 & 15 \\ 15 & 40 & 15 \\ 15 & 15 & 45 \end{pmatrix}$	$\begin{pmatrix} 30 & -10 & -10 \\ -10 & 40 & -10 \\ -10 & -10 & 25 \end{pmatrix}$	$\begin{pmatrix} 60 & -5 & -5 \\ -5 & 40 & -5 \\ -5 & -5 & 50 \end{pmatrix}$	$\begin{pmatrix} 45 & 5 & 5 \\ 5 & 40 & 5 \\ 5 & 5 & 50 \end{pmatrix}$

5×10^{-5} represents a baseline constant intensity. We assume that the three sub-types have intensity functions:

$$\begin{aligned}
\lambda_1(\mathrm{x}) &= \epsilon \qquad\qquad\qquad\qquad\quad + \theta_2\phi_3(\mathrm{x};\mu_2,\Sigma_2) + \theta_3\phi_3(\mathrm{x};\mu_3,\Sigma_3) \qquad\qquad\qquad\qquad\quad , \\
\lambda_2(\mathrm{x}) &= \epsilon \qquad\qquad\qquad\qquad\quad + \theta_2\phi_3(\mathrm{x};\mu_2,\Sigma_2) \qquad\qquad\qquad\qquad\quad + \theta_4\phi_3(\mathrm{x};\mu_4,\Sigma_4)\,, \\
\lambda_3(\mathrm{x}) &= \epsilon + \theta_1\phi_3(\mathrm{x};\mu_1,\Sigma_1) + \theta_2\phi_3(\mathrm{x};\mu_2,\Sigma_2) + \theta_3\phi_3(\mathrm{x};\mu_3,\Sigma_3) \qquad\qquad\qquad\qquad\quad .
\end{aligned}$$

The true intensity functions for the three sub-types are presented in Figure 1. This shows that the three sub-types share clustering in region 2 and type 1 and type 3 share clustering in region 3. Also, only points from type 2 and type 3 cluster about region 4 and region 1, respectively.

4.2. Posterior simulation. We apply our new algorithm (Alg.1) and the inverse-Lévy algorithm (Alg.2) by [**Kan11**] to a simulated data set from (4.1). In both algorithms, the possible number of jump points for gamma random fields is set to 1,000. The number of distinct jump points in Alg.1, denoted by M^*, is changing in the simulation, where M^* is equal to the number of distinct jump point indices $\{s_{i,j,l}\}$ in (3.6). Initially, $s_{i,j,l}$'s are randomly assigned to one of $\{1,\ldots,M\}$. Thus, $M^* = M = 1{,}000$ in iteration 1 in Alg.1. In Alg.2. the number of jump points is fixed at M. Both algorithms are run for 10,000 iterations in total with a burn-in of 3,000. We diagnose the simulation convergence by visual checking the trace plots of log-likelihood which show that the chains are mixing well in both algorithms after burn-in.

Figure 1 presents the estimated posterior mean intensity functions. We can see that the two algorithms provide very similar intensity estimates and they are both very closed to the true intensities. The intensity function estimates from Alg.2 are a bit more diffuse than those estimated by Alg.1. This is due to the difference in the number of jump points in the gamma random fields between the two algorithms, i.e. M^* is much smaller than M in the posterior simulation. The number of distinct jump points, M^*, has a posterior mean of 221.5 with a range of $[218, 226]$.

4.3. Performance. We simulate $K = 100$ data sets according to the model specifications in the previous section and fit each data sets by Alg.1 and Alg.2.

We measure the algorithm performance based on the sub-type average integrated mean square error (IMSE) and integrated weighted mean square error

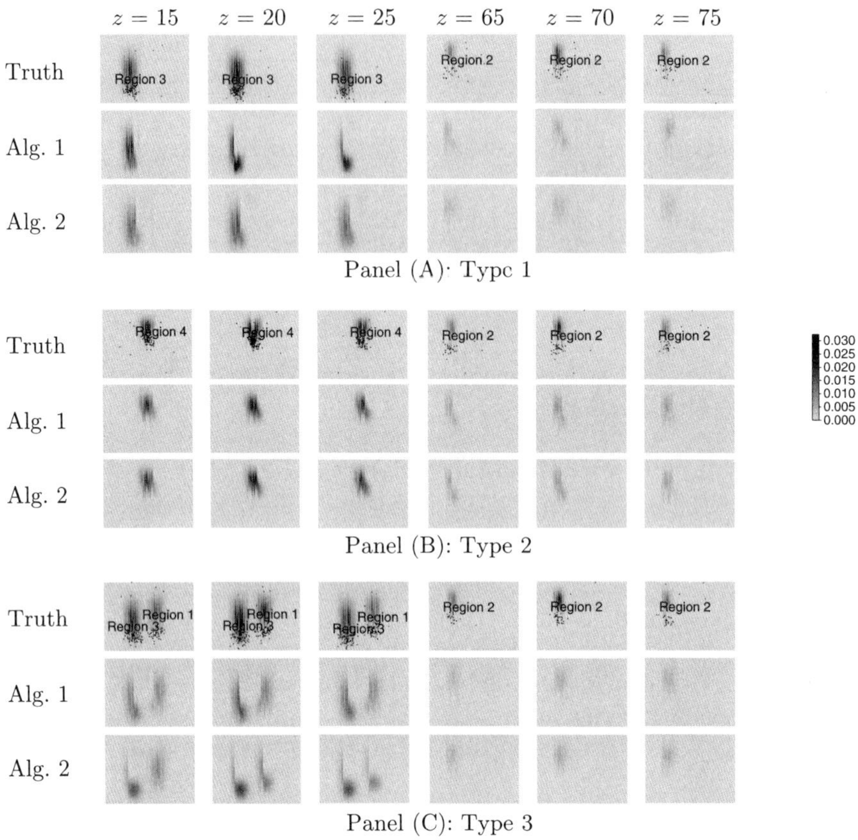

FIGURE 1. Panel (A), (B) and (C) respectively show intensities for type 1, 2 and 3 spatial point patterns in the simulation. In each panel, six x-y slices ($z = 15, 20, 25, 65, 70, 75$) are shown cutting through the high intensity regions (regions 1–4). The top row shows the true intensities along with the simulated point patterns. The middle and bottom row present the posterior mean intensities estimated by algorithms 1 and 2, respectively, where Alg.1 is the proposed algorithm and Alg.2 is the inverse-Lévy algorithm proposed by [**Kan11**].

(IWMSE), i.e.,

$$\text{IMSE} = \frac{1}{JK}\sum_{j=1}^{J}\sum_{k=1}^{K}\int_A [\tilde{\lambda}_{jk}(\mathrm{x}) - \lambda_j(\mathrm{x})]^2 d\mathrm{x}$$

$$\text{IWMSE} = \frac{1}{JK}\sum_{j=1}^{J}\sum_{k=1}^{K}\int_A [\tilde{\lambda}_{jk}(\mathrm{x}) - \lambda_j(\mathrm{x})]^2 \lambda_j(\mathrm{x}) d\mathrm{x}$$

where $\tilde{\lambda}_{jk}(\mathrm{x})$ is the estimated posterior mean intensity function for type j in simulation k and $\lambda_j(\mathrm{x})$ is the true intensity function. IWMSE gives more weight to regions with a large true intensity. The results are summarized in Table 2 which

TABLE 2. Comparison between two algorithms for the simulated data: IMSE, IWMSE and CPU times.

	Alg.1 (s.d.)	Alg.2 (s.d.)	Alg.1/Alg.2 (s.d.)
IMSE ($\times 10^{-7}$)	4.146 (0.825)	3.839 (0.724)	1.103 (0.225)
IWMSE ($\times 10^{-7}$)	0.028 (0.005)	0.028 (0.008)	1.076 (0.288)
CPU[a] times (hours)	0.521 (0.092)	3.112 (0.531)	0.171 (0.041)

[a]Produced by a Mac Pro with 2.7 GHz intel Core i7 processor and 16 GB of memory.

shows that Alg.1 is a bit less accurate than Alg.2. Alg.1 has around 10% inflation in both IMSE and IWMSE. However, Alg.1 is very computationally efficient where the computational cost is a mere 17% of Alg.2.

5. Meta-analysis of functional neuroimaging data

In this section, we apply our proposed algorithm to a functional neuroimaging meta-analysis of emotions first reported by [**KBJ$^+$08**], and further analyzed in [**KJNW11**] and [**Kan11**]. This data set consists of 164 fMRI and PET studies designed to determine brain activation elicited by five emotions (sad, happy, anger, fear, disgust). The goal is to identify consistent regions of activation across the different studies and types of emotion. In [**KJNW11**], we address this problem using a Bayesian hierarchical independent cluster process (BHICP) model. However, both the MKDA and the BHICP can only analyze a single group of spatial point patterns. Those models focus on finding consistent activation over a particular type of emotion. It is also of great interest to identify activation regions shared between different emotions. To this end, [**Kan11**] proposed the HPGRF model to account for the correlation between types and identifies consistent activations regions that are shared by the different types of emotion. In particular, a subset of the meta-analysis dataset including 219 studies and 1393 foci of the five emotions: sad, happy, anger, fear and disgust have been fitted by the HPGRF using the inverse-Lévy algorithm (Alg.2). The posterior estimates are based on 120,000 iterations of simulation with a burn-in of 20,000. The number of jump points in the model was set at 10,000. In the proposed algorithm, we set the initial number of distinct jump points to $M^* = 10{,}000$. We compare the intensity estimates by Alg.1 and Alg.2 in Figure 2. The two algorithms provide qualitatively similar intensity estimates. Although the intensity estimates obtained from Alg.2 are a little more spatially diffuse than those obtained from Alg.1. This is consistent with our simulation study results. However, the computational time of Alg.1 is 2.3 hours which is much smaller than the 33.1 hours needed by Alg.2. Both algorithms are written in C++ and implemented on a MacBook Pro with 16 GB of memory and a 2.7 GHz Intel Core i7 processor. Both our simulation study and real data analysis show that the proposed algorithm is more computationally efficient (about 15 times faster for the emotions data set) than, and is nearly as accurate as, the inverse-Lévy algorithm.

6. Discussion

In this work, we proposed a computationally efficient algorithm for the posterior simulation of the HPGRF model. The key is the stick breaking construction of the

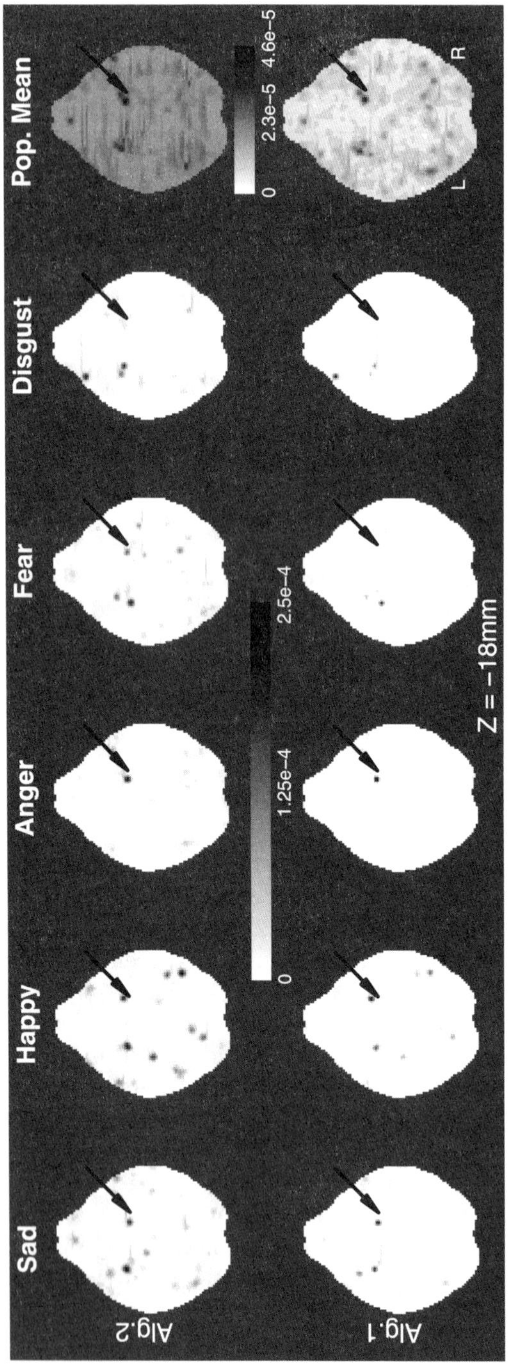

FIGURE 2. Posterior mean intensity estimates (top) by Alg.2 compared with the Alg.1 (bottom). Here we only show one axial slice (Z = −18 mm) of the fully 3D results for the five emotions (sad, happy, anger, fear and disgust). The arrows point to the right amygdala. All intensity functions have units of expected foci/mm^3.

HPGRF model and the introduction of a set of auxiliary variables to truncate the weights of the Gamma process. This algorithm is in a similar fashion with the slice sampler for the mixture of Dirichlet processes proposed by [**Wal07**]. It is remarkably computationally efficient compared to our previous algorithm based on the inverse-Lévy construction [**Kan11**]. Furthermore, our simulation study provides evidence that it produces good estimates of the intensity functions for different sub-types as well as the population level intensity function.

One possible extension is to modify the priors for the weights in the stick-breaking construction to make the simulation more efficient as suggested by [**KGW11**] for the slice sampling algorithm for a mixture of Dirichlet processes. To further improve computational efficiency, another direction is to develop a parallel algorithm using general purpose graphical process unit (GPGPU) techniques. [**SWC$^+$10**] proposed to use GPGPU algorithm for massive Bayesian mixture models. The HPGRF model has this structure and the GPGPU techniques can be readily applied.

Another future direction is to model more flexible dependence between different types of spatial point patterns. Potentially, this can be achieved by constructing dependent Poisson Gamma random fields using an analogous method proposed by MacEachern for dependent Dirichlet processes [**Mac99**]. This framework may provide better fitting of multi-type neuroimaging meta-analysis data that are heterogeneous in their covariance structure.

Acknowledgements. The authors would like to thank the United States National Institutes of Health for their support. Professor Johnson is partially supported by the NINDS grant number 5-R01-NS-075066-02. The work presented in this manuscript represents the views of the authors and not necessarily that of the NINDS. The authors would also like to thank Professor Tor Wager, University of Colorado, and Professor Lisa Feldman Barrett, Northeastern University, for their generous sharing of the emotions meta-analysis data set.

References

[BTMH05] A. Baddeley, R. Turner, J. Møller, and M. Hazelton, *Residual analysis for spatial point processes. With discussion and a reply by the authors*, J. R. Stat. Soc. Ser. B Stat. Methodol. **67** (2005), no. 5, 617–666, DOI 10.1111/j.1467-9868.2005.00519.x. MR2210685

[Bad10] A. Baddeley, *Multivariate and marked point processes*, Handbook of spatial statistics, Chapman & Hall/CRC Handb. Mod. Stat. Methods, CRC Press, Boca Raton, FL, 2010, pp. 371–402, DOI 10.1201/9781420072884-c21. MR2730956

[BIW00] N. G. Best, K. Ickstadt, and R. L. Wolpert, *Spatial Poisson regression for health and exposure data measured at disparate resolutions*, J. Amer. Statist. Assoc. **95** (2000), no. 452, 1076–1088, DOI 10.2307/2669744. MR1821716

[BIW$^+$02] N. G. Best, K. Ickstadt, R. L. Wolpert, S. Cockings, P. Elliott, J. Bennett, A. Bottle, and S. Reed, *Modeling the impact of traffic-related air pollution on childhood respiratory illness. With a discussion by F. Dominici and J. Samet and a rejoinder by the authors*, Case studies in Bayesian statistics, Vol. V (Pittsburgh, PA, 1999), Lecture Notes in Statist., vol. 162, Springer, New York, 2002, pp. 183–259, DOI 10.1007/978-1-4613-0035-9_3. MR1931868

[BM01] A. Brix and J. Møller, *Space-time multi type log Gaussian Cox processes with a view to modelling weeds*, Scand. J. Statist. **28** (2001), no. 3, 471–488, DOI 10.1111/1467-9469.00249. MR1858412 (2002g:60072)

[BM96] C. A. Bush and S. N. MacEachern, *A semiparametric Bayesian model for randomised block designs*, Biometrika **83** (1996), no. 2, 275–285, DOI 10.1093/biomet/83.2.275.

[Cox55] D. R. Cox, *Some statistical methods connected with series of events*, J. Roy. Statist. Soc. Ser. B. **17** (1955), 129–157; discussion, 157–164. MR0092301 (19,1094d)

[Dig81] P. J. Diggle, *Some graphical methods in the analysis of spatial point patterns*, Interpreting Multivariate Data (Sheffield, 1980) (V. Barnett, ed.), Wiley Ser. Probab. Math. Statist., Wiley, Chichester, 1981, pp. 55–73.

[Dig83] P. J. Diggle, *Statistical analysis of spatial point patterns*, Mathematics in Biology, Academic Press Inc. [Harcourt Brace Jovanovich Publishers], London, 1983. MR743593 (85m:62205)

[DM83] P. J. Diggle and R. K. Milne, *Bivariate Cox processes: some models for bivariate spatial point patterns*, J. Roy. Statist. Soc. Ser. B **45** (1983), no. 1, 11–21. MR701070 (84g:62140)

[DR94] P. J. Diggle and B. S. Rowlingson, *A conditional approach to point process modelling of elevated risk*, J. Roy. Statist. Soc. Ser. A **157** (1994), no. 3, 433–440.

[ELG$^+$09] S. B. Eickhoff, A. R. Laird, C. Grefkes, L. E. Wang, K. Zilles, and P. T. Fox, *Coordinate-based activation likelihood estimation meta-analysis of neuroimaging data: a random-effects approach based on empirical estimates of spatial uncertainty*, Human Brain Mapping **30** (2009), no. 9, 2907–2926, DOI 10.1002/hbm.20718.

[Esc95] M. D. Escobar, *Nonparametric Bayesian methods in hierarchical models*, J. Statist. Plann. Inference **43** (1995), no. 1-2, 97–106, DOI 10.1016/0378-3758(94)00011-J. MR1314130

[EW95] M. D. Escobar and M. West, *Bayesian density estimation and inference using mixtures*, J. Amer. Statist. Assoc. **90** (1995), no. 430, 577–588. MR1340510 (96d:62054)

[Fer73] T. S. Ferguson, *A Bayesian analysis of some nonparametric problems*, Ann. Statist. **1** (1973), 209–230. MR0350949 (50 #3441)

[IW99] K. Ickstadt and R. L. Wolpert, *Spatial regression for marked point processes*, Bayesian statistics, 6 (Alcoceber, 1998), Oxford Univ. Press, New York, 1999, pp. 323–341. MR1723503

[IJ01] H. Ishwaran and L. F. James, *Gibbs sampling methods for stick-breaking priors*, J. Amer. Statist. Assoc. **96** (2001), no. 453, 161–173, DOI 10.1198/016214501750332758. MR1952729 (2004d:62018)

[JN04] S. Jain and R. M. Neal, *A split-merge Markov chain Monte Carlo procedure for the Dirichlet process mixture model*, J. Comput. Graph. Statist. **13** (2004), no. 1, 158–182, DOI 10.1198/1061860043001. MR2044876

[KGW11] M. Kalli, J. E. Griffin, and S. G. Walker, *Slice sampling mixture models*, Stat. Comput. **21** (2011), no. 1, 93–105, DOI 10.1007/s11222-009-9150-y. MR2746606

[Kan11] J. Kang, *Some Novel Spatial Stochastic Models for Functional Neuroimaging Analysis*, ProQuest LLC, Ann Arbor, MI, 2011. Thesis (Ph.D.)–University of Michigan. MR2995940

[KJNW11] J. Kang, T. D. Johnson, T. E. Nichols, and T. D. Wager, *Meta analysis of functional neuroimaging data via Bayesian spatial point processes*, J. Amer. Statist. Assoc. **106** (2011), no. 493, 124–134, DOI 10.1198/jasa.2011.ap09735. MR2816707 (2012d:62372)

[KBJ$^+$08] H. Kober, L. F. Barrett, J. Joseph, E. Bliss-Moreau, K. Lindquist, and T. D. Wager, *Functional grouping and cortical–subcortical interactions in emotion: A meta-analysis of neuroimaging studies*, NeuroImage **42** (2008), no. 2, 998–1031, DOI 10.1016/j.neuroimage.2008.03.059.

[LBB$^+$00] A. B. Lawson, A. B. Biggeri, D. Boehning, E. Lesaffre, J-F. Viel, A. Clark, P. Schlattmann, and F. Divino, *Disease mapping models: an empirical evaluation*, Stat. Med. **19** (2000), no. 17-18, 2217–2241, DOI 10.1002/1097-0258(20000915/30)19:17/18<2217::AID-SIM565>3.0.CO;2-E.

[Mac99] S. N. MacEachern, *Dependent nonparametric processes*, 1999 Proceedings of the Section on Bayesian Statistical Science (Baltimore, MD, 1999), Amer. Statist. Assoc., Alexandria, VA, 1999, pp. 50–55.

[Nea00] R. M. Neal, *Markov chain sampling methods for Dirichlet process mixture models*, J. Comput. Graph. Statist. **9** (2000), no. 2, 249–265, DOI 10.2307/1390653. MR1823804

[NF10] A. Niemi and C. Fernández, *Bayesian spatial point process modeling of line transect data*, J. Agric. Biol. Environ. Stat. **15** (2010), no. 3, 327–345, DOI 10.1007/s13253-010-0024-8. MR2787262

[SP00] D. Stoyan and A. Penttinen, *Recent applications of point process methods in forestry statistics*, Statist. Sci. **15** (2000), no. 1, 61–78, DOI 10.1214/ss/1009212674. MR1842237

[SWC+10] M. A. Suchard, Q. Wang, C. Chan, J. Frelinger, A. Cron, and M. West, *Understanding GPU programming for statistical computation: studies in massively parallel massive mixtures*, J. Comput. Graph. Statist. **19** (2010), no. 2, 419–438, DOI 10.1198/jcgs.2010.10016. MR2758309

[TJBB06] Y. W. Teh, M. I. Jordan, M. J. Beal, and D. M. Blei, *Hierarchical Dirichlet processes*, J. Amer. Statist. Assoc. **101** (2006), no. 476, 1566–1581, DOI 10.1198/016214506000000302. MR2279480

[vLB99] M. N. M. van Lieshout and A. J. Baddeley, *Indices of dependence between types in multivariate point patterns*, Scand. J. Statist. **26** (1999), no. 4, 511–532, DOI 10.1111/1467-9469.00165. MR1734259 (2001a:62123)

[WLN+09] T. D. Wager, M. A. Lindquist, T. E. Nichols, H. Kober, and J. X. Van Snellenberg, *Evaluating the consistency and specificity of neuroimaging data using meta-analysis*, NeuroImage **45** (2009), no. 1, Suppl. 1, S210–S221, DOI 10.1016/j.neuroimage.2008.10.061.

[Wal07] S. G. Walker, *Sampling the Dirichlet mixture model with slices*, Comm. Statist. Simulation Comput. **36** (2007), no. 1-3, 45–54, DOI 10.1080/03610910601096262. MR2370888

[WI98] R. L. Wolpert and K. Ickstadt, *Poisson/gamma random field models for spatial statistics*, Biometrika **85** (1998), no. 2, 251–267, DOI 10.1093/biomet/85.2.251. MR1649114 (99k:62110)

[WWO10] D. B. Woodard, R. L. Wolpert, and M. A. O'Connell, *Spatial inference of nitrate concentrations in groundwater*, J. Agric. Biol. Environ. Stat. **15** (2010), no. 2, 209–227, DOI 10.1007/s13253-009-0006-x. MR2787272

Department of Biostatistics and Bioinformatics, Emory University, Atlanta, Georgia 30322

E-mail address: jian.kang@emory.edu

Department of Biostatistics, University of Michigan, Ann Arbor, Michigan 48109

E-mail address: tdjtdj@umich.edu

Contemporary Mathematics
Volume **622**, 2014
http://dx.doi.org/10.1090/conm/622/12435

A new penalized quasi-likelihood approach for estimating the number of states in a hidden Markov model

Annaliza McGillivray and Abbas Khalili

ABSTRACT. In this paper, we propose a new penalized quasi-likelihood method for estimating the number of hidden states (or order) of a hidden Markov model (HMM). Our method makes use of the fact that the marginal distribution of the HMM observations is a finite mixture model. The new method starts with a HMM with a large number of states, and obtains a model of lower order by clustering and combining similar states through two penalty functions. The method is computationally efficient, and unlike most of the existing methods such as AIC and BIC, it obtains the order of a HMM in a single optimization procedure without examining several submodels. We study some theoretical properties of the method, and assess its performance via extensive simulation studies. Two well-known data sets are also analyzed to demonstrate the use of the new method.

1. Introduction

Hidden Markov models (HMMs) have been widely used for modeling heterogeneous and dependent data. They have been applied in fields such as engineering [**LRS**], biology [**C2**], medicine [**A2**] and finance [**RTÅ**]. The problem of parameter estimation in HMMs has been extensively studied in the case where the number of hidden states (or order) of the model is known. The two influential papers by [**L1**] and [**BRR**] studied the consistency and asymptotic normality of the maximum likelihood estimators.

While in some applications, the number of states of a HMM to be fitted is clear from the background of the problem, we may not always have sufficient scientific understanding to determine the number of states. For example, in the analysis of epileptic seizure count data [**A2**, **ZM**], a model of order 2 seems the most biologically plausible to represent the periods of high and low seizure activity undergone by a patient afflicted with epilepsy. However, in the analysis of DNA sequences [**C2**, **PZ**], the order of the model, which corresponds to the number of segments reflecting distinct regions of base composition on a DNA molecule, is not immediately apparent. In this case, it must be estimated from the data, which may consist

2010 *Mathematics Subject Classification.* Primary 62M05; Secondary 65F22.
Key words and phrases. Hidden Markov model; penalized quasi-likelihood; mixture model.
For this research the first author was supported by the Fonds de recherche du Québec – Nature et technologies (FRQNT).
The second author was supported by the Natural Sciences and Engineering Research Council of Canada (NSERC), and the Fonds de recherche du Québec – Nature et technologies (FRQNT).

of thousands of base-pairs on a DNA molecule. In general, a HMM with a large number of states is likely to provide an adequate fit to the data, but it may lack interpretability. Simple models are often favoured in applications to avoid overfitting the data. Order estimation in HMMs, where the order is the minimal number of hidden states, is thus a problem of significant practical and theoretical interest.

The order estimation methods that have been proposed thus far have utilized either the full-model likelihood function, or a so-called quasi-likelihood function which is based on the marginal mixture distribution of a HMM and neglects the Markov dependence of the observations. Information-theoretic approaches, such as the Akaike information criterion (AIC) [**A1**] and the Bayesian information criterion (BIC) [**S**], have been used to estimate the order of a HMM. [**R**] applied AIC and BIC based on the full-model likelihood and showed that they do not underestimate the true order asymptotically. [**GB**] also investigated penalized likelihood estimators of the order based on the full-model likelihood. On the other hand, [**PZ**] used AIC and BIC based on the quasi-likelihood. They showed that BIC based on quasi-likelihood inference provides a consistent estimator of the true order. Another method that has been considered is based on a penalized distance measure between the empirical distribution function of the observed data and the fitted cumulative distribution function, proposed in [**M**]. [**DH**] approached this problem from the point of view of hypothesis-testing on the order of a HMM. They proposed testing the hypothesis of two states by extending the modified likelihood ratio test of [**CCK2**] for testing two components in a finite mixture model. Working within a Bayesian framework, [**RRT**] proposed a reversible jump Markov chain Monte Carlo algorithm, and [**FS**] selected the number of states by comparing marginal likelihoods. While all of the existing methods are advantageous in certain aspects, none of them are proved to be optimal in all situations. Some methods might have computational advantages, while others might have a higher probability in selecting the most suitable model.

In this paper, we propose a new method for order estimation in HMMs that is an extension of the modified smoothly clipped absolute deviation (MSCAD) procedure of [**CK2**] for estimating the number of components in a finite mixture model. Through the use of a quasi-likelihood function, we reduce the problem of estimating the number of states of a HMM to estimating the number of components of its marginal mixture distribution. We therefore call the new method MSCAD_Q. The new method is based on the maximization of a penalized quasi-likelihood, which incorporates the penalty function from the modified likelihood ratio test of [**CCK2**] as well as the smoothly clipped absolute deviation (SCAD) penalty proposed by [**FL**] for variable selection in multiple regression models. The penalty from the modified likelihood ratio test clusters the states of the fitted model around the states of the true model. The SCAD penalty will then combine the states in each cluster into a single state. Therefore, the method starts off with a HMM with a large number of states and obtains a model of lower order by clustering and combining similar states. The advantage of using the quasi-likelihood in our method is that it simplifies the optimization problem since the objective function is only a function of the parameters of the marginal mixture distribution of the HMM. Additionally, there are computational gains that arise from using the quasi-likelihood. Given a sample of size n, the complexity of the efficient evaluation of the full-model likelihood of a K-state HMM is $O(nK^2)$, whereas that of the quasi-likelihood is $O(nK)$.

The rest of the paper is organized as follows. In Section 2, we formally define HMMs. In Section 3, we introduce the new method MSCAD_Q. We present some of the asymptotic properties of the method in Section 4. In Section 5, we evaluate the performance of the method against AIC and BIC, based on both the full-model likelihood and the quasi-likelihood. We then demonstrate the use of the method with the analysis of two well-known data sets in Section 6. Finally, in Section 7, we provide a discussion and conclusion.

2. Hidden Markov models

A hidden Markov model (HMM) is a bivariate stochastic process $\{Y_t, Z_t; t = 1, 2, \dots\}$, where $\{Y_t; t = 1, 2, \dots\}$ is an observed process and $\{Z_t; t = 1, 2, \dots\}$ is an unobserved process such that (i) $\{Z_t; t = 1, 2, \dots\}$ follows a finite-state Markov chain, and (ii) given $\{Z_t; t = 1, 2, \dots\}$, the sequence of random variables $\{Y_t; t = 1, 2, \dots\}$ are conditionally independent with the conditional distribution of Y_t depending only on Z_t. We let $\{Z_t; t = 1, 2, \dots\}$ be stationary with discrete state space $\{1, 2, \dots, K\}$, homogeneous transition probabilities $p_{ij} = P(Z_t = j \mid Z_{t-1} = i)$ and initial distribution $\boldsymbol{\alpha} = (\alpha_1, \alpha_2, \dots, \alpha_K)$, which is the stationary distribution.

The conditional distributions of Y_t given Z_t are called state-dependent distributions; we assume that their corresponding (probability) density functions belong to some parametric family $\Gamma = \{g(y; \theta) : \theta \in \Theta\}$, where $\Theta \subseteq \mathbb{R}$. Since $\{Z_t\}$ is stationary, it is well-known that the marginal distribution of Y_t is a finite mixture with density function given by

$$f(y_t; G) = \sum_{k=1}^{K} \alpha_k g(y_t; \theta_k), \tag{2.1}$$

where $G(\cdot)$ is called the mixing distribution and is given by

$$G(\theta) = \sum_{k=1}^{K} \alpha_k I(\theta_k \leq \theta). \tag{2.2}$$

Here $I(\cdot)$ is an indicator function, $\theta_1, \theta_2, \dots, \theta_K \in \Theta$ are atoms or support points of G, and $\alpha_1, \alpha_2, \dots, \alpha_K$ satisfy $\alpha_k > 0$ and $\sum_{k=1}^{K} \alpha_k = 1$. In contrast to the case of a finite mixture model, $Y_1, Y_2, \dots, Y_n$ are marginally dependent. A HMM can thus be viewed as a generalization of a finite mixture model.

Let K_0 be the smallest number of states such that the stationary mixing proportions α_k are positive and the state-dependent parameters θ_k are distinct. We call K_0 the true order of the HMM.

One difficulty associated with HMMs is their identifiability. In general, a parametric model is said to be identifiable if different values of the parameter generate different probability distributions. HMMs are not identifiable in the strict sense since the labels of the hidden states can be permuted without changing the distribution of $\{Y_t\}$. To ensure model identifiability, we shall assume that $\{Z_t\}$ is ergodic, which would imply that its stationary distribution $\boldsymbol{\alpha} = (\alpha_1, \alpha_2, \dots, \alpha_K)$ is uniquely determined from the transition matrix $\mathbb{P} = (p_{ij})$. We shall also assume that the family of finite mixtures based on $\Gamma = \{g(y; \theta) : \theta \in \Theta\}$ is identifiable, i.e., equality of the density functions $f(y; G) = f(y; G')$, for any y, implies that $G = G'$ up to a permutation of the labels of the hidden states. This condition is satisfied for mixture models based on well-known families of distributions such as normal, Poisson and binomial.

3. The new method for order estimation: $\mathbf{MSCAD_Q}$

Let $y_1, y_2, \ldots, y_n$ be a sample of observations from a HMM. Given a known upper bound K for the order of the HMM, the quasi-likelihood function of the marginal mixing distribution (2.1) is given by

$$\mathcal{L}_n^{\mathrm{Q}}(G) = \prod_{t=1}^{n} f(y_t; G), \tag{3.1}$$

which neglects the dependence between the observations y_t. Note that, given K, the quasi-likelihood in (3.1) is only a function of the parameters $(\alpha_k, \theta_k), k = 1, 2, \ldots, K$, and for simplicity in notation we show this dependency by writing it as a function of G.

While the true order K_0 may be unknown, we assume that some information is available on an upper bound K of K_0. Typically researchers or analysts have some idea of possible orders. In the analysis of DNA sequences [**C3**], for example, scientists may suspect that a particular gene has two distinct regions of dinucleotide composition. In this case, a possible upper bound on the order of the HMM is four. In our simulations and data analysis, we use an upper bound $K = 15$. Since the new method does not require the fitting of each of the candidate submodels, such a large upper bound can be used in practice.

By maximizing the quasi-likelihood function in (3.1), the resulting fitted marginal mixture distribution may suffer from two types of overfitting, namely (i) estimated values of some α_k that are near 0 and/or (ii) some state-dependent parameters θ_k that are too close to one another. To prevent these two types of overfitting, we maximize an objective function that incorporates a penalty on the mixing proportions α_k as well as a penalty on the differences in parameters θ_k of the marginal mixture distribution.

We shall assume that $\theta_1 \leq \theta_2 \leq \cdots \leq \theta_K$ and set $\beta_k = \theta_{k+1} - \theta_k$ for $k = 1, 2, \ldots, K-1$. Our proposed method estimates the number of states by maximizing the penalized quasi-log-likelihood function

$$\tilde{\ell}_n^{\mathrm{Q}}(G) = \ell_n^{\mathrm{Q}}(G) + C_K \sum_{k=1}^{K} \log \alpha_k - \sum_{k=1}^{K-1} p_n(\beta_k), \tag{3.2}$$

where $\ell_n^{\mathrm{Q}}(G) = \log \mathcal{L}_n^{\mathrm{Q}}(G)$; for some positive constant C_K and the SCAD penalty function $p_n(\beta_k)$ which is often defined through its first derivative

$$p_n'(\beta) = \left\{ \gamma_n \sqrt{n} I\{\sqrt{n}|\beta| \leq \gamma_n\} + \frac{\sqrt{n}(a\gamma_n - \sqrt{n}|\beta|)_+}{(a-1)} I(\sqrt{n}|\beta| > \gamma_n) \right\} \operatorname{sign}(\beta)$$

for some tuning parameters $\gamma_n > 0$ and $a > 2$, where $I(\cdot)$ is an indicator function and $(\cdot)_+$ denotes the positive part of a quantity. The first penalty function in (3.2) forces the estimated values of α_k away from the boundary point 0 to prevent the first type of overfitting. As in the multiple regression context, the SCAD penalty shrinks the values of the differences β_k that are close to 0 to exactly 0, preventing the second type of overfitting. [**FL**] showed that the value $a = 3.7$ minimizes a Bayes risk criterion for regression parameter β, and they argued that this choice of a provides good practical performance in various variable selection problems. In this paper, we used the SCAD penalty with $a = 3.7$, and it worked well in our simulation studies. The choice of γ_n is discussed later on in this paper.

We denote the maximizer of the penalized quasi-log-likelihood function $\tilde{\ell}_n^{\mathrm{Q}}(G)$ in (3.2) by $\widehat{G}_n$, which we refer to as the maximum penalized quasi-likelihood estimator (MPQLE) of G. We use the modified EM algorithm proposed in [**CK2**] for our numerical computations. With $\beta_k = \theta_{k+1} - \theta_k$ for $k = 1, 2, \ldots, K-1$, the constraints $\theta_1 \le \theta_2 \le \cdots \le \theta_K$ are incorporated in the modified EM algorithm by restricting β_k to be non-negative, which is done by taking $\max(0, \hat{\beta}_k)$ as the estimate of β_k.

When some $\hat{\beta}_k$ are 0, then $\hat{\theta}_k = \hat{\theta}_{k+1}$. Thus, the actual order of the fitted HMM will be smaller than the upper bound K since the number of distinct $\hat{\theta}_k$ is taken to be the estimated order of the HMM, according to the definition of the order in Section 2. For illustrative purposes, we applied $\mathrm{MSCAD_Q}$ with $K = 15$ to a sample of 100 observations, generated from the 2-state normal HMM with $\boldsymbol{\theta} = (\theta_1, \theta_2) = (0, 3)$, common standard deviation $\sigma = 1$, and dependence structure

$$\mathbb{P} = \begin{pmatrix} 0.90 & 0.10 \\ 0.20 & 0.80 \end{pmatrix}, \quad \boldsymbol{\alpha} = \left(\tfrac{2}{3}, \tfrac{1}{3}\right).$$

In Figure 1, we plot the state index against the estimates of the state-dependent parameters θ_k to demonstrate the role of the penalty functions in the modified EM algorithm at different iterations. For the simulated sample, the algorithm converged in 17 iterations. A HMM was initialized with distinct θ_k's, but as the

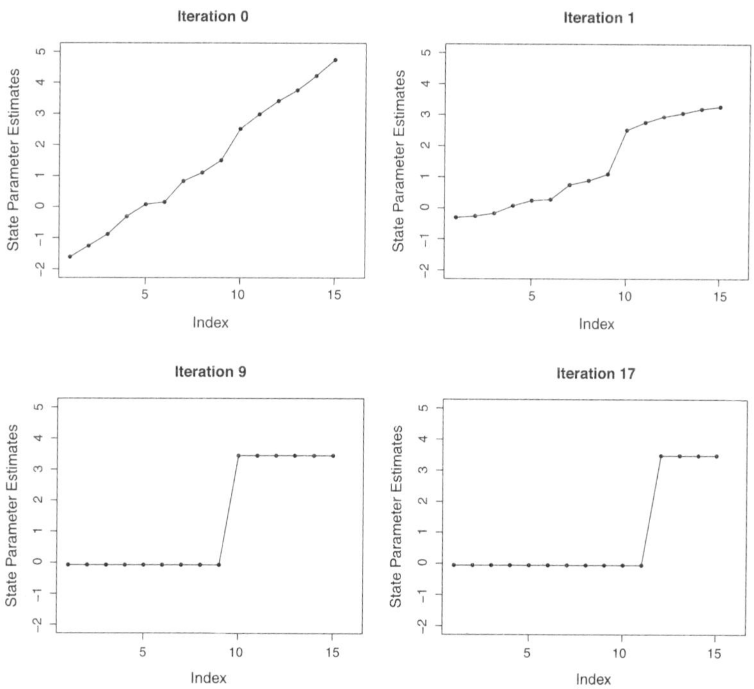

FIGURE 1. The initial values and estimates of the state-dependent parameters θ_k at different iterations of the modified EM algorithm used in $\mathrm{MSCAD_Q}$.

number of iterations increased, the estimates of θ_k were clustered; in particular, 11 initial states were combined into a single state with mean close to 0, and the other 4 states were combined into a single state with mean close to 3. The first penalty kept the mixing proportions away from 0, forming clusters of states with means around either 0 or 3. The second penalty combined clusters of similar states, resulting in a HMM with two states.

4. Asymptotic study

To study some of the asymptotic properties of the MPQLE, we must first establish some notation. Let the class of all finite mixing distributions with at most K support points be given by

$$\mathcal{M}_K = \left\{ G(\theta) = \sum_{k=1}^{K} \alpha_k I(\theta_k \le \theta) : \theta_1 \le \theta_2 \le \cdots \le \theta_K, \sum_{k=1}^{K} \alpha_k = 1, \alpha_k > 0 \right\}.$$

Note that $\mathcal{M}_1 \subseteq \mathcal{M}_2 \subseteq \cdots \subseteq \mathcal{M}_{K-1} \subseteq \mathcal{M}_K$ since the atoms θ_k are allowed to be equal with positive mixing proportions.

The true mixing distribution G_0 is given by

$$G_0(\theta) = \sum_{k=1}^{K_0} \alpha_{0k} I(\theta_{0k} \le \theta),$$

where $\theta_{01} < \cdots < \theta_{0K_0}$ are K_0 distinct interior points of Θ and $(\alpha_{01}, \ldots, \alpha_{0K_0})$ are the true stationary mixing proportions.

We let $\widehat{G}_n(\theta) = \sum_{j=1}^{K} \hat{\alpha}_j I(\hat{\theta}_j \le \theta)$. Following [**CK2**], we then define $I_k = \{j : \theta_{0,k-1} + \theta_{0,k} < 2\hat{\theta}_j \le \theta_{0,k} + \theta_{0,k+1}\}$ for $k = 1, 2, \ldots, K_0$ with $\theta_{0,0} = -\infty$ and $\theta_{0,K_0+1} = \infty$, and

$$\widehat{H}_k(\theta) = \frac{\sum_{j \in I_k} \hat{\alpha}_j I(\hat{\theta}_j \le \theta)}{\sum_{j \in I_k} \hat{\alpha}_j}$$

so that with $\hat{\pi}_k = \sum_{j \in I_k} \hat{\alpha}_j$,

$$\widehat{G}_n(\theta) = \sum_{k=1}^{K_0} \hat{\pi}_k \widehat{H}_k(\theta).$$

Note that $\hat{\pi}_1$ is the probability assigned to the support points $\hat{\theta}_j \le (\theta_{01} + \theta_{02})/2$, $\hat{\pi}_2$ is the probability assigned to the support points $(\theta_{01} + \theta_{02})/2 < \hat{\theta}_j \le (\theta_{02} + \theta_{03})/2$, and so on.

The aim of our proposed method is to cluster the atoms of $\widehat{H}_k$ into a small neighbourhood of θ_{0k} using the penalty on the mixing proportions α_k and then to merge the cluster of atoms into a single atom using the SCAD penalty. In Theorem 4.1, we show that the MPQLE $\widehat{G}_n$ is a consistent estimator of the true marginal mixture distribution G_0. The proof is given in Appendix A.

THEOREM 4.1. *Let $Y_1, Y_2, \ldots, Y_n$ be a sample from a homogeneous, stationary and identifiable HMM satisfying Assumptions* A.1 *to* A.6 *in Appendix* A. *Let $f(y; G_0)$ denote the true density function of the marginal mixture distribution of Y_t. Suppose that we apply the SCAD penalty with $\gamma_n = n^{1/4} \log n$. Then*

(a) *$\widehat{G}_n$ is a consistent estimator of G_0*

(b) *all support points of* $\widehat{H}_k$ *converge in probability to* θ_{0k} *for each* $k = 1, 2, \ldots, K_0$.

Note that Theorem 4.1 does not show that the MPQLE is consistent in estimating the true order K_0. The order may still be overestimated if $\widehat{H}_k$ has more than one atom. We have yet to be able to prove the consistency of MSCAD_Q in estimating the true order of the HMM and must therefore defer this task to future work. The proof requires the extension of the results of [**C1**] to HMMs, which is currently under investigation. The results of Theorem 4.1 can be applied to HMMs based on distributions such as Poisson, normal with known σ, and binomial.

5. Simulation studies

We compared the performance of MSCAD_Q with the two information criteria AIC and BIC, based on the full-model likelihood, and AIC_Q and BIC_Q, based on the quasi-likelihood, for order estimation in normal and Poisson HMMs. We report the percentage of times out of 500 replications that the estimated order equals a given value of K with sample sizes $n = 100, 400$. To select the tuning parameter γ_n, we used cross-validation for samples of size $n = 100$ and 5-fold cross-validation for samples of size $n = 400$. Based on trial runs, we took [0.2, 1.5] as the range of $\gamma_n/\sqrt{n}$ for the normal HMMs and [0.4, 1.6] for the Poisson HMMs. These ranges are in accordance with the theory, which is currently being investigated, on the order of γ_n for the sample sizes under consideration and worked well in our simulations. We let $C_K = \log 25$, following the suggestion of [**CCK1**], and we use the upper bound $K = 15$ for all models in this study. The initial stationary distribution is taken to be $\boldsymbol{\alpha}^{(0)} = (\alpha_1^{(0)}, \alpha_2^{(0)}, \ldots, \alpha_K^{(0)}) = (1/K, 1/K, \ldots, 1/K)$. Twenty initial values are used for $\boldsymbol{\theta} = (\theta_1, \theta_2, \ldots, \theta_K)$, which are obtained by taking the fifteen sample quantiles of bootstrap samples generated from each simulated data set.

To compute the AIC and BIC values based on the full-model likelihood, we used the R software code provided in [**ZM**] for fitting a stationary HMM by direct numerical maximization of the likelihood function. Their program is for the case of state-dependent Poisson distributions, but can be easily altered for the case of state-dependent normal distributions. Maximum quasi-likelihood inference was done with the EM algorithm.

The performance of the methods are assessed for HMMs of order 2, 3, 4 and 6 with the dependence structures S1–S10 displayed in Tables 6 and 7 in Appendix B. Notice that from S1 to S3 in Table 6, the stationary distribution $\boldsymbol{\alpha}$ is the same, but the transition matrices $\mathbb{P}$ are different. For the 2-state HMM with dependence structure S1, there is no expected correlation between the observations. For the 2-state HMMs with dependence structures S2 and S3, on the other hand, correlation is expected between the observations. Therefore, we will be able to observe the influence of the different dependence structures on the performance of the methods under study.

5.1. Simulation study 1. For the normal HMMs under consideration, the marginal distribution of the observations is given by the finite mixture

$$f(y; G) = \sum_{k=1}^{K_0} \frac{\alpha_k}{\sigma} \phi\left(\frac{y - \theta_k}{\sigma}\right),$$

where $\phi(\cdot)$ is the standard normal density function. We consider the scale parameter σ unknown. We generated data from ten normal HMMs with $\sigma = 1$ and dependence structures S1-S10. The simulation results are reported in Tables 8–10.

We let $\boldsymbol{\theta} = (\theta_1, \theta_2, \ldots, \theta_{K_0})$. Models 1–3 have $K_0 = 2$, $\boldsymbol{\theta} = (0, 3)$ and dependence structures S1–S3 given in Table 6. The parameters of the marginal mixture distributions are the same as those in the simulation studies of [**IJS**] for 2-component normal mixtures. For sample size $n = 100$, MSCAD$_\text{Q}$ performs reasonably well in estimating the true order of the HMM (see Table 8). It outperforms AIC and BIC$_\text{Q}$ in all cases. It also appears that the performance of MSCAD$_\text{Q}$ is not significantly affected by the dependence structure. Once dependence within the observations is introduced, the number of times MSCAD$_\text{Q}$ selects the true order decreases, but only slightly.

For models 4 and 5, which have $K_0 = 3$, $\boldsymbol{\theta} = (0, 3, 6)$ and dependence structures S4 and S5 found in Table 6, MSCAD$_\text{Q}$ outperforms the other methods in both cases (see Table 8).

Models 6 and 7 have $K_0 = 4$, $\boldsymbol{\theta} = (0, 3, 6, 9)$ and dependence structures S6 and S7 given in Table 6. For sample sizes $n = 100, 400$, MSCAD$_\text{Q}$ is indisputably the best. BIC and BIC$_\text{Q}$ both have the tendency to underfit, often selecting models of order 2 (see Table 9).

Under models 8–10, which have $K_0 = 6$, $\boldsymbol{\theta} = (0, 3, 6, 9, 12, 15)$ and dependence structures S8-S10 found in Table 7, AIC, AIC$_\text{Q}$ and MSCAD$_\text{Q}$ all perform reasonably well. BIC and BIC$_\text{Q}$ often select models of order 2 or 3 (see Table 10).

5.2. Simulation study 2. Here we consider simulations with Poisson HMMs. The marginal density of the HMM observations is given by

$$f(y; G) = \sum_{k=1}^{K_0} \alpha_k \left(\frac{e^{-\theta_k} \theta_k^y}{y!} \right).$$

We simulated data from ten Poisson HMMs with dependence structures S1-S10. The simulation results can be found in Tables 11–13.

We let $\boldsymbol{\theta} = (\theta_1, \theta_2, \ldots, \theta_{K_0})$. Models 1–3 have $K_0 = 2$, $\boldsymbol{\theta} = (1, 9)$ and dependence structures S1–S3 found in Table 6. For sample size $n = 100$, BIC performs the best, BIC$_\text{Q}$ performs the second best, and MSCAD$_\text{Q}$ performs better than AIC in all cases in estimating the true order of the HMM (see Table 11). It also appears that MSCAD$_\text{Q}$ is not affected by the dependence structure.

For models 4 and 5, which have $K_0 = 3$, $\boldsymbol{\theta} = (1, 5, 10)$ and dependence structures S4 and S5 given in Table 6, MSCAD$_\text{Q}$ is the best in both cases. Both BIC and BIC$_\text{Q}$ have the tendency to underestimate the order of the HMM (see Table 11).

Models 6 and 7 have $K_0 = 4$, $\boldsymbol{\theta} = (1, 5, 10, 15)$ and dependence structures S6 and S7 given in Table 6. For both sample sizes $n = 100, 400$, MSCAD$_\text{Q}$ is almost always the best. The other methods often select a model of order 3 (see Table 12).

Finally, models 8–10 have $K_0 = 6$, $\boldsymbol{\theta} = (1, 5, 10, 15, 20, 25)$ and dependence structures S8-S10 given in Table 7. For sample size $n = 400$, none of the methods perform satisfactory (see Table 13). For the three models, both BIC and BIC$_\text{Q}$ never select the true order out of the 500 simulated data sets. MSCAD$_\text{Q}$, however, provides an estimate closer to the true order most often.

Our simulation studies demonstrate that MSCAD$_\text{Q}$ is an appealing alternative to the information criteria AIC and BIC, based on both the full-model likelihood and the quasi-likelihood. While BIC and BIC_Q performed very well in detecting the true order when it was low, i.e., $K_0 = 2$, they had strong tendencies to underfit models of higher order, i.e., $K_0 = 3, 4$ and 6. MSCAD_Q, on the other hand, had higher success rates than the other methods when the true order was high, especially when $K_0 = 4$ and $K_0 = 6$.

We also compare the time taken by MSCAD_Q and the information criteria to complete the analysis of 500 simulated data sets for the most difficult normal HMM with 6 states and sample size $n = 400$. When MSCAD_Q is used coupled with 5-fold CV, the number of models fitted is 5 times the number of tuning parameters, which is 5 in our simulations. On a typical Unix machine, MSCAD_Q with 5-fold CV for tuning parameter selection took about 13 minutes. The computation of AIC and BIC values for models of order 1 to 9, on the other hand, took over 2 hours. While our code for the computation of AIC_Q and BIC_Q values for models of order 1 to 9 took only about 2 minutes to run, one should also consider that time is taken to modify the code after every model fit. Unlike MSCAD_Q, using AIC and BIC, based on either the full-model likelihood or the quasi-likelihood, requires the fitting of K different models. It is also evident that using a quasi-likelihood instead of the full-model likelihood greatly reduces the computational effort in estimating the order. Methods based on the quasi-likelihood have the advantage of not requiring the estimation of the transition matrix $\mathbb{P}$.

6. Empirical illustrations

EXAMPLE 6.1 (Poisson HMMs for movement counts by a fetal lamb). We consider a time series of overdispersed count data, originally analyzed in [**LP**]. The data set consists of the numbers of movements by a fetal lamb observed through ultrasound in 240 consecutive 5-second intervals. We plot the data in Figure 2.

Let $y_1, y_2, \ldots, y_{240}$ denote the observations. [**LP**] suggest fitting Poisson HMMs to this data for two reasons. The first reason is to accommodate the overdispersion that is present (the sample variance $s^2 = 0.658$ is larger than the sample mean $\bar{y} = 0.358$). The second reason is to capture the serial dependence in the observations, which can be seen from the sample autocorrelation function found in Figure 3.

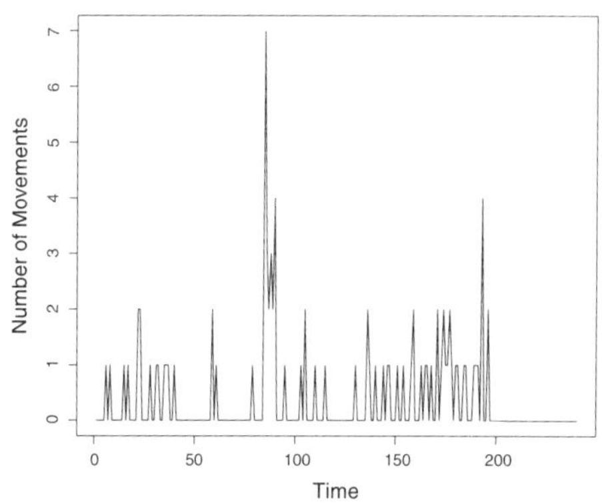

FIGURE 2. Number of movements by a fetal lamb in one of 240 consecutive 5-second intervals.

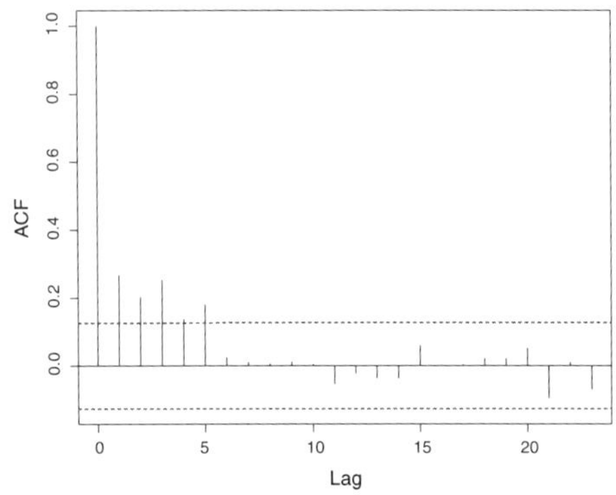

FIGURE 3. Sample ACF for the fetal lamb movement count data.

TABLE 1. AIC and BIC values, based on the quasi-likelihood, for the fetal lamb movement count data.

K	Quasi-log-likelihood	AIC_Q	BIC_Q
1	−201.044	404.087	407.568
2	−186.990	**379.979**	**390.421**
3	−185.796	381.592	398.995
4	−185.793	385.587	409.951
5	−185.790	389.581	420.907

To select the number of states of the Poisson HMM, [**LP**] had used the information criteria AIC and BIC, based on the full-model likelihood. They found that AIC selects a model of 3 states, whereas BIC selects a model of 2 states. In Table 1, we present the AIC and BIC values based on the quasi-likelihood, where we see that both AIC_Q and BIC_Q agree on a 2-state model. [**FS**] had used a Bayesian approach to select the number of states. Through a comparison of marginal likelihoods, she had found that the 3-state model was favoured over the 2-state model.

We applied our proposed method to this data set. Starting from $K = 8$ states, we found that MSCAD_Q, using CV, AIC and BIC to select the tuning parameter, all favour a 2-state model. Figure 4 shows the estimates of the state-dependent parameters θ_k at different iterations of the modified EM algorithm used by MSCAD_Q, which leads to a 2-state model after convergence at the 100th iteration.

Based on the methods discussed above, Poisson HMMs of order 2 or 3 are recommended for this data set and so we do a further analysis of these two possible models. In Table 2, we present the maximum likelihood estimates of the state-dependent parameters θ_k, transition matrix $\mathbb{P}$, and initial distribution $\boldsymbol{\alpha}$ for models of order 2 and 3 obtained using direct numerical maximization of the full-model likelihood. We notice that the estimates of the state-dependent parameters θ_k resulting from MSCAD_Q differ from the maximum likelihood estimates based on the full-model likelihood, which demonstrates the effect of the transition matrix. While the 2-state and 3-state models both provide adequate fits, the 2-state model has the additional advantage of being easily interpreted. As pointed out by [**LP**],

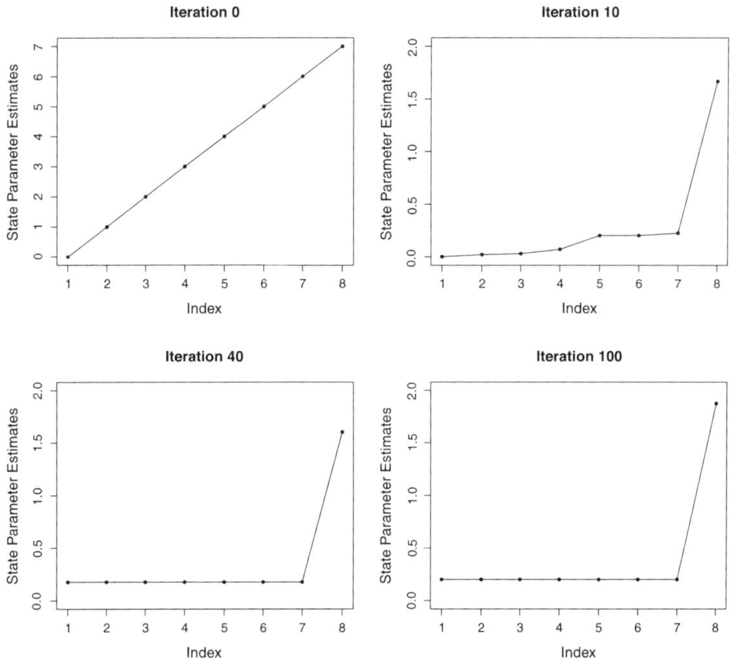

FIGURE 4. The initial values and estimates of the state-dependent parameters θ_k at different iterations of the modified EM algorithm used in $\mathrm{MSCAD_Q}$ for the analysis of fetal lamb movement count data.

TABLE 2. Parameter estimates for Poisson HMMs of order 2 and 3 fitted to the fetal lamb movement count data.

K	$\widehat{\boldsymbol{\theta}}$	$\widehat{\mathbb{P}}$	$\widehat{\boldsymbol{\pi}}$
2	$(0.256, 3.115)$	$\begin{pmatrix} 0.989 & 0.011 \\ 0.310 & 0.690 \end{pmatrix}$	$(0.965, 0.035)$
3	$(0.041, 0.495, 3.413)$	$\begin{pmatrix} 0.950 & 0.040 & 0.010 \\ 0.041 & 0.959 & 0 \\ 0.188 & 0 & 0.812 \end{pmatrix}$	$(0.490, 0.483, 0.027)$

the states may correspond to a relaxed state with regular levels of fetal activity and an excited state with higher levels of fetal activity, which are possibly triggered by physical factors such as the development of the central nervous system or empty space within the uterus.

To assess the marginal properties of both the 2-state and 3-state models, we compared the observed numbers of movement counts to those expected under each of the models, which are displayed in Table 3. As a measure of the goodness-of-fit,

TABLE 3. Observed numbers of movement counts, compared with those expected under models of order 2 and 3.

# of Movements	Observed Frequency	Expected Frequency	
		$K = 2$	$K = 3$
0	182	179.587	183.911
1	41	47.108	40.292
2	12	7.702	9.969
3	2	2.385	2.814
4	2	1.497	1.358
5	0	0.914	0.823
6	0	0.474	0.460
7	1	0.211	0.224

we used Pearson's chi-squared test statistic

$$\chi^2 = \sum_{i=1}^{N} \frac{(O_i - E_i)^2}{E_i},$$

where O_i is the observed frequency, E_i is the expected frequency and $N = 8$. The test statistic is asymptotically chi-squared distributed with $N - p - 1$ degrees of freedom, where p is the number of parameters estimated from the data. The expected frequencies were computed using $E_i = 240 \times \sum_{k=1}^{K} \hat{\alpha}_k f(y_i; \hat{\theta}_k)$. We found $\chi^2 = 7.796$ and $\chi^2 = 4.965$ for models of order 2 and 3, respectively. Using the 0.05 level of significance, the critical values are $\chi^2_{0.95,4} = 9.488$ for the 2-state model and $\chi^2_{0.95,2} = 5.991$ for the 3-state model. Since in both cases the test statistic is smaller than the critical value, it appears that the two models fit the data well. The 2-state model may be preferable since it has a biological interpretation.

EXAMPLE 6.2 (Normal HMMs for waiting times of the Old Faithful geyser). In this example, we consider a time series relating to eruptions of the Old Faithful geyser in Yellowstone National Park in the U.S. state of Wyoming. The data set, which was originally presented in [**AB**], consists of 299 observations of continuous measurement from August 1st to August 15th, 1985. The observations are times between the starts of successive eruptions. From the sample autocorrelation function displayed in Figure 5, we see that there is strong serial dependence in the behaviour of the geyser.

This data set has been previously analyzed by [**ZM**], who fit a series of normal HMMs with unequal variances. Since the variances for their final 3-state model do not differ substantially across states, we decided to fit a series of normal HMMs with equal variances σ^2 and consider the problem of estimating the number of states. In Table 4, we compare models of order 1 to 5 on the basis of AIC and BIC, based on both the full-model likelihood and the quasi-likelihood. We see that AIC selects a 4-state model, while its quasi-likelihood counterpart selects a 3-state model. We also see that BIC selects a 3-state model, while its quasi-likelihood counterpart selects a 2-state model.

We applied MSCAD_Q, starting with an upper bound $K = 15$. MSCAD_Q, using CV and BIC to select the tuning parameter, decided on a 3-state model. Figure 6 shows the estimates of the state-dependent parameters θ_k at different iterations of

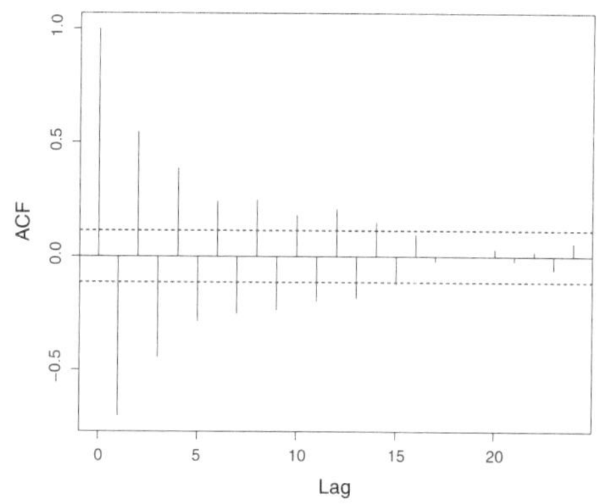

FIGURE 5. Sample ACF for the waiting times of the Old Faithful geyser.

TABLE 4. AIC and BIC values, based on both the full-model likelihood, $\ell_n(\boldsymbol{\Psi})$, and the quasi-likelihood, $\ell_n^{\mathrm{Q}}(\boldsymbol{\Psi})$, for the Old Faithful waiting times.

K	$\ell_n(\boldsymbol{\Psi})$	AIC	BIC	$\ell_n^{\mathrm{Q}}(\boldsymbol{\Psi})$	$\mathrm{AIC_Q}$	$\mathrm{BIC_Q}$
1	−1210.488	2424.977	2432.378	−1210.488	2424.977	2432.378
2	−1099.632	2209.264	2227.766	−1161.709	2331.419	**2346.220**
3	−1053.391	2126.783	**2163.787**	−1158.522	**2329.044**	2351.246
4	−1046.130	**2126.261**	2189.169	−1157.288	2330.575	2360.179
5	−1034.787	2121.574	2217.786	−1157.288	2330.575	2360.179

TABLE 5. Parameter estimates for normal HMMs of order 3 and 4 fitted to the waiting times of the Old Faithful geyser.

K	2	3	4
$\widehat{\boldsymbol{\theta}}$	(57.206, 81.921)	(54.764, 75.414, 85.091)	(53.168, 62.544, 75.978, 85.091)
$\widehat{\sigma}$	6.867	5.287	4.933
$\widehat{\mathbb{P}}$	$\begin{pmatrix} 0.000 & 1.000 \\ 0.638 & 0.362 \end{pmatrix}$	$\begin{pmatrix} 0.000 & 0.000 & 1.000 \\ 0.251 & 0.635 & 0.114 \\ 0.667 & 0.296 & 0.037 \end{pmatrix}$	$\begin{pmatrix} 0.000 & 0.000 & 0.000 & 1.000 \\ 0.000 & 0.000 & 0.000 & 1.000 \\ 0.105 & 0.235 & 0.621 & 0.039 \\ 0.605 & 0.072 & 0.280 & 0.043 \end{pmatrix}$
$\widehat{\boldsymbol{\pi}}$	(0.390, 0.610)	(0.325, 0.302, 0.373)	(0.256, 0.092, 0.277, 0.375)

the modified EM algorithm used by $\mathrm{MSCAD_Q}$, which leads to a 3-state model after convergence at the 130th iteration.

The estimates of the state-dependent parameters θ_k, transition matrix $\mathbb{P}$, and initial distribution $\boldsymbol{\alpha}$ for models of order 2, 3 and 4 are displayed in Table 5. They were obtained using direct numerical maximization of the full-model likelihood. In the left panel of Figure 7, we plot the fitted densities of models with 2, 3 and 4 states on the histogram of waiting times. We also plot the density of the fitted normal HMMs with unequal variances in the right panel of Figure 7. From the left panel of Figure 7, it appears that the 4-state model does not result in a substantial improvement in fit over the 3-state model. Our preference is thus with the normal

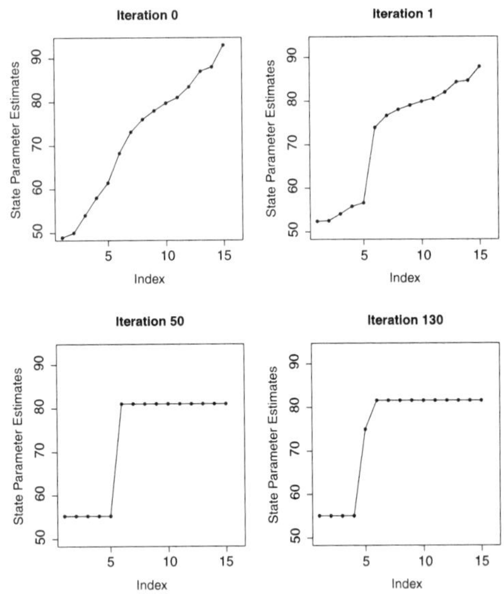

FIGURE 6. The initial values and estimates of the state-dependent parameters θ_k at different iterations of the modified EM algorithm used in MSCAD_Q for the analysis of waiting times of the Old Faithful geyser.

HMM of order 3. We also see that the models with unequal variances provide slightly better fits than the models with equal variances. This is expected since, in general, the addition of more parameters should improve the overall fit of the model.

7. Discussion and conclusion

In this paper, we proposed a new method, MSCAD_Q, for estimating the number of hidden states (order) of a HMM. The method is based on the penalization of a so-called quasi-likelihood, which is constructed from the marginal mixture distribution of the HMM observations. After initializing a HMM with a large number of states, the two penalty functions (SCAD and the log-penalty on the initial distribution) work simultaneously to cluster and combine similar states so that a model with an appropriate lower order is obtained. Given the estimated order, the state-dependent parameters, transition matrix, and initial distribution can then be estimated by maximizing the full-model likelihood.

We evaluated the performance of the proposed method through simulation, comparing it to the performance of the information criteria AIC and BIC, based on both the full-model likelihood and the quasi-likelihood. Our simulation results indicate that when the true order K_0 is small, MSCAD_Q is on par with these methods, but when K_0 is large, its success rates in detecting the true order are generally higher than those of the information criteria.

Unlike AIC and BIC, one of the main advantages of MSCAD_Q is that it does not require the fitting of candidate submodels. Furthermore, using the quasi-likelihood instead of the full-model likelihood simplifies the optimization problem since the

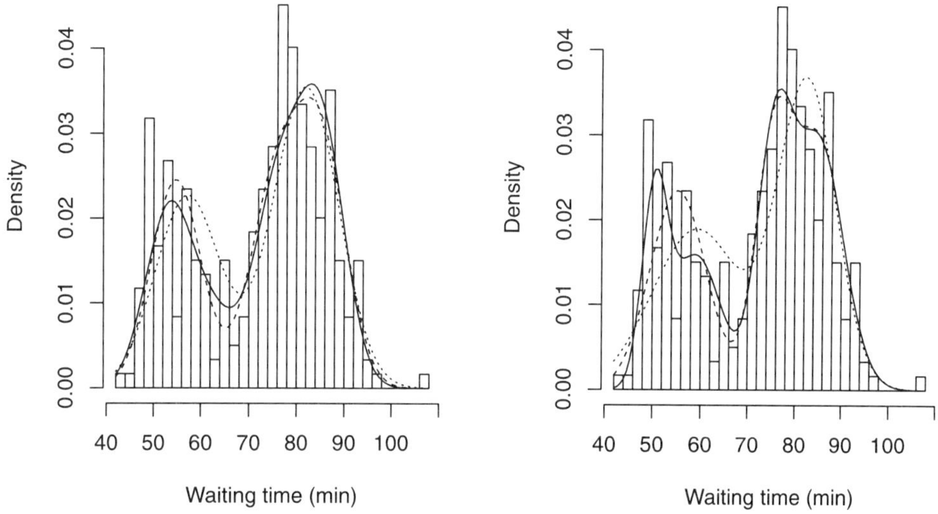

FIGURE 7. Normal HMMs of order 2 (dotted), 3 (dashed) and 4 (solid) with equal variances (left) and unequal variances (right) fitted to the waiting times of the Old Faithful geyser.

objective function does not involve the transition matrix $\mathbb{P} = (p_{ij})$. Methods based on the full-likelihood therefore require the estimation of a larger number of parameters and are more sensitive to the transition matrix than those based on the quasi-likelihood. MSCAD_Q should not be as sensitive to the transition matrix as AIC and BIC based on the full-likelihood, which can be seen, for example, in Table 13.

Several aspects of the MSCAD_Q method require further investigation. First, how to select a suitable value of the tuning parameter for the SCAD penalty in a computationally efficient manner is still unclear. Also, while MSCAD_Q was found to perform well in simulation, its theoretical properties have yet to be fully established. Our current theoretical results could be generalized to the multi-parameter setting. However, for the implementation of the new method, the ordering of the state-dependent parameters θ_k is needed. In multivariate space, the ordering of the θ_k's is not well-defined at the moment and remains a subject of ongoing research.

Appendix A. Proofs

In what follows, the expectations are under the true mixing distribution G_0. Assumptions A.3 to A.6 correspond to those in [**PZ**], who show that the sequence $(1/n)\ell_n^{\text{Q}}(G)$ converges to $\mathbb{E}[\log f(Y; G)]$ almost surely and uniformly over the compact space of G.

ASSUMPTION A.1. The parameter space Θ is compact. Furthermore, the following two conditions hold.

(1) $\mathbb{E}[|\log g(Y_1; \theta)|] < \infty \; \forall \theta \in \Theta$.
(2) There exists $\epsilon > 0$ such that for each $\theta \in \Theta$, $g(y; \theta)$ is measurable and $\mathbb{E}[|\log g(Y_1; \theta, \epsilon)|] < \infty$, where $g(y; \theta, \epsilon) = 1 + \sup_{|\theta - \theta'| \leq \epsilon} g(y; \theta')$.

ASSUMPTION A.2. The family $\{g(y;\theta);\theta \in \Theta\}$ is strongly identifiable in the sense that for K distinct $\theta_1, \theta_2, \dots, \theta_K$,

$$\sum_{j=1}^{K} \{a_j g(y;\theta_j) + b_j g'(y;\theta_j) + c_j g''(y;\theta_j)\} = 0$$

for all y implies that $a_j = b_j = c_j = 0$ for $j = 1, 2, \dots, K$.

ASSUMPTION A.3. The density $g(y;\theta)$ is differentiable with respect to $\theta \in \Theta$ and y, and three times continuously differentiable with respect to θ.

ASSUMPTION A.4. There exists a continuous function $h(y)$ such that $g(y;\theta) \le h(y)$ and $\mathbb{E}[|\log h(y)|] < \infty$.

ASSUMPTION A.5. First- and second-order partial derivatives of $\log g(y;\theta)$ satisfy

$$\left|\frac{\partial \log g(y;\theta)}{\partial \theta_u}\right| < h_u(y), \qquad \left|\frac{\partial \log g(y;\theta)}{\partial \theta_u \partial \theta_v}\right| < h_{u,v}(y),$$

where $\mathbb{E}[h(y)] < \infty$ and $\mathbb{E}[h_{u,v}(y)] < \infty$.

ASSUMPTION A.6. There exists $\delta > 0$ such that $\int \|y\|^{2+\delta} g(y;\theta)\, dy < \infty$ for all $\theta \in \Theta$, where $\|\cdot\|$ is Euclidean distance.

In Lemma A.7, we show that the MPQLE $\widehat{G}_n$ satisfies $0 < \hat{\alpha}_k < 1$ for $k = 1, 2, \dots, K$ in probability as $n \to \infty$. We use similar proof techniques to those in [**CK2**].

LEMMA A.7. *Let $Y_1, Y_2, \dots, Y_n$ be a sample from a homogeneous, stationary and identifiable HMM satisfying Assumptions* A.1 *to* A.6. *Let $f(y;G_0)$ denote the true density function of the marginal mixture distribution of Y_t. Then, as $n \to \infty$, the MPQLE $\widehat{G}_n$ has the property*

$$\sum_{k=1}^{K} \log \hat{\alpha}_k = O_p(1).$$

PROOF. In what follows, the expectations are taken with respect to the true marginal mixture distribution G_0. Note that by Jensen's inequality,

$$\mathbb{E}\left[-\log\left\{\frac{f(Y;G)}{f(Y;G_0)}\right\}\right] \ge -\log\left\{\mathbb{E}\left[\frac{f(Y;G)}{f(Y;G_0)}\right]\right\} = -\log 1 = 0$$

under identifiability and Assumption A.1 for the existence of the expectation. Therefore, for any $G \neq G_0$, we have that

$$\mathbb{E}[\log f(Y;G)] - \mathbb{E}[\log f(Y;G_0)] < 0. \tag{A.1}$$

Now, as shown by [**PZ**], since Y_t is stationary and ergodic,

$$\frac{1}{n}\ell_n^{\mathrm{Q}}(G) \xrightarrow{\text{a.s.}} \mathbb{E}[\log f(Y;G)] \quad \text{as } n \to \infty.$$

Therefore,

$$\lim_{n\to\infty}\left\{\frac{1}{n}\ell_n^{\mathrm{Q}}(G) - \frac{1}{n}\ell_n^{\mathrm{Q}}(G_0)\right\} = \mathbb{E}[\log f(Y;G)] - \mathbb{E}[\log f(Y;G_0)] \tag{A.2}$$

almost surely. [**PZ**] and references therein also show that this convergence is uniform.

Now combining results (A.1) and (A.2), we obtain

$$\ell_n^{Q}(G) - \ell_n^{Q}(G_0) < -M \times n$$

almost surely for any $G \neq G_0$ with some constant $M > 0$. Due to the compactness of the space of G from Assumption A.1, we can use a finite open coverage result in topology to strengthen the inequality to

$$\sup_{G \in N} \{\ell_n^{Q}(G) - \ell_n^{Q}(G_0)\} < -M \times n$$

for any compact neighbourhood N not containing G_0. In other words, the difference $\ell_n^{Q}(G) - \ell_n^{Q}(G_0)$ is negative in the order of n, uniformly for any G outside a neighbourhood of G_0.

Now when $\gamma_n = n^{1/4} \log n$, it is straightforward to check that $\lim_{n\to\infty} p_n(\beta)/n = 0$. Due to this property, the SCAD penalty $p_n(\cdot)$ for any $a > 2$ satisfies

$$\sum_{k=1}^{K-1} p_n(\beta_k) - \sum_{k=1}^{K_0-1} p_n(\beta_{0k}) = o(n)$$

when $\gamma_n = n^{1/4} \log n$. Thus, since the addition of the SCAD penalty to $\ell_n^{Q}(G)$ does not change the order assessment and the term $C_K \sum_{k=1}^{K} \log \alpha_k - C_{K_0} \sum_{k=1}^{K_0} \log \alpha_{0k}$ is constant with respect to n, we have that

$$\sup_{G \in N} \left\{\tilde{\ell}_n^{Q}(G) - \tilde{\ell}_n^{Q}(G_0)\right\} \le -M \times n.$$

Therefore, since $\widehat{G}_n$ is the maximizer of $\tilde{\ell}_n^{Q}(G)$, it must be in a small neighbourhood of G_0. That is, $\widehat{G}_n \xrightarrow{P} G_0$ and so it has at least K_0 distinct support points.

Since each β_{0k} is positive and approximated by one of the estimated differences in atoms $\hat{\beta}_k$, we must have that $p_n(\hat{\beta}_k) = p_n(\beta_{0k})$ in probability due to the fact that the SCAD penalty is constant outside a neighbourhood of 0. Therefore,

$$\sum_{k=1}^{K-1} p_n(\hat{\beta}_k) - \sum_{k=1}^{K_0-1} p_n(\beta_{0k}) \ge 0.$$

in probability.

Now let $\overline{G}_n$ be the maximum quasi-likelihood estimator (MQLE) of G_0 so that $\overline{G}_n$ has at most K support points. Then, by definition,

$$\begin{aligned}
0 &\le \tilde{\ell}_n^{Q}(\widehat{G}_n) - \tilde{\ell}_n^{Q}(G_0) \\
&= \{\ell_n^{Q}(\widehat{G}_n) - \ell_n^{Q}(G_0)\} - \left\{\sum_{k=1}^{K-1} p_n(\hat{\beta}_k) - \sum_{k=1}^{K_0-1} p_n(\beta_{0k})\right\} \\
&\qquad + \left\{C_K \sum_{k=1}^{K} \log \hat{\alpha}_k - C_{K_0} \sum_{k=1}^{K_0} \log \alpha_{0k}\right\} \\
&\le \{\ell_n^{Q}(\widehat{G}_n) - \ell_n^{Q}(G_0)\} + \left\{C_K \sum_{k=1}^{K} \log \hat{\alpha}_k - C_{K_0} \sum_{k=1}^{K_0} \log \alpha_{0k}\right\} \\
&\le \{\ell_n^{Q}(\overline{G}_n) - \ell_n^{Q}(G_0)\} + \left\{C_K \sum_{k=1}^{K} \log \hat{\alpha}_k - C_{K_0} \sum_{k=1}^{K_0} \log \alpha_{0k}\right\}.
\end{aligned}$$

Now under the assumptions of stationarity and ergodicity of Y_t, [**LL**] show that

$$\ell_n^{\mathrm{Q}}(\overline{G}_n) - \ell_n^{\mathrm{Q}}(G_0) = O_p(1).$$

Also, since the quantity $C_{K_0}\sum_{k=1}^{K_0}\log\alpha_{0k}$ is constant with respect to n, we have that

$$C_K\sum_{k=1}^{K}\log\hat{\alpha}_k \geq -\{\ell_n^{\mathrm{Q}}(\overline{G}_n) - \ell_n^{\mathrm{Q}}(G_0)\} + C_{K_0}\sum_{k=1}^{K_0}\log\alpha_{0k} = O_p(1). \qquad \square$$

In Theorem 4.1, we show that the MPQLE $\widehat{G}_n$ is a consistent estimator of the true mixing distribution G_0 of the marginal distribution.

PROOF (Theorem 4.1). In combination with the results from Lemma A.7, the proof follows from [**KM**] and [**CK2**]. $\square$

Appendix B. Simulations

TABLE 6. Transition matrices and corresponding stationary distributions in simulation studies for HMMs of order 2, 3 and 4 (S1–S7).

Structure	Transition matrix	Initial distribution
S1-S3: $K_0 = 2$	$\begin{pmatrix}0.50 & 0.50\\0.50 & 0.50\end{pmatrix}, \begin{pmatrix}0.25 & 0.75\\0.75 & 0.25\end{pmatrix}, \begin{pmatrix}0.75 & 0.25\\0.25 & 0.75\end{pmatrix}$	(0.50, 0.50)
S4: $K_0 = 3$	$\begin{pmatrix}\frac{1}{3} & \frac{1}{3} & \frac{1}{3}\\\frac{1}{3} & \frac{1}{3} & \frac{1}{3}\\\frac{1}{3} & \frac{1}{3} & \frac{1}{3}\end{pmatrix}$	$(\frac{1}{3}, \frac{1}{3}, \frac{1}{3})$
S5: $K_0 = 3$	$\begin{pmatrix}0.10 & 0.20 & 0.70\\0.20 & 0.60 & 0.20\\0.80 & 0.10 & 0.10\end{pmatrix}$	$\approx (0.374, 0.275, 0.352)$
S6: $K_0 = 4$	$\begin{pmatrix}\frac{1}{4} & \frac{1}{4} & \frac{1}{4} & \frac{1}{4}\\\frac{1}{4} & \frac{1}{4} & \frac{1}{4} & \frac{1}{4}\\\frac{1}{4} & \frac{1}{4} & \frac{1}{4} & \frac{1}{4}\\\frac{1}{4} & \frac{1}{4} & \frac{1}{4} & \frac{1}{4}\end{pmatrix}$	$(\frac{1}{4}, \frac{1}{4}, \frac{1}{4}, \frac{1}{4})$
S7: $K_0 = 4$	$\begin{pmatrix}0.70 & 0 & 0.10 & 0.20\\0 & 0.80 & 0.20 & 0\\0.10 & 0 & 0.70 & 0.20\\0.10 & 0.10 & 0 & 0.80\end{pmatrix}$	$(0.20, 0.20, 0.20, 0.40)$

TABLE 7. Transition matrices and corresponding stationary distributions in simulation studies for HMMs of order 6 (S8–S10).

Structure	Transition matrix	Initial distribution
S8: $K_0 = 6$	$\begin{pmatrix} \frac{1}{6} & \frac{1}{6} & \frac{1}{6} & \frac{1}{6} & \frac{1}{6} & \frac{1}{6} \\ \frac{1}{6} & \frac{1}{6} & \frac{1}{6} & \frac{1}{6} & \frac{1}{6} & \frac{1}{6} \\ \frac{1}{6} & \frac{1}{6} & \frac{1}{6} & \frac{1}{6} & \frac{1}{6} & \frac{1}{6} \\ \frac{1}{6} & \frac{1}{6} & \frac{1}{6} & \frac{1}{6} & \frac{1}{6} & \frac{1}{6} \\ \frac{1}{6} & \frac{1}{6} & \frac{1}{6} & \frac{1}{6} & \frac{1}{6} & \frac{1}{6} \\ \frac{1}{6} & \frac{1}{6} & \frac{1}{6} & \frac{1}{6} & \frac{1}{6} & \frac{1}{6} \end{pmatrix}$	$(\frac{1}{6}, \frac{1}{6}, \frac{1}{6}, \frac{1}{6}, \frac{1}{6}, \frac{1}{6})$
S9: $K_0 = 6$	$\begin{pmatrix} 0.2 & 0.4 & 0.1 & 0.1 & 0.1 & 0.1 \\ 0.1 & 0.4 & 0.1 & 0.1 & 0.2 & 0.1 \\ 0.2 & 0.2 & 0.1 & 0.1 & 0.2 & 0.2 \\ 0.1 & 0.2 & 0.1 & 0.2 & 0.1 & 0.3 \\ 0.1 & 0.2 & 0.1 & 0.2 & 0.3 & 0.1 \\ 0.1 & 0.2 & 0.1 & 0.2 & 0.1 & 0.3 \end{pmatrix}$	$\approx (0.122, 0.281, 0.100, 0.150, 0.173, 0.175)$
S10: $K_0 = 6$	$\begin{pmatrix} 0.1 & 0.2 & 0.1 & 0.1 & 0.1 & 0.4 \\ 0.1 & 0.2 & 0.1 & 0.1 & 0.4 & 0.1 \\ 0.2 & 0.1 & 0.1 & 0.4 & 0.1 & 0.1 \\ 0.1 & 0.1 & 0.4 & 0.1 & 0.1 & 0.2 \\ 0.1 & 0.4 & 0.1 & 0.1 & 0.2 & 0.1 \\ 0.4 & 0.2 & 0.1 & 0.1 & 0.1 & 0.1 \end{pmatrix}$	$\approx (0.163, 0.208, 0.143, 0.143, 0.180, 0.163)$

TABLE 8. Simulation results for normal HMMs of order K_0 ($n = 100$). Percentage of times out of 500 replications that the estimated order equals a given value of K.

Model	K_0	K	AIC	BIC	AIC_Q	BIC_Q	MSCAD_Q
1	2	1	4.8	41.0	1.8	19.8	6.2
		2	**79.2**	**58.6**	**92.8**	**80.2**	**93.4**
		3	12.2	0.4	5.4	0.0	0.4
		4	3.8	0.0	0.0	0.0	0.0
2	2	1	2.5	3.2	2.6	22.0	5.2
		2	**87.8**	**96.6**	**90.4**	**76.8**	**89.8**
		3	9.4	0.2	7.0	1.2	4.8
		4	0.3	0.0	0.0	0.0	0.2
3	2	1	0.0	3.8	2.8	20.4	4.2
		2	**89.2**	**96.2**	**90.8**	**79.2**	**90.6**
		3	8.2	0.0	6.0	0.4	5.2
		4	2.6	0.0	0.4	0.0	0.0
4	3	1	1.0	43.2	0.4	16.4	0.0
		2	31.0	**53.0**	13.6	40.0	0.4
		3	**58.0**	3.8	**80.0**	**43.6**	**83.6**
		4	10.0	0.0	6.0	0.0	16.0
5	3	1	0.0	0.2	0.8	8.2	0.0
		2	2.4	**41.8**	13.6	40.4	1.8
		3	**50.0**	28.6	**78.0**	**51.0**	**82.2**
		4	47.6	29.4	7.6	0.4	15.6
		5	0.0	0.0	0.0	0.0	0.4

TABLE 9. Simulation results for normal HMMs of order K_0 ($n = 100, 400$). Percentage of times out of 500 replications that the estimated order equals a given value of K.

Model	K_0	K	AIC	BIC	AIC_Q	BIC_Q	$MSCAD_Q$
6	4				$n = 100$		
		1	0.8	33.6	0.0	8.6	0.0
		2	**40.4**	**64.8**	12.8	**62.2**	0.0
		3	23.4	1.6	18.0	13.2	2.4
		4	30.2	0.0	**62.4**	15.8	**70.4**
		5	2.8	0.0	6.0	0.2	26.4
		6	2.4	0.0	0.8	0.0	0.8
					$n = 400$		
		1	0.0	0.0	0.0	0.0	0.0
		2	0.8	**84.4**	0.0	4.6	0.0
		3	2.8	12.6	0.2	7.2	2.0
		4	**80.6**	3.0	**94.2**	**88.0**	**95.2**
		5	13.4	0.0	4.2	0.2	2.8
		6	2.4	0.0	1.4	0.0	0.0
7	4				$n = 100$		
		1	0.0	0.0	0.2	3.4	0.0
		2	0.0	5.0	13.4	**52.2**	0.0
		3	17.8	**48.4**	34.4	32.6	2.4
		4	**67.0**	46.6	**37.8**	11.0	**67.2**
		5	13.2	0.0	12.0	0.8	29.2
		6	2.0	0.0	2.2	0.0	1.2
					$n = 400$		
		1	0.0	0.0	0.0	0.0	0.0
		2	0.0	0.0	0.0	4.8	0.0
		3	0.0	1.4	2.4	23.0	0.4
		4	**76.0**	**77.8**	32.2	**46.4**	**87.6**
		5	19.8	18.2	**65.4**	25.8	12.0
		6	4.2	2.6	0.0	0.0	0.0

TABLE 10. Simulation results for normal HMMs of order K_0 ($n = 400$). Percentage of times out of 500 replications that the estimated order equals a given value of K.

Model	K_0	K	AIC	BIC	AIC_Q	BIC_Q	$MSCAD_Q$
8	6	1	0.0	0.0	0.0	0.0	0.0
		2	0.2	**64.8**	0.0	2.6	0.0
		3	14.6	35.0	0.0	36.8	0.0
		4	24.0	0.2	1.6	18.6	0.0
		5	12.8	0.0	0.6	3.0	6.6
		6	**40.0**	0.0	**93.4**	**39.0**	**89.2**
		7	8.4	0.0	4.4	0.0	4.2
9	6	1	0.0	0.0	0.0	0.0	0.0
		2	0.0	**63.6**	0.0	12.2	0.0
		3	4.4	36.4	1.0	**51.6**	0.0
		4	2.2	0.0	0.2	1.6	1.6
		5	16.2	0.0	6.6	8.0	13.2
		6	**66.8**	0.0	**81.0**	26.4	**83.2**
		7	10.4	0.0	11.2	0.2	2.0
10	6	1	0.0	0.0	0.0	0.0	0.0
		2	0.0	21.4	0.0	23.8	0.0
		3	0.0	**67.6**	0.4	**35.2**	0.0
		4	0.0	9.8	0.2	7.8	0.0
		5	1.0	1.2	1.6	3.4	7.0
		6	**94.4**	0.0	**92.2**	29.8	**87.8**
		7	4.6	0.0	5.6	0.0	5.2

TABLE 11. Simulation results for Poisson HMMs of order K_0 ($n = 100$). Percentage of times out of 500 replications that the estimated order equals a given value of K.

Model	K_0	K	AIC	BIC	AIC_Q	BIC_Q	$MSCAD_Q$
1	2	1	0.0	0.0	0.0	0.0	3.2
		2	**93.8**	**100**	**95.6**	**99.4**	**96.8**
		3	6.0	0.0	4.4	0.6	0.0
		4	0.2	0.0	0.0	0.0	0.0
2	2	1	0.0	0.0	0.0	0.0	0.016
		2	**95.6**	**100**	**95.2**	**99.4**	**98.4**
		3	3.8	0.0	4.8	0.6	0.0
		4	0.6	0.0	0.0	0.0	0.0
3	2	1	0.0	0.0	0.0	0.0	0.0
		2	**94.4**	**100**	**97.0**	**99.8**	**94.8**
		3	5.2	0.0	3.0	0.2	5.2
		4	0.4	0.0	0.0	0.0	0.0
4	3	1	0.0	0.0	0.0	0.0	0.0
		2	44.0	**96.2**	31.4	**69.0**	2.8
		3	**53.0**	3.8	**68.4**	30.8	**75.2**
		4	3.0	0.0	0.2	0.2	21.2
		5	0.0	0.0	0.0	0.0	0.8
5	3	1	0.0	0.0	0.0	0.0	0.0
		2	32.8	**88.6**	39.4	**79.2**	24.4
		3	**66.0**	11.4	**60.4**	20.8	**73.8**
		4	1.2	0.0	0.2	0.0	1.8
		5	0.0	0.0	0.0	0.0	0.0

TABLE 12. Simulation results for Poisson HMMs ($n = 100, 400$). Percentage of times out of 500 replications that the estimated order equals a given value of K.

Model	K_0	K	AIC	BIC	AIC_Q	BIC_Q	$MSCAD_Q$
6	4				$n = 100$		
		1	0.0	0.0	0.0	0.0	0.0
		2	5.0	**52.0**	1.0	11.0	0.0
		3	**84.6**	47.8	**93.2**	**88.8**	30.2
		4	10.4	0.2	5.8	0.2	**52.6**
		5	0.0	0.0	0.0	0.0	16.6
		6	0.0	0.0	0.0	0.0	0.6
					$n = 400$		
		1	0.0	0.0	0.0	0.0	0.0
		2	0.0	0.0	0.0	0.0	0.0
		3	**73.6**	**100**	**65.2**	**97.8**	8.2
		4	25.8	0.0	34.8	2.2	**61.0**
		5	0.6	0.0	0.0	0.0	26.2
		6	0.0	0.0	0.0	0.0	4.6
7	4				$n = 100$		
		1	0.0	0.0	0.0	0.0	0.0
		2	4.0	24.0	5.4	24.8	0.6
		3	**88.8**	**75.8**	**89.8**	**75.0**	36.2
		4	7.2	0.2	4.8	0.2	**50.8**
		5	0.0	0.0	0.0	0.0	11.8
		6	0.0	0.0	0.0	0.0	0.6
					$n = 400$		
		1	0.0	0.0	0.0	0.0	0.0
		2	0.0	0.0	0.0	0.0	0.0
		3	30.0	**89.4**	**76.4**	**99.0**	15.8
		4	**66.0**	10.6	23.6	1.0	**60.6**
		5	4.0	0.0	0.0	0.0	21.0
		6	0.0	0.0	0.0	0.0	2.6

TABLE 13. Simulation results for Poisson HMMs of order K_0 ($n = 400$). Percentage of times out of 500 replications that the estimated order equals a given value of K.

Model	K_0	K	AIC	BIC	AIC_Q	BIC_Q	MSCAD_Q
8	6	1	0.0	0.0	0.0	0.0	0.0
		2	0.0	0.0	0.0	0.0	0.0
		3	1.2	**81.6**	0.2	5.8	0.0
		4	**85.0**	18.4	**83.2**	**94.0**	9.8
		5	13.0	0.0	16.6	0.2	**41.0**
		6	0.8	0.0	0.0	0.0	32.6
		7	0.0	0.0	0.0	0.0	13.8
9	6	1	0.0	0.0	0.0	0.0	0.0
		2	0.0	0.0	0.0	0.0	0.0
		3	1.2	**63.2**	0.0	3.8	0.0
		4	**87.6**	36.8	**87.0**	**95.8**	13.2
		5	11.0	0.0	13.0	0.4	**43.6**
		6	0.2	0.0	0.0	0.0	31.2
		7	0.0	0.0	0.0	0.0	10.0
10	6	1	0.0	0.0	0.0	0.0	0.0
		2	0.0	0.0	0.0	0.0	0.0
		3	0.6	48.2	0.2	4.6	0.0
		4	**70.8**	**51.8**	**83.4**	**95.2**	9.8
		5	28.0	0.0	16.2	0.2	**45.2**
		6	0.6	0.0	0.2	0.0	27.2
		7	0.0	0.0	0.0	0.0	15.2

References

[A1] H. Akaike, *Information theory and an extension of the maximum likelihood principle*, Second International Symposium on Information Theory (Tsahkadsor, 1971), Akadémiai Kiadó, Budapest, 1973, pp. 267–281. MR0483125 (58 #3144)

[A2] P. S. Albert, *A two-state Markov mixture model for a time series of epileptic seizure counts*, Biometrics **47** (1991), no. 4, 1371–1381.

[AB] A. Azzalini and A. W. Bowman, *A look at some data on the Old Faithful geyser*, J. Roy Statist. Soc. Ser. C. **39** (1990), 357–365.

[BRR] P. J. Bickel, Y. Ritov, and T. Rydén, *Asymptotic normality of the maximum-likelihood estimator for general hidden Markov models*, Ann. Statist. **26** (1998), no. 4, 1614–1635, DOI 10.1214/aos/1024691255. MR1647705 (99j:62120)

[C1] J. H. Chen, *Optimal rate of convergence for finite mixture models*, Ann. Statist. **23** (1995), no. 1, 221–233, DOI 10.1214/aos/1176324464. MR1331665 (96c:62047)

[C2] G. A. Churchill, *Stochastic models for heterogeneous DNA sequences*, Bull. Math. Biol. **51** (1989), no. 1, 79–94, DOI 10.1016/S0092-8240(89)80049-7. MR978904 (89k:92027)

[C3] G. A. Churchill, *Hidden Markov chains and the analysis of genome structure*, Computers & Chemistry **16** (1992), no. 2, 107–115.

[CCK1] H. Chen, J. Chen, and J. D. Kalbfleisch, *A modified likelihood ratio test for homogeneity in finite mixture models*, J. R. Stat. Soc. Ser. B Stat. Methodol. **63** (2001), no. 1, 19–29, DOI 10.1111/1467-9868.00273. MR1811988 (2001k:62023)

[CCK2] H. Chen, J. Chen, and J. D. Kalbfleisch, *Testing for a finite mixture model with two components*, J. R. Stat. Soc. Ser. B Stat. Methodol. **66** (2004), no. 1, 95–115, DOI 10.1111/j.1467-9868.2004.00434.x. MR2035761 (2004k:62051)

[CK1] J. Chen and J. D. Kalbfleisch, *Penalized minimum-distance estimates in finite mixture models* (English, with English and French summaries), Canad. J. Statist. **24** (1996), no. 2, 167–175, DOI 10.2307/3315623. MR1406173 (97g:62031)

[CK2] J. Chen and A. Khalili, *Order selection in finite mixture models with a non-smooth penalty*, J. Amer. Statist. Assoc. **103** (2008), no. 484, 1674–1683, DOI 10.1198/016214508000001075. MR2722574

[DH] J. Dannemann and H. Holzmann, *Testing for two states in a hidden Markov model* (English, with English and French summaries), Canad. J. Statist. **36** (2008), no. 4, 505–520, DOI 10.1002/cjs.5550360402. MR2532249 (2011b:62030)

[FL] J. Fan and R. Li, *Variable selection via nonconcave penalized likelihood and its oracle properties*, J. Amer. Statist. Assoc. **96** (2001), no. 456, 1348–1360, DOI 10.1198/016214501753382273. MR1946581 (2003k:62160)

[FS] S. Frühwirth-Schnatter, *Finite mixture and Markov switching models*, Springer Series in Statistics, Springer, New York, 2006. MR2265601 (2007e:62005)

[GB] E. Gassiat and S. Boucheron, *Optimal error exponents in hidden Markov models order estimation*, IEEE Trans. Inform. Theory **49** (2003), no. 4, 964–980, DOI 10.1109/TIT.2003.809574. MR1984482 (2004j:62069)

[IJS] H. Ishwaran, L. F. James, and J. Sun, *Bayesian model selection in finite mixtures by marginal density decompositions*, J. Amer. Statist. Assoc. **96** (2001), no. 456, 1316–1332, DOI 10.1198/016214501753382255. MR1946579 (2003k:62013)

[KM] A. Khalili Mahmoudabadi, *Order selection in classical finite mixture models and variable selection and inference in finite mixture of regression models*, ProQuest LLC, Ann Arbor, MI, 2006. Thesis (Ph.D.)–University of Waterloo (Canada). MR2709100

[L1] B. G. Leroux, *Maximum-likelihood estimation for hidden Markov models*, Stochastic Process. Appl. **40** (1992), no. 1, 127–143, DOI 10.1016/0304-4149(92)90141-C. MR1145463 (93d:62154)

[L2] G. Lindgren, *Markov regime models for mixed distributions and switching regressions*, Scand. J. Statist. **5** (1978), no. 2, 81–91. MR497061 (80c:62032)

[LL] S. Lee and T. Lee, *Robust estimation for order of hidden Markov models based on density power divergences*, J. Stat. Comput. Simul. **80** (2010), no. 5-6, 503–512, DOI 10.1080/00949650902725155. MR2758143

[LP] B. G. Leroux and M. L. Puterman, *Maximum-penalized likelihood estimation for independent and Markov-dependent mixture models*, Biometrics **48** (1992), no. 2, 545–558.

[LRS] S. E. Levinson, L. R. Rabiner, and M. M. Sondhi, *An introduction to the application of the theory of probabilistic functions of a Markov process to automatic speech recognition*, Bell System Tech. J. **62** (1983), no. 4, 1035–1074, DOI 10.1002/j.1538-7305.1983.tb03114.x. MR702893 (84k:68054)

[M] R. J. MacKay, *Estimating the order of a hidden Markov model* (English, with English and French summaries), Canad. J. Statist. **30** (2002), no. 4, 573–589, DOI 10.2307/3316097. MR1964428 (2004b:62189)

[PZ] D. S. Poskitt and J. Zhang, *Estimating components in finite mixtures and hidden Markov models*, Aust. N. Z. J. Stat. **47** (2005), no. 3, 269–286, DOI 10.1111/j.1467-842X.2005.00393.x. MR2169527 (2006f:62078)

[R] T. Rydén, *Estimating the order of hidden Markov models*, Statistics **26** (1995), no. 4, 345–354, DOI 10.1080/02331889508802501. MR1365683 (96m:62156)

[RRT] C. P. Robert, T. Rydén, and D. M. Titterington, *Bayesian inference in hidden Markov models through the reversible jump Markov chain Monte Carlo method*, J. R. Stat. Soc. Ser. B Stat. Methodol. **62** (2000), no. 1, 57–75, DOI 10.1111/1467-9868.00219. MR1747395 (2001b:62018)

[RTÅ] T. Rydén, T. Teräsvirta, and S. Åsbrink, *Stylized facts of daily return series and the hidden Markov model*, J. Appl. Econometrics **13** (1998), 217–244.

[S] G. Schwarz, *Estimating the dimension of a model*, Ann. Statist. **6** (1978), no. 2, 461–464. MR0468014 (57 #7855)

[ZM] W. Zucchini and I. L. MacDonald, *Hidden Markov models for time series*, Monographs on Statistics and Applied Probability, vol. 110, CRC Press, Boca Raton, FL, 2009. An introduction using R. MR2523850 (2010k:62001)

Department of Mathematics and Statistics, McGill University, Montréal, QC, H3A 0B9, Canada

E-mail address: mcgillivray@math.mcgill.ca

Department of Mathematics and Statistics, McGill University, Montréal, QC, H3A 0B9, Canada

E-mail address, A. Khalili: khalili@math.mcgill.ca

Contemporary Mathematics
Volume **622**, 2014
http://dx.doi.org/10.1090/conm/622/12441

Efficient adaptive estimation strategies in high-dimensional partially linear regression models

Xiaoli Gao and S. Ejaz Ahmed

ABSTRACT. We consider estimation of regression coefficients in a partially linear model with a diverging number of predictors. In the arena of high-dimensional (HD) data analysis, it is often assumed that the model is sparse for simultaneous variable selection and estimation. In other words, it is assumed that the signals are well separated from zero in the model at hand. However, in many studies this assumption may not hold, resulting in producing biased parameters estimation. Our aim is to improve the estimation of the regression parameters in a realistic scenario when the model consists of weak signals as well as strong signals. We suggest an improved estimation strategy which incorporates the effect of weak signals. More specifically, we propose a high-dimensional shrinkage estimation strategy to improve the prediction performance of the model. We develop the asymptotic properties of the shrinkage estimator and assess its relative performances with respect to the full model and submodel estimators using the quadratic loss function. We show both analytically and numerically that the proposed high-dimensional shrinkage estimator performs better than the full model estimator. Interestingly, the simulation study reveals that suggested strategy significantly improves the prediction accuracy of any candidate submodel estimates generated from available LASSO-type variable selection methods. Finally, we apply the suggested estimation strategy to real data to illustrate its practical usefulness. Our data analysis strongly corroborates the analytical and simulated findings that the performance of the shrinkage strategy in comparison with other estimators is favorable.

1. Introduction

We consider a partially linear model (PLM)

$$y_i = \mathbf{x}_i'\boldsymbol{\beta} + g(u_i) + \varepsilon_i, \quad 1 \leq i \leq n, \tag{1.1}$$

where y_is are responses, $\mathbf{x}_i = (x_{i1}, \dots, x_{ip})'$ are predictors, $\boldsymbol{\beta} = (\beta_1, \dots, \beta_p)'$ is a $p \times 1$ unknown vector, $g(\cdot)$ is an unknown smooth function, u_is take values in a compact interval, and the ϵ_is are independent and identically distributed random

2010 *Mathematics Subject Classification.* Primary 62J07.

Key words and phrases. High-dimensional shrinkage estimation, sparse model, asymptotic distributional risk, partially linear model, penalty estimators.

The research of the first author was supported by the OU-Beaumont Multidisciplinary Research Award.

The research of the second author was supported by the Natural Sciences and Engineering Research Council of Canada.

errors with mean 0 and finite variance σ^2. In this paper, we assume that p grows with n but $p < n$. For simplicity, we consider $u_i \in [0,1]$ for $1 \le i \le n$ and the intercept is not included. Otherwise, we can either standardize, or center the data.

In this model, the response variable y_i relates to the covariates $\mathbf{x}_i$ linearly, while its relation with u_i is not specified up to any finite number of parameters. Comparing with the multiple linear regression model, PLM has a flexible relationship between y_i and u_i. Comparing with the the nonparametric regression, PLM provides more interpretation of the relationship between the response and covariates vector $\mathbf{x}_i$.

Generally speaking, the study of PLM includes two cases: fixed p and diverging p. When the number of parametric effects p is fixed, Wahba (**1990**), Engle et al. (**1986**) and Heckman (**1986**) proposed a partial spline estimator to estimate $\boldsymbol{\beta}$ and g. On the other hand, Robinson (**1988**), Speckman (**1988**) and Chen (**1988**) proposed partial residual estimators.

Recently, a lot of work is reported for sparse models in high-dimensional settings where p is not fixed. When p increases with the sample size n, sparsity of the true model is often assumed. Sparsity means that some covariates do not contribute to the response variable in the sense that some β_js in (1.1) are exactly 0. Under such a sparsity assumption, Xie and Huang (**2009**) estimated $\boldsymbol{\beta}$ and g using polynomial regression splines and the SCAD penalty. Ni et al. (**2009**) investigated a penalized PLM using partially smoothing splines. In addition, previous studies assume the important (nonzero) β_j's are assumed big enough to be separable from those non-important ones (zero β_j's). Even though the true model can be sparse, it is too strict to assume no effect on the response variable for any other unimportant covariates. In the real world, many covariates still have an effect to the response variable, but may not be as large as the important ones. Suppose the regression parameter vector $\boldsymbol{\beta}$ can be partitioned into two sub-vectors, $\boldsymbol{\beta}_1$ and $\boldsymbol{\beta}_2$, respectively. For example, $\boldsymbol{\beta}_1$ includes only the coefficients of main effects (e.g., treatment effects, genetic effects) and $\boldsymbol{\beta}_2$ is only associated with some "nuisance" effects (e.g., age, laboratory). It is worth noting that ignoring nuisance covariates completely or enforcing the nuisance parameters to be zero, however, may reduce the overall prediction performance in many experimental situations, for example in microarray data analysis. We suggest that a more reasonable assumption on the parametric effects is to permit the co-existence of both zero and weak effects, besides those coefficients with relatively large effects.

To formulate the problem, let $p = p_1 + p_2$ and $\mathbf{X} = (\mathbf{x}_1, \ldots, \mathbf{x}_p)$. We can partition $\mathbf{X} = (\mathbf{X}_1, \mathbf{X}_2)$, where $\mathbf{X}_1, \mathbf{X}_2$ are of order $n \times p_1$ and $n \times p_2$, respectively. Thus, $\boldsymbol{\beta}'_n = (\boldsymbol{\beta}'_1, \boldsymbol{\beta}'_2)$, where $\boldsymbol{\beta}_1$ consists of strong and moderate effects, and $\boldsymbol{\beta}_2$ consists of coefficients with no effects. It is commonly assumed that those covariates with strong effects can be easily detected by some prior information or by existing variable selection methods. However, a relatively aggressive variable selection method will often ignore covariates with weak coefficients completely. Interestingly, a Steinian shrinkage estimation strategy suggests that if a subset model is predefined from some prior information, then inferences about the main effects may benefit from moving the full model estimate in the direction of the submodel estimate. In an essence, the Stein-type or shrinkage estimator combines estimation problems by shrinking a base estimator to a plausible alternative estimator. When p is fixed, **Ahmed et al. (2007)**, **Fallahpour et al. (2012)** and

Raheem et al. (2012), investigated properties of the shrinkage estimator of regression parameters in PLM. As we know that, the assumption of fixed p is often violated in real applications.

In this paper, we are primarily interested in improving the estimation of $\boldsymbol{\beta}_1$ under a predefined subset PLM when the number of parametric coefficients p grows with the sample size n. In a possibly ultrahigh dimension ($p \gg n$) estimation problem, we make the assumption that a successful reduction to a problem with much lower dimensions ($p = o(n)$) exists. In any event, we shrink a full estimator towards the submodel estimator using available *uncertain prior information* (UPI) that some variables already included in the model may not be useful for the prediction purpose. However, there are situations when the UPI may not be readily available. In this case we suggest to use existing variable selection methods to obtain a subset of variables. Thus we obtain *auxiliary information* (AI) to build a subset model, which can be used in our proposed estimation strategy to improve the prediction performance.

The proposed shrinkage strategy has the following advantages:

(1) Using prior information, it significantly improves the submodel estimators based on only the predefined covariates.
(2) It permits the number of parameter coefficients, p, to grow with n.
(3) It can be easily combined with existing sparse variable selection methods to improve estimation and prediction.

To end the section, we give the well-known penalized regularization approach for obtaining some auxiliary information as follows. To explicitly make the covariates and linear coefficients depend on n, we write $p = p_n$ and $p_k = p_{kn}$ to address that p and p_k grow with n. Corresponding subscripts are also applied to other notations, such as $\mathbf{X}_{kn}$ and $\boldsymbol{\beta}_{kn}$, for $k = 1, 2$. A penalized least squares estimation can be obtained by minimizing

$$\widehat{\boldsymbol{\beta}}_n^{\mathrm{PEN}} = \arg\min\left\{\sum_{i=1}^{n}\left(y_i - \sum_{j=1}^{p_n} x_{ij}\beta_j\right)^2 + \sum_{j=1}^{p_n} p_\lambda(\beta_j)\right\},$$

where $p_\lambda(\beta_j)$ is the penalty function on β_j, λ is the penalized tuning parameter. Specifically, if $p_\lambda(\beta_j) = \lambda|\beta_j|$, then $\widehat{\boldsymbol{\beta}}_n^{\mathrm{PEN}}$ is the famous LASSO estimator (**Tibshirani, 1996**), $\widehat{\boldsymbol{\beta}}_n^{\mathrm{LASSO}}$. Similarly, an Adaptive LASSO estimator (**Zou, 2006**), $\widehat{\boldsymbol{\beta}}_n^{\mathrm{ALASSO}}$, is obtained if $p_\lambda(\beta_j) = \lambda w_j|\beta_j|$, where w_j can be chosen to be a proxy of the inverse of an initial estimator of β_j. A SCAD estimator, $\widehat{\boldsymbol{\beta}}_n^{\mathrm{SCAD}}$ (**Fan and Li, 2001**) is obtained if $p_\lambda(t) = \lambda|t|$, $-(t^2 - 2\alpha\lambda|t| + \lambda^2)/[2(\alpha-1)]$, $(\alpha+1)\lambda^2/2$ when $|t| < \lambda$, $\lambda < |t| < \alpha\lambda$, and $|t| > \alpha\lambda$, where $\alpha > 2$ is a regularization parameter. An MCP estimator (**Zhang, 2010**), $\widehat{\boldsymbol{\beta}}_n^{\mathrm{MCP}}$, is obtained if $p_\lambda(t) = \int_0^t (\lambda - x/\gamma)_+ \,\mathrm{d}x$, where $\gamma > 0$ is a regularization parameter and $c_+ = \max\{c, 0\}$.

The remainder of this paper is organized as follows. In Section 2, we extend Steinian shrinkage methods in the high-dimensional-PLM (HD-PLM) model. The asymptotic properties of the proposed shrinkage strategy in the HD-PLM are developed in Section 3. More specifically, we investigate the asymptotic normality and asymptotic distributional risks of the HD shrinkage estimators. Some numerical studies are given in Section 4. Section 5 showcases a published data example. We summarize the paper with some discussions in the final section. All proofs are given in the Appendix.

2. High-dimensional shrinkage strategy

In this section, we present an estimation strategy to improve the performance of a subset candidate model in a HD-PLM model. Roughly speaking, we first approximate the non-parametric part and transform the PLM in (1.1) into a linear model with weighted matrix. Then we design a *high-dimensional shrinkage* (HDS) strategy to improve the estimation efficiency of the subset partial linear model where only a subset of covariates is included in the original PLM model.

Xie and Huang (2009) studied the approximation of the nonparametric function g in (1.1) using B-spline. Consider a partition of $[0,1]$: $0 = \eta_0 < \eta_1 < \cdots < \eta_{M_n} < \eta_{M_n+1} = 1$. Here M_n grows with n at a certain rate, but slower than n. We will provide more details in Section 3. Let $S_m(\boldsymbol{\eta})$ be the space of polynomial splines of order m and knots $\boldsymbol{\eta}$, where $\boldsymbol{\eta} = (\eta_0, \eta_1, \ldots, \eta_{M_n}, \eta_{M_n+1})'$. Suppose the largest mesh size $\bar{\tau}_n \equiv \max_{0\le k\le M_n} \eta_{k+1} - \eta_k = \bar{\tau}_n = O(M_n^{-1})$. Then $S_m(\boldsymbol{\eta})$ has dimension $q_n = M_n + m$. For any function $s \in S_m(\boldsymbol{\eta})$ with $m > 2$, it is $m-2$ differentiable polynomial of order m at interval $[\eta_k, \eta_{k+1})$. From **Schumaker (1981**, Corollary 4.10), there always exists local basis $\{B_{nw}, 1 \le w \le q_n\}$ for $S_m(\boldsymbol{\eta})$. Let $\mathbf{z}(u, \bar{\tau}_n) \equiv \big(B_{n1}(u), \ldots, B_{nq_n}(u)\big)'$ and $\mathbf{Z}_n \equiv \mathbf{Z}_n(\mathbf{u}, \bar{\tau}_n)$ be the $n \times q_n$ matrix with ith row being $\mathbf{z}(u_i, \bar{\tau}_n)'$, where $\mathbf{u} = (u_1, \ldots, u_n)$. Thus under reasonable smoothness conditions, g can be well approximated by an element in $S_m(\boldsymbol{\eta})$, that is, there exists a $q_n \times 1$ vector $\mathbf{a}_n = (a_1, \ldots, a_{q_n})'$ such that $g(\mathbf{u}) \asymp \mathbf{z}(\mathbf{u}, \bar{\tau}_n)'\mathbf{a}_n$, with "$\asymp$" meaning an approximation.

Denote $\mathbf{y} = (y_1, \ldots, y_n)'$. Thus a PLM estimator $(\widehat{\boldsymbol{\beta}}_n, \widehat{\boldsymbol{a}}_n)$ of model (1.1) is to minimize the following least squares objective function,

$$Q(\boldsymbol{\beta}_n, \mathbf{a}_n; \bar{\tau}_n, m) = \arg\min\{\|\mathbf{y} - \mathbf{X}_n\boldsymbol{\beta}_n - \mathbf{Z}_n\mathbf{a}_n\|_2^2\}, \tag{2.1}$$

where m and $\bar{\tau}_n$ are the degrees of freedom and the largest mesh size of the polynomial spline. Throughout this paper, $\widehat{\boldsymbol{\beta}}_n$ is considered to be a full estimation of the parameter part in PLM. For any given $\boldsymbol{\beta}_n$, an estimator of a_n satisfies

$$\mathbf{Z}_n'\mathbf{Z}_n\mathbf{a}_n = \mathbf{Z}'(\mathbf{y} - \mathbf{X}_n\boldsymbol{\beta}_n).$$

Therefore, the profile objective function of the parametric part,

$$\widetilde{Q}(\boldsymbol{\beta}_n, \bar{\tau}_n, m) = \arg\min\{\|(\mathbf{I}-\mathbf{P}_{\mathbf{Z}_n})(\mathbf{y}-\mathbf{X}_n\boldsymbol{\beta}_n)\|_2^2\} = \arg\min\{\|\widetilde{\mathbf{y}}-\widetilde{\mathbf{X}}_n\boldsymbol{\beta}_n\|_2^2\}, \tag{2.2}$$

where $\mathbf{P}_{\mathbf{Z}_n} = \mathbf{Z}_n(\mathbf{Z}_n'\mathbf{Z}_n)^{-1}\mathbf{Z}_n'$ is the projection matrix of $\mathbf{Z}_n$, $\widetilde{\mathbf{X}}_n = (\mathbf{I}_n - \mathbf{P}_{\mathbf{Z}_n})\mathbf{X}$ and $\widetilde{\mathbf{y}} = (\mathbf{I}_n - \mathbf{P}_{\mathbf{Z}_n})\mathbf{y}$. Based on the partition of $\mathbf{X}_n = (\mathbf{X}_{1n}, \mathbf{X}_{2n})$, we assume the true linear coefficients vector $\boldsymbol{\beta}_0 = (\boldsymbol{\beta}_{10}', \boldsymbol{\beta}_{20}')'$, where $\boldsymbol{\beta}_{10}$ associated with $\mathbf{X}_{1n}$ includes moderate or strong coefficients, and $\boldsymbol{\beta}_{20}$ associated with $\mathbf{X}_{2n}$ includes very weak but nonzero coefficients. Some UPI or AI may only let moderate or strong signals in $\boldsymbol{\beta}_{10}$ stand out, that is,

$$\text{UPI or AI:}\quad \boldsymbol{\beta}_{20} = \mathbf{0}_{p_{2n}}, \tag{2.3}$$

where $\mathbf{0}_q$ represents a q-dimensional vector with all zeros.

Thus under UPI or AI, we can only consider the following subset candidate partial linear model (SM) in a finite dimension p_1,

$$y_i = x_{i1}\beta_1 + \cdots + x_{ip_1}\beta_{p_1} + g(t_i) + \varepsilon_i, \quad i = 1, \ldots, n. \tag{2.4}$$

Similar to the above, we can use the B-spline approximation to approximate the nonparametric part. Then we obtain a submodel PLM estimator of $\boldsymbol{\beta}_{10}$ in (2.4) as follows:

$$\widehat{\boldsymbol{\beta}}_{1n}^{\text{SM}} = \arg\min\{\|(\mathbf{I}_n - \mathbf{P}_{\mathbf{Z}_n})(\mathbf{y} - \mathbf{X}_{1n}\boldsymbol{\beta}_{1n})\|_2^2\} = \arg\min\{\|\widetilde{\mathbf{y}} - \widetilde{\mathbf{X}}_{1n}\boldsymbol{\beta}_{1n}\|_2^2\}. \tag{2.5}$$

The submodel estimate $\widehat{\boldsymbol{\beta}}_{1n}^{\text{SM}}$ performs well when $\boldsymbol{\beta}_{20} = \mathbf{0}$ or is close to $\mathbf{0}$. However, when $\boldsymbol{\beta}_{20}$ is staying away from $\mathbf{0}$, $\widehat{\boldsymbol{\beta}}_{1n}^{\text{SM}}$ can be biased and inefficient. Our interest is to improve the estimation of $\boldsymbol{\beta}_{1n}$ when it is plausible that $\boldsymbol{\beta}_{2n}$ includes the nuisance coefficient vectors under UPI or AI.

Suppose $\widehat{\boldsymbol{\beta}}_n = (\widehat{\boldsymbol{\beta}}_{1n}', \widehat{\boldsymbol{\beta}}_{2n}')'$ and $\widehat{\boldsymbol{a}}$ minimize the objective function in (2.1). Let $\boldsymbol{\Upsilon}_n$ consists of any ι different rows of an p_{2n}-dimensional identity matrix $\mathbf{I}_{p_{2n}}$, where ι is a constant independent of n. Thus $\boldsymbol{\Upsilon}_n$ is a ι-dimensional full rank matrix. Define a ι-dimensional vector $\boldsymbol{\vartheta} = \boldsymbol{\Upsilon}_n\boldsymbol{\beta}_{2n}$ and $\widehat{\boldsymbol{\vartheta}} = \boldsymbol{\Upsilon}_n\widehat{\boldsymbol{\beta}}_{2n}$. We want to improve the estimation of $\boldsymbol{\beta}_{1n}$ by borrowing some information from the following statistic,

$$T_n \equiv \frac{n}{\sigma^2}\widehat{\boldsymbol{\vartheta}}'(\boldsymbol{\Upsilon}_n(\widetilde{\mathbf{X}}_{2n}'\mathbf{P}_{\widetilde{\mathbf{X}}_{1n}}^{\perp}\widetilde{\mathbf{X}}_{2n})^{-1}\boldsymbol{\Upsilon}_n')^{-1}\widehat{\boldsymbol{\vartheta}}, \tag{2.6}$$

where $\mathbf{P}_{\widetilde{\mathbf{X}}_{1n}}^{\perp} = \mathbf{I}_n - \widetilde{\mathbf{X}}_{1n}(\widetilde{\mathbf{X}}_{1n}'\widetilde{\mathbf{X}}_{1n})^{-1}\widetilde{\mathbf{X}}_{1n}$. Thus T_n depends on $\boldsymbol{\Upsilon}_n$. If σ^2 is unknown, we can consider an unbiased estimation $\widehat{\sigma}_n^2 = n^{-1}\|\mathbf{y} - \mathbf{X}_n\widehat{\boldsymbol{\beta}}_n - \mathbf{Z}_n\widehat{\boldsymbol{a}}_n\|_2^2$ if p is small. However, if p is large, we suggest to estimating σ alternatively based on a subset model in (2.5). We shall later see that under (2.3) the statistic T_n follows a chi-square distribution with ι degrees of freedom for large n under some sufficient conditions. We define a shrinkage estimator of $\boldsymbol{\beta}_{1n}$ as:

$$\widehat{\boldsymbol{\beta}}_{1n}^{\text{S}} = \widehat{\boldsymbol{\beta}}_{1n}^{\text{SM}} + \left[1 - \frac{(\iota-2)}{T_n}\right](\widehat{\boldsymbol{\beta}}_{1n} - \widehat{\boldsymbol{\beta}}_{1n}^{\text{SM}}), \quad \iota \geq 3. \tag{2.7}$$

This estimator is the weighted average of a full and a submodel estimator. By including the matrix $\boldsymbol{\Upsilon}_n$ in T_n, we restrict the information we want to borrow from $\widehat{\boldsymbol{\beta}}_{2n}$ to a fixed ι-dimensional subspace even though p_{2n} increases with n. We expect to improve the risk of the estimation of $\boldsymbol{\beta}_1$ by borrowing some information from the estimation of ι elements in $\widehat{\boldsymbol{\beta}}_{2n}$.

To avoid over-shrinking where $\widehat{\boldsymbol{\beta}}_{1n}$ has a different sign from $\widehat{\boldsymbol{\beta}}_{1n}^{\text{S}}$, we consider a positive shrinkage estimator given by a convex combination of $\widehat{\boldsymbol{\beta}}_{1n}$ and $\widehat{\boldsymbol{\beta}}_{1n}^{\text{SM}}$,

$$\widehat{\boldsymbol{\beta}}_{1n}^{\text{PS}} = \widehat{\boldsymbol{\beta}}_{1n} + \left[1 - \frac{(\iota-2)}{T_n}\right]^{+}(\widehat{\boldsymbol{\beta}}_{1n} - \widehat{\boldsymbol{\beta}}_{1n}^{\text{SM}}), \quad \iota \geq 3, \tag{2.8}$$

where $z^+ = \max(0, z)$. Again,we emphasize here that $\widehat{\boldsymbol{\beta}}_{1n}^{\text{PS}}$ is particularly important for controlling the over-shrinking problem inherit in the shrinkage estimator.

3. Asymptotic analysis

In this section we investigate the asymptotic properties of the proposed estimators in HD-PLM settings.

3.1. Consistency and asymptotic normality. As a preparation, we first present the asymptotic property of $\widehat{\boldsymbol{\beta}}_n$ when p_n grows with n. Following **Xie and Huang (2009)**, we let $\theta_j(u) = E[x_j \mid U = u]$, $j = 1, \ldots, p_n$. Denote $\mathbf{x}_n = (x_1, \ldots, x_{p_n})'$ and let $\mathbf{e}_n = (e_1, \ldots, e_{p_n})' = \mathbf{x}_n - E[\mathbf{x}_n \mid U]$. Let $\boldsymbol{\Sigma}_n(u) = \text{Var}(\mathbf{e}_n \mid U = u) = \text{Var}(\mathbf{x}_n \mid U = u)$ be the conditional covariance matrix of $(\mathbf{x}_n \mid U = u)$. Let $\boldsymbol{\Gamma} = E[\boldsymbol{\Sigma}_n(U)]$ with the largest and smallest eigenvalues being λ_{1n} and λ_{2n}.

We make the following assumptions.

(C1) There exist constants $\gamma_\theta > 0$ and $\zeta_\theta > 0$ such that

$$\sup_{n\geq 1}\sup_{1\leq j\leq p_n} |\theta_j^{(r_\theta)}(u_2) - \theta_j^{(r_\theta)}(u_1)| \leq \zeta_\theta |u_2 - u_1|^{\gamma_\theta}, \quad 0 \leq u_1,\ u_2 \leq 1,$$

where $r_\theta \leq m-1$ and $r_\theta + \gamma_\theta > \frac{1}{2}$.

(C2) There exist constants $\gamma_g > 0$ and $\zeta_g > 0$ such that

$$|g^{(r_g)}(u_2) - g^{(r_g)}(u_1)| \leq \zeta_g |u_2 - u_1|^{\gamma_g}, \quad 0 \leq u_1,\ u_2 \leq 1,$$

where $r_g \leq m-1$ and $r_g + \gamma_g > 1/2$

(C3) $\sum_{1\leq j \leq p_n} E[e_j^4] \leq \sigma_1$, where σ_1 is independent of n.

(C4) (a) $\lim_{n\to\infty} p_n^2/n = 0$;
(b) $\lim_{n\to\infty} p_n^2 M_n^2/n^2 = 0$;
(c) $\lim_{n\to\infty} p_n/M_n^{r_\theta+\gamma_\theta} = 0$.

(C5) There exists a constant $c_1 > 0$, such that $\liminf_{n\to\infty} \lambda_{2n} = c_1$.

(C6) Suppose for all $u \in [0,1]$, $\mathrm{Tr}\big(\boldsymbol{\Sigma}_n(u)\big) \leq c_2 p_n$ for any $0 \leq u \leq 1$, where c_2 is a positive constant.

Conditions (C1)–(C2) are smooth conditions of g and θ_j's. Both γ_θ and γ_g are Lipschitz condition orders. Condition (C4) indicates that p_n can grow with n at certain rate. (C5) is the regularity condition on the covariance matrix of the parametric part. We can consider a polynomial spline with $\lim_{n\to\infty} n^{1/2}/M_n^{r_\theta+\gamma_\theta+r_g+\gamma_g} = 0$ to approximate the non-parametric part. Let us define

$$\boldsymbol{\Omega}_n = [n^{-1}\mathbf{X}_n'(\mathbf{I}_n - \mathbf{P}_{\mathbf{Z}_n})\mathbf{X}_n]\boldsymbol{\Gamma}_n^{-1}[n^{-1}\mathbf{X}_n'(\mathbf{I}_n - \mathbf{P}_{\mathbf{Z}_n})\mathbf{X}_n]. \tag{3.1}$$

Then we have the following asymptotic results.

THEOREM 3.1. (i) *Under conditions* (C1)–(C5),

$$\|\widehat{\boldsymbol{\beta}}_n - \boldsymbol{\beta}_0\|_2 = O_P\big((p_n/n)^{1/2}\big) + O_P\Big(M_n^{-(r_g+\gamma_g)}\Big).$$

(ii) *Suppose* $\mathbf{A}_n$ *is a fixed* $\iota\times p_n$ *full rank matrix. Under conditions* (C1)–(C6),

$$\sqrt{n}(\mathbf{A}_n\boldsymbol{\Omega}_n^{-1}\mathbf{A}_n')^{-1/2}\mathbf{A}_n(\widehat{\boldsymbol{\beta}}_n - \boldsymbol{\beta}_0) \rightsquigarrow N(\mathbf{0}_\iota, \sigma^2\mathbf{I}_\iota),$$

where "$\rightsquigarrow$" means "converge in distribution."

If we partition the matrix $\boldsymbol{\Omega}_n$ in (3.1) into

$$\begin{pmatrix} \boldsymbol{\Omega}_{11} & \boldsymbol{\Omega}_{12} \\ \boldsymbol{\Omega}_{12}' & \boldsymbol{\Omega}_{22} \end{pmatrix}$$

based on the partition of $\mathbf{X}_n$, we can denote $\boldsymbol{\Omega}_{22.1} = \boldsymbol{\Omega}_{22} - \boldsymbol{\Omega}_{12}'\boldsymbol{\Omega}_{11}'^{-1}\boldsymbol{\Omega}_{12}$ and $\boldsymbol{\Omega}_{11.2} = \boldsymbol{\Omega}_{11} - \boldsymbol{\Omega}_{12}\boldsymbol{\Omega}_{22}'^{-1}\boldsymbol{\Omega}_{12}'$. If we choose $\mathbf{A}_n = \begin{pmatrix}\mathbf{0} & \boldsymbol{\Upsilon}_n\end{pmatrix}$ in Theorem 3.1, then $\sqrt{n}(\boldsymbol{\Upsilon}_n\boldsymbol{\Omega}_{22.1}^{-1}\boldsymbol{\Upsilon}_n')^{-1/2}(\widehat{\boldsymbol{\vartheta}} - \boldsymbol{\vartheta}_0) \rightsquigarrow N(\mathbf{0}_\iota, \sigma^2\mathbf{I}_\iota)$, where $\boldsymbol{\vartheta}_0 = \boldsymbol{\Upsilon}_n\boldsymbol{\beta}_{20}$. From $\boldsymbol{\Omega}_n = \boldsymbol{\Gamma}_n + o_P(1)$ in Lemma A.1 in Appendix, we can further prove that

$$(n/\sigma^2)(\widehat{\boldsymbol{\vartheta}} - \boldsymbol{\vartheta}_0)'(\boldsymbol{\Upsilon}_n(\widetilde{\mathbf{X}}_{2n}'\mathbf{P}_{\widetilde{\mathbf{X}}_{1n}}^{\perp}\widetilde{\mathbf{X}}_{2n})^{-1}\boldsymbol{\Upsilon}_n')^{-1}(\widehat{\boldsymbol{\vartheta}} - \boldsymbol{\vartheta}_0) \rightsquigarrow \chi^2(\iota),$$

where $\chi^2(\iota)$ is a χ^2 distribution with degrees of freedom ι.

3.2. Asymptotic distribution risk. In this section, we will apply the asymptotic result of Theorem 3.1 to investigate the risk reduction for shrinkage estimators of $\boldsymbol{\beta}_{1n}$. For the rest of the paper, we will adopt similar partition and notations to $\boldsymbol{\Omega}_n$ on covariance matrix $\boldsymbol{\Gamma}_n$.

Let $\boldsymbol{\Phi}$ be a p_1-dimensional positive definite matrix. We consider a quadratic loss function of $\boldsymbol{\beta}_1$ using

$$\mathcal{L}(\boldsymbol{\beta}_{1n}^*, \boldsymbol{\beta}_{10}) = n(\boldsymbol{\beta}_{1n}^* - \boldsymbol{\beta}_{10})'\boldsymbol{\Phi}(\boldsymbol{\beta}_{1n}^* - \boldsymbol{\beta}_{10}),$$

where $\boldsymbol{\beta}_{1n}^*$ can be any one of the suggested estimators of $\boldsymbol{\beta}_{1n}$. Recall that $\boldsymbol{\Upsilon}_n$ consists of row vectors of $\mathbf{I}_{p_{2n}}$. If we denote $\mathcal{S} = \{1, \ldots, p_{2n}\}$ and $\mathcal{S}_1$ be corresponding row indexes in $\mathbf{I}_{p_{2n}}$ for $\boldsymbol{\Upsilon}_n$. We consider a PLM with only possible weak signals when $j \in \mathcal{S}_1$ and zero signals when $j \notin \mathcal{S}_1$. In particular, we consider to use the following model K_n to represent the small signals in $\boldsymbol{\vartheta}$, a ι-dimensional sub-vector from $\boldsymbol{\beta}_{20}$,

$$K_n \colon \boldsymbol{\vartheta}_0 = n^{-1/2}\boldsymbol{\delta} \quad \text{and} \quad \|\boldsymbol{\beta}_{20}\| = \|\boldsymbol{\vartheta}_0\|, \tag{3.2}$$

where $\boldsymbol{\delta} = (\delta_1, \delta_2, \ldots, \delta_\iota)' \in \mathfrak{R}^\iota$ is a fixed vector. If we define

$$\Delta_{\boldsymbol{\Upsilon}} = \sigma^{-2}\boldsymbol{\delta}'(\boldsymbol{\Upsilon}_n\boldsymbol{\Gamma}_{22.1}^{-1}\boldsymbol{\Upsilon}_n')^{-1}\boldsymbol{\delta}, \tag{3.3}$$

then $\Delta_{\boldsymbol{\Upsilon}} = 0$ if $\boldsymbol{\beta}_{20} = \mathbf{0}_{p_{2n}}$ or $\boldsymbol{\vartheta}_0 = \mathbf{0}_\iota$.

Under $\{K_n\}$, we can consider the asymptotic distribution function of $\boldsymbol{\beta}_{1n}^*$,

$$F(\mathbf{x}) = \lim_{n\to\infty} P(\sqrt{n}(\boldsymbol{\beta}_{1n}^* - \boldsymbol{\beta}_{10}) \le \mathbf{x} \mid K_n), \tag{3.4}$$

where $F(\mathbf{x})$ is nondegenerate. Thus the asymptotic distributional risk (ADR) of $\boldsymbol{\beta}_{1n}^*$ is defined as

(3.5)

$$\begin{aligned}\mathrm{ADR}(\boldsymbol{\beta}_{1n}^*) &= E\Big[\lim_{n\to\infty} n(\boldsymbol{\beta}_{1n}^* - \boldsymbol{\beta}_{10})'\boldsymbol{\Phi}(\boldsymbol{\beta}_{1n}^* - \boldsymbol{\beta}_{10})\Big] \\ &= \mathrm{Tr}(\boldsymbol{\Phi}\mathbf{V}_n) + E\Big[\lim_{n\to\infty} n(\boldsymbol{\beta}_{1n}^* - \boldsymbol{\beta}_{10})\Big]'\boldsymbol{\Phi}^{-1}E\Big[\lim_{n\to\infty} n(\boldsymbol{\beta}_{1n}^* - \boldsymbol{\beta}_{10})\Big],\end{aligned}$$

where $\mathbf{V}_n$ is the dispersion matrix for the distribution $F(\mathbf{x})$ given in (3.4). Let $H_v(x; c)$ be the cumulative distribution function of the noncentral chi-square distribution with noncentrality parameter c and v degrees of freedom. Further,

$$E(\chi_v^{-2j}(c)) = \int_0^\infty x^{-2j}\,\mathrm{d}H_v(x; c).$$

We have the following theorem on the asymptotic risk.

THEOREM 3.2. *Suppose $\boldsymbol{\beta}_0$ is satisfying K_n in* (3.2) *and conditions* (C1)–(C7) *hold. Let $\boldsymbol{\Upsilon}_n$ consists of any ι independent rows of $\mathbf{I}_{p_{2n}}$. Then we have*

$$\mathrm{ADR}(\widehat{\boldsymbol{\beta}}_{1n}) = \sigma^2\,\mathrm{Tr}(\boldsymbol{\Phi}\boldsymbol{\Gamma}_{11.2}^{-1}), \tag{3.6a}$$

$$\mathrm{ADR}(\widehat{\boldsymbol{\beta}}_{1n}^{\mathrm{SM}}) = \boldsymbol{\delta}'\boldsymbol{\Psi}\boldsymbol{\delta} + \sigma^2\,\mathrm{Tr}(\boldsymbol{\Phi}\boldsymbol{\Gamma}_{11}^{-1}), \tag{3.6b}$$

$$\begin{aligned}\mathrm{ADR}(\widehat{\boldsymbol{\beta}}_{1n}^{\mathrm{S}}) = \sigma^2\Big[&\mathrm{Tr}(\boldsymbol{\Phi}\boldsymbol{\Gamma}_{11.2}^{-1}) - (\iota - 2)\,\mathrm{Tr}(\boldsymbol{\Psi}\boldsymbol{\Gamma}_{22.1}^{-1}) \\ &\times \Big(2E(\chi_{\iota+2}^{-2}(\Delta_{\boldsymbol{\Upsilon}})) - (\iota - 2)E(\chi_{\iota+2}^{-4}(\Delta_{\boldsymbol{\Upsilon}}))\Big)\Big] \\ &+ (\iota^2 - 4)\boldsymbol{\delta}'\boldsymbol{\Psi}\boldsymbol{\delta}E(\chi_{\iota+4}^{-4}(\Delta_{\boldsymbol{\Upsilon}})),\end{aligned}$$

$$\begin{aligned}\mathrm{ADR}(\widehat{\boldsymbol{\beta}}_{1n}^{\mathrm{PS}}) &= \mathrm{ADR}(\widehat{\boldsymbol{\beta}}_{1}^{\mathrm{S}}) - \mathrm{Tr}(\boldsymbol{\Psi}\boldsymbol{\Gamma}_{22.1}^{-1})H_{(\iota+2)}(\iota-2;\Delta_{\boldsymbol{\Upsilon}})\\&\quad + (\iota-2)\,\mathrm{Tr}(\boldsymbol{\Psi}\boldsymbol{\Gamma}_{11.2}^{-1})\big[2E\big(\chi_{\iota+2}^{-2}(\Delta_{\boldsymbol{\Upsilon}})I(\chi_{\iota+2}^{2}(\Delta_{\boldsymbol{\Upsilon}})\le\iota-2)\big)\\&\qquad - (\iota-2)E\big(\chi_{\iota+2}^{-4}(\Delta_{\boldsymbol{\Upsilon}})I(\chi_{\iota+2}^{2}(\Delta_{\boldsymbol{\Upsilon}})\le\iota-2)\big)\big]\\&\quad + \boldsymbol{\delta}_n'\boldsymbol{\Psi}\boldsymbol{\delta}_n[2H_{\iota+2}(\iota-2;\Delta_{\boldsymbol{\Upsilon}}) - H_{\iota+4}(\iota-2;\Delta_{\boldsymbol{\Upsilon}})]\\&\quad - (\iota-2)\boldsymbol{\delta}'\boldsymbol{\Psi}\boldsymbol{\delta}\big[2E\big(\chi_{\iota+2}^{-2}(\Delta_{\boldsymbol{\Upsilon}})I(\chi_{\iota+2}^{2}(\Delta_{\boldsymbol{\Upsilon}})\le\iota-2)\big)\\&\qquad - 2E\big(\chi_{\iota+4}^{-2}(\Delta_{\boldsymbol{\Upsilon}})I(\chi_{\iota+4}^{2}(\Delta_{\boldsymbol{\Upsilon}})\le\iota-2)\big)\\&\qquad + (\iota-2)E\big(\chi_{\iota+4}^{-4}(\Delta_{\boldsymbol{\Upsilon}})I(\chi_{\iota+4}^{2}(\Delta_{\boldsymbol{\Upsilon}})\le\iota-2)\big)\big],\end{aligned}$$

where $\Delta_{\boldsymbol{\Upsilon}}$ *is defined in* (3.3) *and* $\boldsymbol{\Psi} = \boldsymbol{\Upsilon}_n\boldsymbol{\Gamma}_{21}\boldsymbol{\Gamma}_{11}^{-1}\boldsymbol{\Phi}\boldsymbol{\Gamma}_{11}^{-1}\boldsymbol{\Gamma}_{12}\boldsymbol{\Upsilon}_n'$.

The proof is given in the Appendix. Theorem 3.2 results in the following two results directly. We omit the proof since they are analogue to **Hossain (2008**, Section 4.5.1).

COROLLARY 1. *Under conditions in Theorem* 3.2, *for any* $\boldsymbol{\Phi}$ *such that*

$$\mathrm{Tr}(\boldsymbol{\Phi}\boldsymbol{\Gamma}_{22.1}^{-1})/\lambda_{\max}(\boldsymbol{\Phi}\boldsymbol{\Gamma}_{22.1}^{-1}) \ge (\iota+2)/2,$$

we have

$$\mathrm{ADR}(\widehat{\boldsymbol{\beta}}_{1n}^{\mathrm{PS}}) \le \mathrm{ADR}(\widehat{\boldsymbol{\beta}}_{1n}^{\mathrm{S}}) \le \mathrm{ADR}(\widehat{\boldsymbol{\beta}}_{1n}),$$

where $\lambda_{\max}(\boldsymbol{\Phi}\boldsymbol{\Gamma}_{22.1}^{-1})$ *is the maximum eigenvalue of* $\boldsymbol{\Phi}\boldsymbol{\Gamma}_{22.1}^{-1}$.

Corollary 1 tells us that a shrinkage estimator dominates the full PLM estimator regardlessly whether the submodel is correctly specified or not. In the worst possible scenario where $\boldsymbol{\Delta}_{\boldsymbol{\Upsilon}} \to \infty$, that is, a submodel is grossly missspecified, the ADR of the shrinkage estimator converges to the ADR of the full model estimator.

COROLLARY 2. *If* $\Delta_{\boldsymbol{\Upsilon}} = 0$, *then*

$$\mathrm{ADR}(\widehat{\boldsymbol{\beta}}_{1n}^{\mathrm{SM}}) < \mathrm{ADR}(\widehat{\boldsymbol{\beta}}_{1n}^{\mathrm{PS}}) < \mathrm{ADR}(\widehat{\boldsymbol{\beta}}_{1n}^{\mathrm{S}}) < \mathrm{ADR}(\widehat{\boldsymbol{\beta}}_{1n}).$$

Corollary 2 tells us two facts: (1) if the predefined submodel or a submodel based on AE is correct, then shrinkage estimator introduces more bias and is less efficient than the restricted PLM estimator; (2) even though a submodel is misspecified, if we do not pick those useful (true nonzero) elements from $\boldsymbol{\beta}_{2n}$ to use, there still no improvement for the asymptotic risk for the shrinkage estimator, especially when ι is too small.

REMARK 1. The simultaneous variable selection and parameter estimation strategy may not lead to a good estimation strategy when weak signals are co-exist with zero signals. Even though subset models can be provided by some existing variable selection techniques when $p_n < n$, the prediction performance can be improved by the shrinkage strategy, especially when an under-fitted subset model is selected by an aggressive variable selection procedure.

REMARK 2. When p_{2n} grows with n and more and more covariates make mild contributions to the response variable, it is important for us to decide how to seek help from $\widehat{\boldsymbol{\beta}}_{2n}$ to improve the efficiency of the estimation of $\boldsymbol{\beta}_1$ from the above observations. Corollary 2 encourages us to use more elements from $\widehat{\boldsymbol{\beta}}_{2n}$ in the shrinkage step, while Corollary 1 encourages a smaller number of elements.

4. Numerical studies

We use simulation to examine the quadratic risk performance of the various estimators. Our simulation is based on a linear and non-linear regression model (1.1) with a large number of explanatory variables. In all experiments, the ε_i's are simulated from i.i.d. standard normal random variables, $x_{is} = (\xi^1_{(is)})^2 + \xi^2_{(is)}$, where $\xi^1_{(is)}$ and $\xi^2_{(is)}$, $i = 1, \ldots, n$, $s = 1, \ldots, p_n$ are also independent copies of standard normal random variables.

Throughout, we use $m = 4$, $M_n = 4$ and the sample quantiles of u_i's as knots. Each design is repeated 1000 times, as a further increase in the number of realizations did not significantly change the result.

We will investigate the performance of submodel and shrinkage estimators under different scenarios when the true model may or may not contain the submodel. The performance of an estimator $\boldsymbol{\beta}^{\diamond}_{1n}$ is evaluated by the relative mean squared error (RMSE) criterion with respect to $\widehat{\boldsymbol{\beta}}_{1n}$ as follows:

$$\text{RMSE}(\boldsymbol{\beta}^{\diamond}_{1n}) = \frac{E\|\widehat{\boldsymbol{\beta}}_{1n} - \boldsymbol{\beta}_{10}\|^2}{E\|\boldsymbol{\beta}^{\diamond}_{1n} - \boldsymbol{\beta}_{10}\|^2}. \tag{4.1}$$

Therefore, $\text{RMSE}(\boldsymbol{\beta}^{\diamond}_{1n}) > 1$ means the superiority of $\boldsymbol{\beta}^{\diamond}_{1n}$ over $\widehat{\boldsymbol{\beta}}_{1n}$. Larger RMSE indicates the stronger degree of superiority of the estimate $\boldsymbol{\beta}^{\diamond}_{1n}$ over $\widehat{\boldsymbol{\beta}}_{1n}$.

We consider two different scenarios based on whether a submode is predefined or not.

4.1. When a submodel is predefined. We will use two examples to examine the quadratic risk performance of the proposed shrinkage estimator when a predefined submodel is specified in advance. In the first example, a predefined submodel is fixed, and the true model grows bigger from the predefined subset model. We will compare the behavior of the shrinkage estimation with the submodel estimation.

EXAMPLE 4.1 (Comparison of submodel and shrinkage estimation). In the true model, $p_1 = 3$, $p_n = cn^{\alpha}$ for different sample size n, where α changes from 0.3 to 0.5 with an increment of 0.05. We let $\boldsymbol{\beta}_{10} = (1.5, 3, 2)'$ be fixed for every design. We provide different $\boldsymbol{\beta}_{20}$s in the way that $\Delta^* = \|\boldsymbol{\beta}_{20}\|^2$ varying between 0 and $a > 0$. We choose (n, c) to be either $(100, 3.2)$ or $(200, 4.9)$.

Partial results on RMSE in (4.1) generated from SM and PS are reported in Table 1. In summary, $\widehat{\boldsymbol{\beta}}^{\text{PS}}_{1n}$ always dominates $\widehat{\boldsymbol{\beta}}_{1n}$, indicated by $\text{RMSE}(\widehat{\boldsymbol{\beta}}^{\text{PS}}_{1n})$ being larger than 1. When Δ^* is nearby to 0, $\widehat{\boldsymbol{\beta}}^{\text{SM}}_{1n}$ has higher RMSE than $\widehat{\boldsymbol{\beta}}^{\text{PS}}_{1n}$. However, $\widehat{\boldsymbol{\beta}}^{\text{PS}}_{1n}$ turns to dominate $\widehat{\boldsymbol{\beta}}^{\text{SM}}_{1n}$ when Δ^* is further away from 0. For relatively larger Δ^*, $\widehat{\boldsymbol{\beta}}^{\text{SM}}_{1n}$ loses its efficiency quickly, while $\widehat{\boldsymbol{\beta}}^{\text{PS}}_{1n}$ is less affected. We also plot the trend of RMSEs versus Δ^* in Figure 1, where $\text{RMSE}(\widehat{\boldsymbol{\beta}}^{\text{SM}}_{1n})$ goes to 0 quickly and $\text{RMSE}(\widehat{\boldsymbol{\beta}}^{\text{PS}}_{1n})$ goes to 1 when Δ^* becomes larger.

In the following example, we always choose the predefined submodel to be the true model. We will compare the behavior of the shrinkage estimation with the LASSO-type estimation based on the true model.

EXAMPLE 4.2 (Comparison of shrinkage and LASSO-type estimators). In the true model, we let $\boldsymbol{\beta}_{10} = (1.5, 3, 2, \underbrace{0.1, \ldots, 0.1}_{p_1 - 3})'$, $\boldsymbol{\beta}_{20} = \mathbf{0}_{p_{2n}}$. Thus the true model

TABLE 1. Simulated RMSEs in (4.1) for Example 4.1.

(n,p)	Δ^*	$\widehat{\boldsymbol{\beta}}_{1n}^{\mathrm{SM}}$	$\widehat{\boldsymbol{\beta}}_{1n}^{\mathrm{PS}}$	(n,p)	Δ^*	$\widehat{\boldsymbol{\beta}}_{1n}^{\mathrm{SM}}$	$\widehat{\boldsymbol{\beta}}_{1n}^{\mathrm{PS}}$
	0.00	2.088	1.456		0.00	2.958	2.000
	0.05	1.121	1.146		0.05	1.351	1.542
	0.30	0.300	1.050		0.30	0.334	1.153
(100, 13)	0.5	0.192	1.041	(100, 20)	0.50	0.203	1.105
	1.00	0.092	1.034		1.00	0.127	1.062
	1.50	0.061	1.010		1.50	0.077	1.031
	2.00	0.047	1.006		2.00	0.052	1.020
	0.00	2.374	1.735		0.00	3.281	2.511
	0.05	0.801	1.355		0.05	1.081	1.724
	0.30	0.168	1.154		0.30	0.204	1.313
(200, 33)	0.50	0.102	1.117	(200, 53)	0.50	0.137	1.294
	1.00	0.052	1.037		1.00	0.071	1.170
	1.50	0.036	1.026		1.50	0.035	1.098
	2.00	0.027	1.005		2.00	0.032	1.088

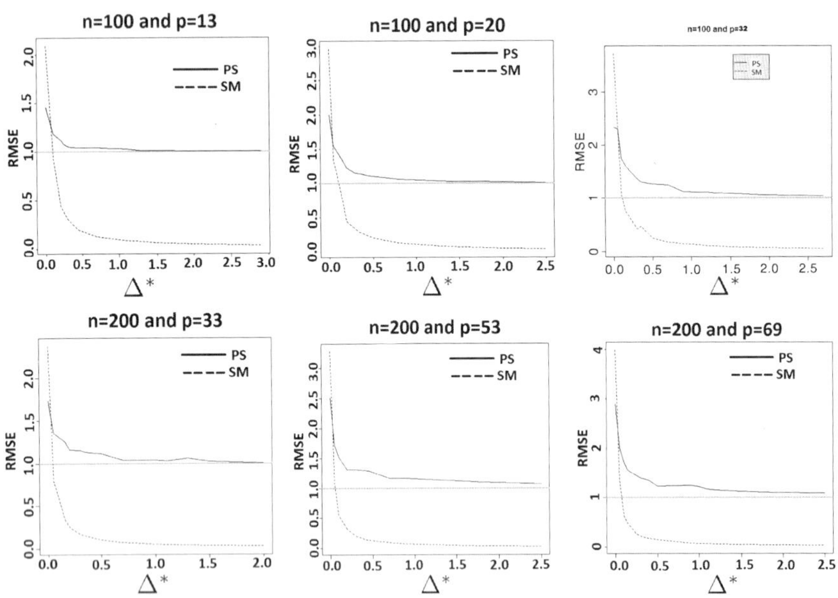

FIGURE 1. RMSEs for SM and PM in Example 4.1. The top three panels are for $n = 100$ and $p_n = 13, 20, 32$ from the left to the right. The bottom panels are for $n = 200$ and $p_n = 33, 53, 69$ from the left to the right. Solid curves: $\mathrm{RMSE}(\widehat{\boldsymbol{\beta}}_{1n}^{\mathrm{PS}})$; Dashed curves: $\mathrm{RMSE}(\widehat{\boldsymbol{\beta}}_{1n}^{\mathrm{SM}})$.

includes some covariates with weak but nonzero effects. We consider $n = 100$ and $p_1 = 3, 4, 8$.

We only report the results of MCP and LASSO since both SCAD and Adaptive LASSO generate similar results to MCP. We choose $\gamma = 3$ for MCP. The optimal tuning parameters are chosen using Schwarz's Bayesian Criterion (BIC, **Schwarz (1978)**) for both LASSO and MCP. Notice that although different variable selection criteria may select different subset models, we still do the comparison based on the same submodel determined by UPI, that is, $\boldsymbol{\beta}_{20} = \mathbf{0}_{p_{2n}}$.

Simulation results are reported in both Table 2 and Figure 2. From the simulation results, the SM dominates all other methods by providing the largest RMSEs. This is not a surprise since the SM is computed based on the true model. We observe that LASSO is less efficient than the PS especially when p_n grows. In addition, PS is comparable to MCP in term of estimation performance when p_1 is

TABLE 2. Simulated RMSEs in (4.1) for $n = 100$ in Example 4.2.

p_1	p_n	$\widehat{\boldsymbol{\beta}}_{1n}^{\mathrm{SM}}$	$\widehat{\boldsymbol{\beta}}_{1n}^{\mathrm{PS}}$	$\widehat{\boldsymbol{\beta}}_{1n}^{\mathrm{MCP}}$	$\widehat{\boldsymbol{\beta}}_{1n}^{\mathrm{LASSO}}$
3	10	1.743	1.265	1.392	1.034
	20	2.318	1.794	1.867	1.081
	32	4.517	2.948	3.480	1.710
	40	5.892	3.837	4.507	1.975
8	10	1.118	1.090	0.938	1.023
	20	1.367	1.231	0.955	1.126
	32	2.138	1.661	1.064	1.293
	40	2.070	1.808	1.261	1.419

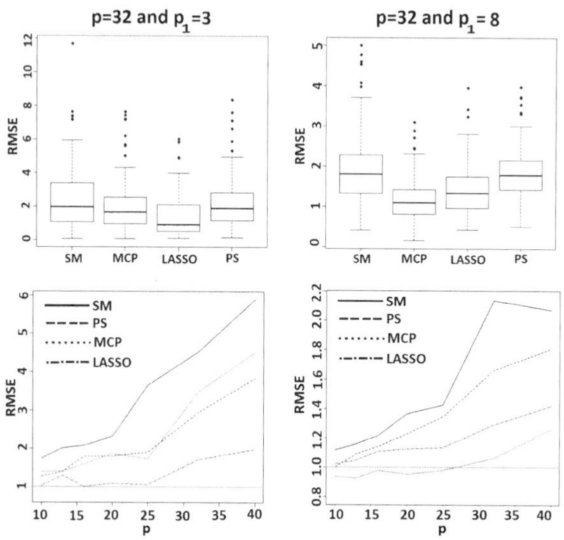

FIGURE 2. RMSEs for PS and MCP or LASSO in Example 4.2, where $\Delta^* = 0$ and $n = 100$. The first two plots are the boxplots of RMSEs out of 1000 iterations; Left panel: $(p_1, p) = (3, 32)$, right panel: $(p_1, p) = (8, 32)$. The last two plots are means of all those RMSEs when p increases from 10 to 40.

small. However, when p_1 gets bigger, and the number of the weaker coefficients is larger PS dominates both MCP and LASSO. This phenomenon can be explained by the existence of covariates with weak effects, which cannot be separated from zero effects using the penalized approach. Overall, shrinkage estimates perform more robustly than the LASSO-type estimates and better than the full model estimates.

4.2. When a submodel is NOT predefined. In Section 4.1, we have demonstrated the performance of the proposed shrinkage estimators when a predefined submodel is specified in advance. In many cases, a subset model is not available. In this case, a LASSO-type method can be extended to do both the variable selection and estimation simultaneously. It is known that both the SCAD and MCP have the oracle properties. In other words, a well-chosen tuning parameter is able to obtain a final procedure with both variable selection consistency and estimation consistency under some sufficient conditions (**Xie and Huang, 2009**; **Zhang, 2010**). However, LASSO does not have variable selection consistency unless crucial Sparse Riesz condition condition and certain other regularity conditions are satisfied (**Zhang and Huang, 2008**). A small λ in LASSO model turns to overfit the data and a big λ turns to shrink too many weak coefficients to be zero. In the following example, we first use either the LASSO or MCP to choose a submodel (defined as $\widehat{\mathcal{M}}$) and then apply the proposed shrinkage methods on $\widehat{\mathcal{M}}$. Here $\widehat{\mathcal{M}}$ may change all the time due to the different tuning parameters. We use BIC to choose the optimal tuning parameter in the final model $\widehat{\mathcal{M}}$.

EXAMPLE 4.3. To begin our study, in the true model, we let

(a) $\boldsymbol{\beta}_{10} = (5, 5, 5, \underbrace{0.01, \ldots, 0.01}_{10})'$, $\boldsymbol{\beta}_{20} = \mathbf{0}'_{p_{2n}}$.

(b) $\boldsymbol{\beta}_{10} = (\underbrace{0.5, \ldots, 0.5}_{13}, \underbrace{0.1, \ldots, 0.1}_{10})'$, $\boldsymbol{\beta}_{20} = \mathbf{0}'_{p_{2n}}$.

(c) $\boldsymbol{\beta}_{10} = (5, 5, 5, \underbrace{0.2, \ldots, 0.2}_{10})'$, $\boldsymbol{\beta}_{20} = \mathbf{0}'_{p_{2n}}$.

All nonzero coefficients are randomly assigned to be either positive or negative. Thus the true model includes some covariates with weak but nonzero effects. We consider $n = 200$ and $p_{2n} = n^{\alpha}$, where α increases from 0.7 to 0.9 with an increment of 0.05.

The relative risk of the proposed shrinkage estimator and the estimator based on corresponding variable selection approaches, LASSO/MCP (based on $\mathcal{M}$) are compared. In Table 3, we report the simulation results when LASSO is used to generate the subset model. We report both the average and standard deviation (in the parenthesis) of RMSEs generated from each method out of 1000 iterations. The mean and standard deviations of the number of linear covariates in $\widehat{\mathcal{M}}$ ($\#\widehat{\mathcal{M}}$) are also reported in Table 3. In Figure 3, plots in the top/middle/bottom panels are for (a)/(b)/(c) in Example 4.3 while LASSO is adopted to select the submodel in all iterations. In each panel, the first two plots are boxplots of RMSEs from 1000 iterations, the third plot gives the 10% trimmed mean values from RMSEs generated from submodel, shrinkage and LASSO estimates, respectively.

In Table 4 and Figure 4, we also report corresponding simulation results while MCP is adopted to choose the final submodel.

From the simulation results, we obtained the following observations.

TABLE 3. Simulated RMSEs in (4.1) in Example 4.3. Subset model is chosen from LASSO.

Case	p_n	$\widehat{\beta}_{1n}^{\text{SM}}$	$\widehat{\beta}_{1n}^{\text{PS}}$	$\widehat{\beta}_{1n}^{\text{LASSO}}$	$\#\widehat{\mathcal{M}}$
(a)	41	0.897 (1.085)	1.128(0.245)	0.127 (0.180)	7.100 (2.558)
	53	0.880 (0.576)	1.156 (0.322)	0.176 (0.177)	7.787 (2.349)
	69	1.083 (1.114)	1.293 (0.541)	0.157 (0.198)	7.112 (2.223)
	90	1.243 (1.135)	1.366 (0.821)	0.186 (0.229)	7.000 (2.149)
	118	1.691 (1.415)	1.566 (0.962)	0.222 (0.330)	6.338 (1.883)
(b)	41	1.128 (0.131)	1.131 (0.171)	0.934 (0.470)	19.61 (2.913)
	53	1.187 (0.184)	1.201 (.204)	1.040 (0.535)	20.97 (3.040)
	69	1.344 (0.306)	1.329 (0.295)	0.939 (0.449)	21.11 (3.387)
	90	1.511 (0.415)	1.501 (0.378)	1.046 (0.528)	23.03 (4.681)
	118	1.882 (0.687)	1.769 (0.471)	1.396 (0.738)	23.73 (4.194)
(c)	41	1.060 (0.321)	1.131 (0.221)	0.495 (0.323)	14.037 (3.783)
	53	1.084 (0.422)	1.191 (0.262)	0.537 (0.558)	14.037 (4.465)
	69	1.225 (0.556)	1.304 (0.404)	0.435 (0.431)	12.162 (4.373)
	90	1.318 (4.493)	1.366 (0.569)	0.381 (0.403)	10.688 (3.605)
	118	1.727 (1.073)	1.610 (0.468)	0.623 (0.595)	11.287 (3.782)

TABLE 4. Simulated RMSEs in (4.1) in Example 4.3. Subset model is chosen from MCP.

Case	p_n	$\widehat{\beta}_{1n}^{\text{SM}}$	$\widehat{\beta}_{1n}^{\text{PS}}$	$\widehat{\beta}_{1n}^{\text{MCP}}$	$\#\widehat{\mathcal{M}}$
(a)	41	0.928 (0.534)	1.074 (0.255)	0.788 (0.722)	8.725 (3.652)
	53	0.924 (0.581)	1.091 (0.475)	0.809 (0.699)	8.725 (3.466)
	69	1.018 (0.692)	1.167 (0.387)	0.913 (0.714)	7.475 (3.210)
	90	1.105 (1.881)	1.253 (0.831)	1.049 (0.926)	6.825 (3.420)
	118	1.448 (1.352)	1.467 (1.416)	1.514 (1.520)	7.075 (3.563)
(b)	41	1.276 (0.319)	1.231 (0.219)	1.183 (0.440)	14.012 (0.479)
	53	1.332 (0.437)	1.293 (0.315)	1.249 (0.488)	14 (0.525)
	69	1.498 (0.688)	1.426 (0.467)	1.498 (0.886)	14 (0.371)
	90	1.91 (0.879)	1.744 (0.557)	1.833 (0.960)	14 (0.351)
	118	2.378 (1.698)	2.088 (0.870)	2.499 (2.196)	14.125 (0.813)
(c)	41	1.072 (0.414)	1.094 (0.24)	1.034 (0.59)	13.025 (4.262)
	53	1.033 (0.533)	1.113 (0.253)	0.945 (0.641)	11.463 (4.589)
	69	1.134 (0.611)	1.214 (0.371)	0.987 (0.661)	10.787 (5.067)
	90	1.203 (1.822)	1.336 (0.759)	1.212 (1.193)	9.387 (5.196)
	118	1.548 (1.055)	1.509 (0.572)	1.543 (1.133)	10.625 (5.582)

(i) When data includes both strong signals and weak signals, both MCP and LASSO tend to ignore those weak covariates. In this case, the shrinkage estimate dominates LASSO/MCP estimates in terms of the risk performance. See results for Case (a) or (c), where $\#\widehat{\mathcal{M}}$ is much smaller or smaller than 14, the number of parametric terms in the true model. This is because shrinkage estimates can recover

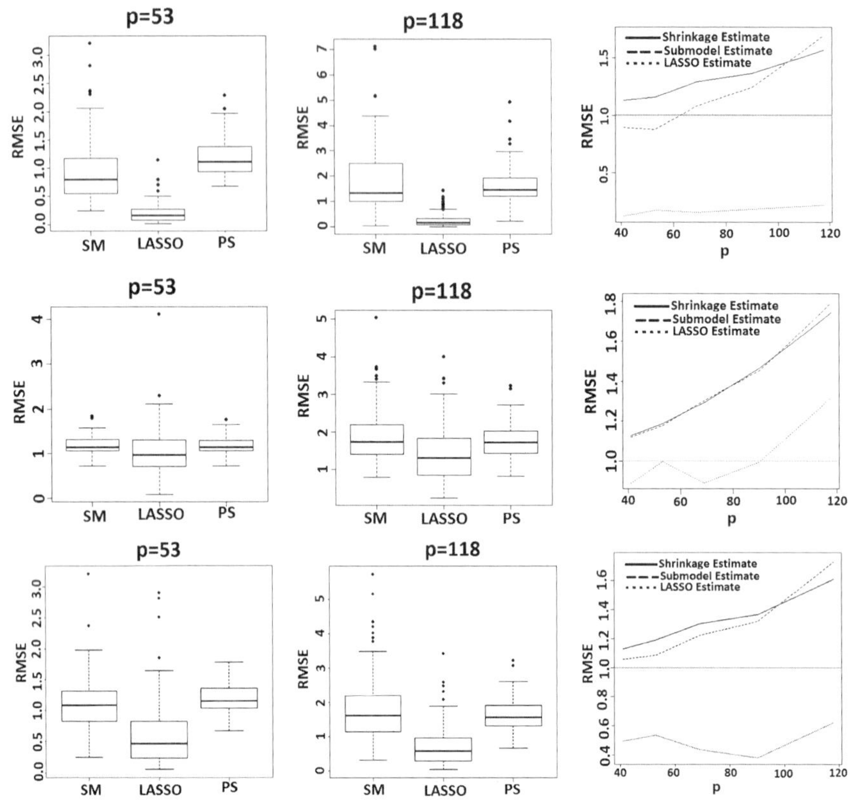

FIGURE 3. RMSEs for SM, PS and LASSO in Example 4.3. Plots in the top/middle/bottom panels are for (a)/(b)/(c). In each panel, the first two boxplots are RMSEs from 1000 iterations, the third plot gives the 10% trimmed mean values from RMSEs generated for SM, PS and LASSO, respectively.

some information ignored by LASSO/MCP when they underfit the model. When p grows, MCP tends to work better than the shrinkage estimator. For example, RMSE of MCP is larger than one from the shrinkage estimate when $p = 118$. This is because shrinkage estimation tends to borrow too much adundant information from large amounts of unimportant covariates when p is large.

(ii) When data includes all moderate signals, LASSO tends to overfit the model by including more linear covariates into the model, while MCP has much better results in terms of variable selection. See $\#\widehat{\mathcal{M}}$ in Case (b). The simple explanation is that LASSO does not have the oracle properties. In many cases, LASSO cannot separate the weak and zero covariates. Therefore, a more aggressive tuning parameter tends to sacrifice the estimation performance by generating a sparser solution. However, MCP has a better performance due to its better performance of variable selection. When p grows at a faster rate, it can outperform the shrinkage estimation.

(iii) Shrinkage estimates are more stable than LASSO/MCP in terms of risk performance. They are not seriously affected by a heavily underfitted model.

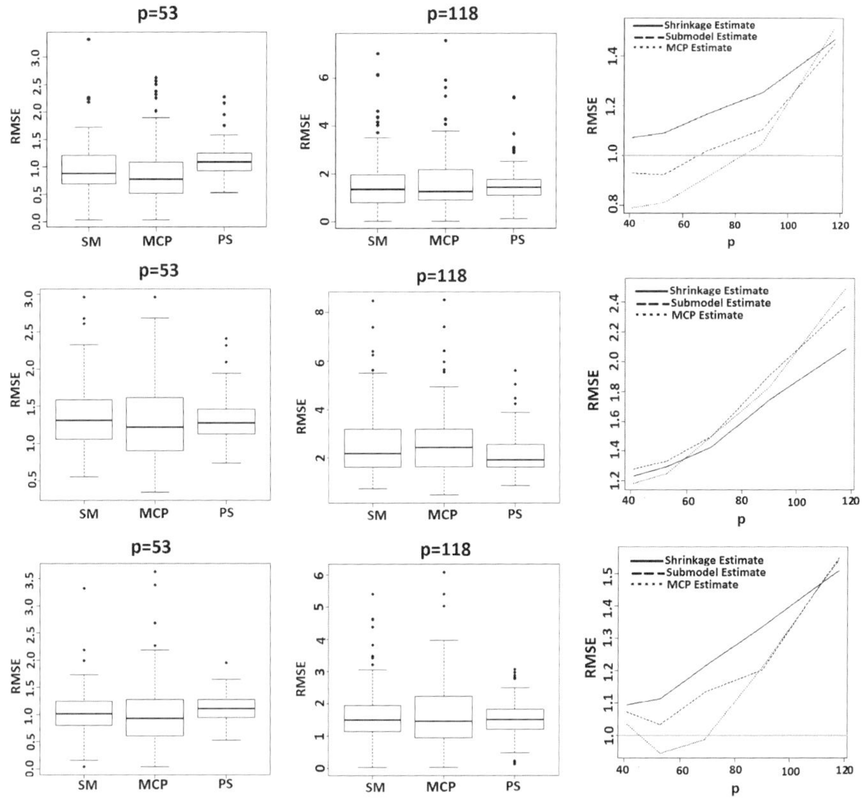

FIGURE 4. RMSEs for SM, PS and LASSO in Example 4.3. Plots in the top/middle/bottom panels are for (a)/(b)/(c). In each panel, the first two boxplots are RMSEs from 1000 iterations, the third plot gives the 10% trimmed mean values for RMSEs generated from SM, PS and MCP, respectively.

(iv) When facing a high-dimensional problem where p is much larger than n, we suggest using two step variable selection approaches. In the first screening step, MCP or any other penalized approach reduces the dimension to a relatively lower dimension, $o(n)$. When a variable selection approach is used in the second step, shrinkage strategy should be applied to improve the estimation performance.

5. Real data analysis

We use the data set reported in **Berndt (1991)** to illustrate the application and the performance of the HD shrinkage and penalty estimators. In this dataset, the wage information of 534 workers and their education, living region, gender, race, occupation, marriage status and years of experience (T) are collected. Assuming a nonlinear relationship between the years of experience and wage level, **Xie and Huang (2009)** suggested a partial linear regression model,

$$y_i = \sum_{j=1}^{14} x_{ij}\beta_j + g(T_i) + \varepsilon_i, \quad i = 1, \ldots, 534,$$

where y_i is the log-value of the worker's wage, T_i is his or her years of experience, x_{ij} are other variables for the ith worker.

The Adaptive LASSO and LASSO methods only provide non-zero estimates on a subset of all 14 covariates. We apply both LASSO and Adaptive LASSO and find 10 and 8 important covariates besides the years of experiences, respectively. Here we compute both shrinkage estimates in the direction of those two predefined submodels. The estimates based on different methods are reported in Table 5.

The ratio of the mean estimation error estimator $\boldsymbol{\beta}^*$ and full model estimator $\widehat{\boldsymbol{\beta}}$ is computed as follows

$$\text{PMSE}(\boldsymbol{\beta}^*_{\mathcal{J}}, \widehat{\boldsymbol{\beta}}_{\mathcal{J}}) = \frac{\sum_{i=1}^n \|\mathbf{y} - \sum_{j\in\mathcal{J}} \mathbf{X}_{\mathcal{J}} \widehat{\boldsymbol{\beta}}_{\mathcal{J}} - \widehat{g}(t)\|^2}{\sum_{i=1}^n \|\mathbf{y} - \sum_{j\in\mathcal{J}} \mathbf{X}_{\mathcal{J}} \boldsymbol{\beta}^*_{\mathcal{J}} - g^*(t)\|^2}, \tag{5.1}$$

where $\mathcal{J}$ is the index of the subset model. Larger PMSE indicates the supremacy of $\boldsymbol{\beta}^*$ over the full model estimator. We compute the prediction errors for positive shrinkage estimates under both subsets chosen by Adaptive LASSO and LASSO, respectively. The prediction errors are computed using cross validation following 300 random partitions of the data set. In each partition, the training set consists of $\frac{2}{3}$ observations and the test set consists of the remaining $\frac{1}{3}$ observations. The results are reported in Figure 5, where the first panel gives the boxplot of 300 PMSEs between Adaptive LASSO estimates and corresponding positive shrinkage estimates. The second panel gives boxplot of 300 PMSEs between positive shrinkage estimates and Adaptive LASSO estimates. It is observed that the positive shrinkage estimates generated from two predefined subsets produce a much smaller prediction error than the corresponding penalized estimates only.

6. Concluding remarks and future outlook

For a partial linear model where the dimension of the regression parameter vector increases with the sample size, we suggest a high dimensional shrinkage

TABLE 5. Positive shrinkage estimates based on the Adaptive LASSO (ALASSO) and LASSO.

Variable	Description	ALASSO	PS ALASSO	LASSO	PS LASSO
X1	Number of years of education	0.0654	0.0639	0.0624	0.0636
X2	1 = southern region, 0 = other	0	—	−0.0724	−0.0795
X3	1 = Female, 0 =Male	−0.2025	−0.2144	−0.2112	−0.2118
X4	1 = Union member, 0 = nonmember	0.1991	0.2121	0.2059	0.2073
X5	1 = Caucasian, 0 = other	0	—	0	—
X6	1 = Hispanic, 0 = other	0	—	0	—
X7	1 = Management, 0 = other	0.2222	0.2714	0.2665	0.2702
X8	1 = sales, 0 = other	−0.0074	−0.0707	−0.0603	−0.0683
X9	1 = clerical, 0 = other	0	—	0	—
X10	1 = service, 0 = other	−0.1108	−0.1056	−0.0965	−0.1053
X11	1 = professional, 0 = other	0.1811	0.2211	0.2185	0.2196
X12	1 = manufacturing, 0 = other	0.0726	0.1210	0.1163	0.1209
X13	1 = construction, 0 = other	0	—	0.1040	0.1186
X14	1 = married, 0 = other	0	—	0	—

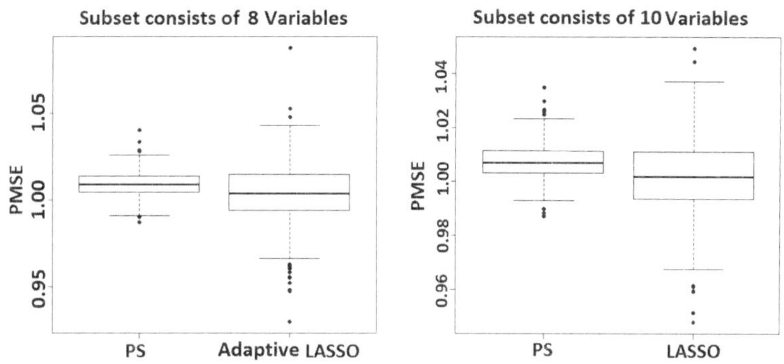

FIGURE 5. Prediction error ratio computed using cross validation following 300 random partitions of the data set. In each partition, the training set consists of $\frac{2}{3}$ observations and the test set consists of the remaining $\frac{1}{3}$ observations. The left panel shows the comparison between positive shrinkage and the Adaptive LASSO estimates; the right panel shows the comparison between positive shrinkage and the LASSO estimates.

estimation strategy. The asymptotic risk properties of the HD shrinkage estimator are developed and are assessed with the risk of the full model, submodel and LASSO-type estimators. We conclude that the suggested estimation strategy is very competitive with LASSO-type estimators and in many cases performs better. Particularly, it performs very well when there are weak signals in terms of regression coefficients. The bulk of current literature on high-dimensional data assumed that the signals and noises (regression coefficients zero) are well separated for simultaneous variable section and post estimation. An exception is **Yang et al. (2014)** in this volume. Realistically, zero coefficients may not be the case in most experiments. For this reason, we suggested a more realistic approach for parameter estimation when the model is not completely sparse, and there are many regression coefficients with weak effects. The proposed strategy is intuitively appealing and can be easily realized. Theoretical and simulation results demonstrated that the HD shrinkage estimator has favorable performance and is a good alternative to LASSO-type estimators. It can save the loss of efficiency of LASSO due to the effect of variable selection.

When p increases with n rapidly then it is reasonable to suspect the co-existence of sparsity and nuisance coefficients. In other words, many predictors in the model may not contribute individually, but at the same time, these semi-active predictors may jointly make some small but not trivial contribution towards the prediction accuracy. Indeed, penalized regularization approaches have the advantage of generating a parsimonious sparse model, but are not able to separate those small contributors and non-contributors. In this situation, the proposed HD shrinkage estimator has a superior prediction performance over other penalty type estimators. The penalty estimators do not permit other covariates to contribute once a sparse model or submodel is generated. The proposed HD shrinkage estimators inherit the advantage of the Stein-type estimators and take into account possible contributions of other nuisance parameters. We acknowledge the importance of penalty type estimates, but at the same time, we should not depend completely on the these type of estimates. A LASSO-type estimator can be the start, but not the end. We can still make prediction improvement by using the proposed HD shrinkage estimation approach. This paper contributes to the investigation of post penalty estimation analysis.

In summary we established the asymptotic normality of the full model estimator when p increases with n. Further, the relative performance of the proposed HD shrinkage estimator to its competitors is appraised analytically and numerically. We show that the performance of the shrinkage strategy is favorable relative to the class of estimators examined in this paper.

In this paper we only investigated the high-dimensional shrinkage estimation in the least squares framework, where the random error is often assumed to be normal. We are also interested in the extension of the proposed shrinkage strategy in more general cases in future studies.

Appendix A

PROOF OF THEOREM 3.1. Define

$$\mathbf{w} = n^{-1/2}[\mathbf{X}'(\mathbf{I}_n - \mathbf{P}_{\mathbf{Z}_n})\mathbf{X}]^{1/2}\mathbf{X}'(\mathbf{I}_n - \mathbf{P}_{\mathbf{Z}_n})\big(\boldsymbol{\varepsilon} + g(\mathbf{U})\big).$$

First, we have the following lemmas from **Xie and Huang (2009)**.

LEMMA A.1. *Under* (C1)–(C2) *and* (C4),

$$\|n^{-1}\mathbf{X}_n'(\mathbf{I} - \mathbf{P}_{\mathbf{Z}})\mathbf{X}_n - \boldsymbol{\Gamma}_n\|_2 = o_P(1).$$

LEMMA A.2. *Under* (C1)–(C2) *and* (C4),

$$\|\mathbf{w}_n\|_2^2 = O_P(p_n/n + M_n^{-2(r_g+\gamma_g)}).$$

Since $\widehat{\boldsymbol{\beta}}_n$ minimizes Q_n in (2.1), $Q_n(\widehat{\boldsymbol{\beta}}_n) \leq Q_n(\boldsymbol{\beta}_n)$. It means that

$$\|(\mathbf{I}_n - \mathbf{P}_{\mathbf{Z}_n})\mathbf{X}_n(\widehat{\boldsymbol{\beta}}_n - \boldsymbol{\beta})\|^2 \leq 2(\boldsymbol{\varepsilon} + g(\mathbf{U}))'(\mathbf{I}_n - \mathbf{P}_{\mathbf{Z}_n})\mathbf{X}_n(\widehat{\boldsymbol{\beta}}_n - \boldsymbol{\beta}).$$

Define $\mathbf{d} = n^{-1/2}[\mathbf{X}'(\mathbf{I}_n - \mathbf{P}_{\mathbf{Z}_n})\mathbf{X}]^{1/2}(\widehat{\boldsymbol{\beta}}_n - \boldsymbol{\beta})$. Therefore, $\|\mathbf{d}_n\|^2 \leq 4\|\mathbf{w}\|_2^2$. Thus, from Lemma A.1 and (C5),

$$c_1\|\widehat{\boldsymbol{\beta}} - \boldsymbol{\beta}\|^2 \leq 4\|\mathbf{w}\|_2^2 \leq O_P\Big(p_n/n + M_n^{-2(r_g+\gamma_g)}\Big).$$

Thus, Theorem 3.1(i) holds.

Let us verify the asymptotic normality. From (2.2),

$$\widehat{\boldsymbol{\beta}}_n - \boldsymbol{\beta}_n = (\mathbf{X}'(\mathbf{I}_n - \mathbf{P}_{\mathbf{Z}_n})\mathbf{X}]^{-1}\mathbf{X}'(\mathbf{I}_n - \mathbf{P}_{\mathbf{Z}_n})(g(\mathbf{U}) + \boldsymbol{\varepsilon}).$$

The rest of the proof is similar to **Xie and Huang (2009)**. □

PROOF OF THEOREM 3.2. For the full estimator, the dispersion matrix for the distribution of $\widehat{\boldsymbol{\beta}}_{1n}$ is $\mathbf{V}_n = \sigma^2\mathbf{D}_{11.2}^{-1}$ from Theorem 3.1. Then (3.6a) holds. From (3.1) and Lemma A.1, we know

$$\boldsymbol{\Omega}_n = n^{-1}\widetilde{\mathbf{X}}_n'\widetilde{\mathbf{X}}_n + o_P(1) = \boldsymbol{\Gamma} + o_P(1).$$

Notice that

$$\widehat{\boldsymbol{\beta}}_{1n}^{\mathrm{SM}} = \widehat{\boldsymbol{\beta}}_{1n} + (\widetilde{\mathbf{X}}_{1n}'\widetilde{\mathbf{X}}_{1n})^{-1}\widetilde{\mathbf{X}}_{1n}'\widetilde{\mathbf{X}}_{2n}\widehat{\boldsymbol{\beta}}_{2n}. \tag{A.1}$$

Then,

$$\begin{aligned}
&\lim_{n\to\infty} \sqrt{n}(\widehat{\boldsymbol{\beta}}_{1n}^{\mathrm{SM}} - \boldsymbol{\beta}_{10}) \\
&\quad = \lim_{n\to\infty} \sqrt{n}(\widehat{\boldsymbol{\beta}}_{1n} - \boldsymbol{\beta}_{10}) + \lim_{n\to\infty} \sqrt{n}\boldsymbol{\Omega}_{11}^{-1}\boldsymbol{\Omega}_{12}\boldsymbol{\beta}_{20} + \lim_{n\to\infty} \sqrt{n}\boldsymbol{\Omega}_{11}^{-1}\boldsymbol{\Omega}_{12}(\widehat{\boldsymbol{\beta}}_{2n} - \boldsymbol{\beta}_{20}) \\
&\quad = \lim_{n\to\infty} \sqrt{n}\left(\mathbf{I}_{p_1} \quad \boldsymbol{\Omega}_{11}^{-1}\boldsymbol{\Omega}_{12}\right)(\widehat{\boldsymbol{\beta}}_n - \boldsymbol{\beta}_0) + \lim_{n\to\infty} \sqrt{n}\boldsymbol{\Omega}_{11}^{-1}\boldsymbol{\Omega}_{12}\boldsymbol{\Upsilon}_n'\boldsymbol{\vartheta}_0 \\
&\quad = \mathbf{W}_1 + \mathbf{W}_2.
\end{aligned}$$

Notice that

$$\boldsymbol{\Gamma}_{11}^{-1}\boldsymbol{\Gamma}_{12}\boldsymbol{\Gamma}_{22.1}^{-1}\boldsymbol{\Gamma}_{21}\boldsymbol{\Gamma}_{11}^{-1} = \boldsymbol{\Gamma}_{11.2}^{-1} - \boldsymbol{\Gamma}_{11}^{-1}.$$

Then from Theorem 3.1(ii), we know that

$$\mathbf{W}_1 \equiv \lim_{n\to\infty} \sqrt{n}\left(\mathbf{I}_{p_1} \quad \boldsymbol{\Omega}_{11}^{-1}\boldsymbol{\Omega}_{12}\right)(\widehat{\boldsymbol{\beta}}_n - \boldsymbol{\beta}_0) \sim N(\mathbf{0}_{p_1}, \boldsymbol{\Gamma}_{11}^{-1}).$$

In addition, under K_n,

$$\lim_{n\to\infty} \mathbf{W}_2'\boldsymbol{\Phi}\mathbf{W}_2 = \boldsymbol{\delta}'\boldsymbol{\Upsilon}_n\boldsymbol{\Gamma}_{21}\boldsymbol{\Gamma}_{11}^{-1}\boldsymbol{\Phi}\boldsymbol{\Gamma}_{11}^{-1}\boldsymbol{\Gamma}_{12}\boldsymbol{\Upsilon}_n'\boldsymbol{\delta} = \boldsymbol{\delta}'\boldsymbol{\Psi}\boldsymbol{\delta}.$$

From the definition in (3.5), (3.6a) holds. From the definition in (2.7) and (A.1),

$$\sqrt{n}(\widehat{\boldsymbol{\beta}}_{1n}^{\mathrm{S}} - \boldsymbol{\beta}_{10}) = \sqrt{n}(\widehat{\boldsymbol{\beta}}_{1n} - \boldsymbol{\beta}_{10}) + \sqrt{n}(\widetilde{\mathbf{X}}_{1n}'\widetilde{\mathbf{X}}_{1n})^{-1}\widetilde{\mathbf{X}}_{1n}'\widetilde{\mathbf{X}}_{2n}\widehat{\boldsymbol{\beta}}_{2n}\big((\iota - 2)/T_n\big).$$

In Section 3.1, we have demonstrated that T_n approximates to a χ^2 distribution with degrees of freedom ι and noncentrality $\Delta_{\boldsymbol{\Upsilon}}$. The rest proof just follows **Hossain (2008)**. □

Acknowledgements. We would like to thank the referees for their valuable suggestions in the revision of this paper.

References

Ahmed, S. E., K. A. Doksum, S. Hossain, and J. You. 2007. *Shrinkage, pretest and absolute penalty estimators in partially linear models*, Aust. N. Z. J. Stat. **49**, no. 4, 435–454, DOI 10.1111/j.1467-842X.2007.00493.x. MR2394591 (2009b:62139)

Berndt, E. R. 1991. *The practice of econometrics: classic and contemporary*, Addison-Wesley, Reading, MA.

Chen, H. 1988. *Convergence rates for parametric components in a partly linear model*, Ann. Statist. **16**, no. 1, 136–146, DOI 10.1214/aos/1176350695. MR924861 (89k:62082)

Engle, R. F., C. W. J. Granger, J. Rice, and A. Weiss. 1986. *Semiparametric estimates of the relationship between weather and electricity sales*, J. Amer. Statist. Assoc. **81**, no. 394, 310–320.

Fallahpour, S., S. E. Ahmed, and K. A. Doksum. 2012. *L_1 penalty and shrinkage estimation in partially linear models with random coefficient autoregressive errors*, Appl. Stoch. Models Bus. Ind. **28**, no. 3, 236–250, DOI 10.1002/asmb.933. MR2944145

Fan, J. and R. Li. 2001. *Variable selection via nonconcave penalized likelihood and its oracle properties*, J. Amer. Statist. Assoc. **96**, no. 456, 1348–1360, DOI 10.1198/016214501753382273. MR1946581 (2003k:62160)

Heckman, N. E. 1986. *Spline smoothing in a partly linear model*, J. Roy. Statist. Soc. Ser. B **48**, no. 2, 244–248. MR868002 (88f:62092)

Hossain, Md. S. 2008. *Shrinkage, pretest and LASSO estimators in parametric and semiparametric linear models*, ProQuest LLC, Ann Arbor, MI. Thesis (Ph.D.)–University of Windsor (Canada). MR2712478

Ni, X., H. H. Zhang, and D. Zhang. 2009. *Automatic model selection for partially linear models*, J. Multivariate Anal. **100**, no. 9, 2100–2111, DOI 10.1016/j.jmva.2009.06.009. MR2543089 (2010k:62169)

Raheem, S. M. E., S. E. Ahmed, and K. A. Doksum. 2012. *Absolute penalty and shrinkage estimation in partially linear models*, Comput. Statist. Data Anal. **56**, no. 4, 874–891, DOI 10.1016/j.csda.2011.09.021. MR2888731

Robinson, P. M. 1988. *Root-N-consistent semiparametric regression*, Econometrica **56**, no. 4, 931–954, DOI 10.2307/1912705. MR951762 (89j:62057)

Schumaker, L. L. 1981. *Spline functions: basic theory*, John Wiley & Sons Inc., New York. Pure and Applied Mathematics; A Wiley-Interscience Publication. MR606200 (82j:41001)

Schwarz, G. 1978. *Estimating the dimension of a model*, Ann. Statist. **6**, no. 2, 461–464. MR0468014 (57 #7855)

Speckman, P. 1988. *Kernel smoothing in partial linear models*, J. Roy. Statist. Soc. Ser. B **50**, no. 3, 413–436. MR970977 (90g:62093)

Tibshirani, R. 1996. *Regression shrinkage and selection via the lasso*, J. Roy. Statist. Soc. Ser. B **58**, no. 1, 267–288. MR1379242 (96j:62134)

Xie, H. and J. Huang. 2009. *SCAD-penalized regression in high-dimensional partially linear models*, Ann. Statist. **37**, no. 2, 673–696, DOI 10.1214/07-AOS580. MR2502647 (2010i:62198)

Wahba, G. 1990. *Spline models for observational data*, CBMS-NSF Regional Conference Series in Applied Mathematics, vol. 59, Society for Industrial and Applied Mathematics (SIAM), Philadelphia, PA. MR1045442 (91g:62028)

Yang, F., K. Doksum, and K.-W. Tsui. 2014. *Principal component analysis (PCA) for high-dimensional data. PCA is dead. Long live PCA*, in this book.

Zhang, C.-H. 2010. *Nearly unbiased variable selection under minimax concave penalty*, Ann. Statist. **38**, no. 2, 894–942, DOI 10.1214/09-AOS729. MR2604701 (2011d:62211)

Zhang, C.-H. and J. Huang. 2008. *The sparsity and bias of the LASSO selection in high-dimensional linear regression*, Ann. Statist. **36**, no. 4, 1567–1594, DOI 10.1214/07-AOS520. MR2435448 (2010h:62204)

Zou, H. 2006. *The adaptive lasso and its oracle properties*, J. Amer. Statist. Assoc. **101**, no. 476, 1418–1429, DOI 10.1198/016214506000000735. MR2279469 (2008d:62024)

Department of Mathematics and Statistics, Oakland University, Rochester, Michigan 48309

Current address: Department of Mathematics and Statistics, University of North Carolina at Greensboro, Greensboro, North Carolina 27412

E-mail address: `x_gao2@uncg.edu0`

Faculty of Math & Science, Brock University, 500 Glenridge Ave. St. Catharines, ON L2S 3A1, Canada

E-mail address: `sahmed5@brocku.ca`

Contemporary Mathematics
Volume **622**, 2014
http://dx.doi.org/10.1090/conm/622/12438

Geometry and properties of generalized ridge regression in high dimensions

Hemant Ishwaran and J. Sunil Rao

ABSTRACT. Hoerl and Kennard proposed generalized ridge regression (GRR) over forty years ago as a means to overcome deficiencies in least squares in multicollinear problems. Because high-dimensional regression naturally involves correlated predictors, in part due to the nature of the data and in part due to artifact of the dimensionality, it is reasonable to consider GRR for such problems. We study GRR when the number of predictors p exceeds the sample size n. A novel geometric interpretation for GRR is described in terms of a uniquely defined least squares estimator and lends insight into its properties. It is shown that GRR possesses a shrinkage property useful in correlated settings and that in sparse high-dimensional settings it can have excellent performance but no such guarantees hold in non-sparse settings. We describe a computationally efficient representation for GRR requiring only a linear number of operations in p, thus making GRR computationally applicable to high dimensions.

1. Introduction

Let $\mathbf{Y} = (Y_1, \ldots, Y_n)^T \in \mathbb{R}^n$ be a response vector and $\mathbf{X} = (\mathbf{X}_{(1)}, \ldots, \mathbf{X}_{(p)})$ an $n \times p$ design matrix, where $\mathbf{X}_{(k)} = (x_{k,1}, \ldots, x_{k,n})^T$ denotes the kth column of $\mathbf{X}$. It is assumed that

$$\mathbf{Y} = \mathbf{X}\boldsymbol{\beta} + \boldsymbol{\varepsilon}, \tag{1.1}$$

where $\boldsymbol{\varepsilon} = (\varepsilon_1, \ldots, \varepsilon_n)^T$ is such that $\mathbb{E}(\varepsilon_i) = 0$ and $\mathbb{E}(\varepsilon_i^2) = \sigma_0^2 > 0$. The true value for the coefficient vector $\boldsymbol{\beta} = (\beta_1, \ldots, \beta_p)^T \in \mathbb{R}^p$ is unknown and is denoted by $\boldsymbol{\beta}_0 = (\beta_{0,1}, \ldots, \beta_{0,p})^T$. In this paper, we focus on linear regression models (1.1) in high-dimensional scenarios where $p > n$.

The so-called "big-p small-n problem" poses unique obstacles for estimation of $\boldsymbol{\beta}_0$. One significant concern is multicollinearity. With very large p, the sample correlation between variables can become sizeable as a pure artifact of the dimensionality. Groups of variables become highly correlated with other groups of variables sporadically. These effects can even occur when the population design matrix is orthogonal, i.e., $\mathbb{E}(\mathbf{X}_{(k)}^T \mathbf{X}_{(j)}) = 0$ if $k \neq j$ (see [**1**,**3**] for a discussion of these points). Multicollinearity is further compounded as variables collected in high-dimensional applications are often naturally correlated because of the underlying technology

2010 *Mathematics Subject Classification.* Primary 62G99.

The first author was supported by DMS grant 0705037 from the National Science Foundation.

or science: for example, gene expression values obtained from DNA microarrays, or genotype data collected using SNP arrays in Genome Wide Association Studies (GWAS).

Over 40 years ago, Hoerl and Kennard [**6**, **7**] proposed generalized ridge regression (GRR), a method specifically designed for correlated and ill-conditioned settings. Although it is unlikely they envisioned using GRR in problems where p could be orders of magnitudes larger than n, it is natural to wonder if it can be applied effectively in such contexts.

We recall the definition of GRR. Let $\mathbf{\Lambda} = \text{diag}\{\lambda_k\}_{k=1}^p$ be a $p \times p$ diagonal matrix with diagonal entries $\lambda_k > 0$. The GRR estimator with ridge matrix $\mathbf{\Lambda}$ is

$$\widehat{\boldsymbol{\beta}}_{\text{G}} = (\mathbf{Q} + \mathbf{\Lambda})^{-1}\mathbf{X}^T\mathbf{Y},$$

where $\mathbf{Q} = \mathbf{X}^T\mathbf{X}$. An important property of GRR is that $\widehat{\boldsymbol{\beta}}_{\text{G}}$ is well defined even when $\mathbf{Q}$ is non-invertible. An alternative representation for $\widehat{\boldsymbol{\beta}}_{\text{G}}$ is in terms of ℓ_2-penalization:

$$\widehat{\boldsymbol{\beta}}_{\text{G}} = \underset{\boldsymbol{\beta}\in\mathbb{R}^p}{\arg\min}\left\{\|\mathbf{Y} - \mathbf{X}\boldsymbol{\beta}\|^2 + \sum_{k=1}^p \lambda_k\beta_k^2\right\}, \tag{1.2}$$

where $\|\cdot\|$ is the ℓ_2-norm. Setting $\mathbf{\Lambda} = \lambda\mathbf{I}_p$, where $\mathbf{I}_p$ is the $p \times p$ identity matrix, yields the ridge estimator $\widehat{\boldsymbol{\beta}}_{\text{R}} = (\mathbf{Q} + \lambda\mathbf{I}_p)^{-1}\mathbf{X}^T\mathbf{Y}$ as a special case. The parameter $\lambda > 0$ is referred to as the ridge parameter. Setting $\lambda = 0$ and assuming that $\mathbf{Q}$ is invertible yields the OLS (ordinary least squares) estimator $\widehat{\boldsymbol{\beta}}_{\text{OLS}} = \mathbf{Q}^{-1}\mathbf{X}^T\mathbf{Y}$.

Much of the recent effort to address high-dimensional problems has focused on ℓ_1-penalization (lasso) methods [**15**]. Some of these are similar to GRR in that they allow a unique regularization parameter for each coefficient, although the penalization is in an ℓ_1-sense. Examples include the adaptive lasso for $p < n$ problems [**16**] and for the diverging parameters problem, the recent extension by [**8**]. As well, [**18**] recently introduced the adaptive elastic net, which imposes both an adaptive lasso and a ridge penalty. Similar to the original elastic net [**17**], the additional ridge penalty encourages a grouping effect that can help select groups of correlated variables and stabilizes model predictions.

Unlike lasso-based estimators, GRR imposes parameterwise adaptation using ℓ_2-regularization, which may have benefits in high-dimensional correlated settings. In the classic setting when $n > p$, it is well known that the ridge estimator uniformly shrinks the OLS, thus reducing its squared-length relative to the OLS [**6**]. This variance reduction enables the ridge estimator to outperform the OLS in correlated settings. However, the ridge estimator shrinks all coefficients uniformly to zero, which is effective only when all coefficients are small. On the other hand, GRR generalizes ridge estimation by using a unique parameter λ_k for each coefficient β_k, which allows GRR to achieve non-uniform shrinkage, thus making it feasible to selectively shrink coefficients to zero, similar to the lasso. In the classic settiing $n > p$, it has been shown that GRR estimators can achieve oracle properties [**10**].

We take a geometric approach to study the properties of GRR (Section 2). Because OLS is not uniquely defined when $p > n$, these arguments make use of the *minimum least squares* (MLS) estimator, which is the uniquely defined least squares estimator with minimum squared-length (see Definition 2.3). Using a modified MLS estimator, a novel geometric interpretation for GRR is described that lends insight into its properties. Section 3 lists implications of these findings for GRR in high dimensions. Analogous to ridge estimation in $n > p$ settings, it is shown that the GRR estimator is shrunk relative to MLS in correlated settings and that the GRR predictor has the tendency to shrink towards zero in unfavorable directions in the $\mathbf{X}$-column space. This can lead to both improved estimation and prediction over MLS. However, unlike the classic setting, GRR is constrained to lie in a low-dimensional subspace containing the modified MLS estimator. This implies that for accurate estimation the true parameter vector should be sparse. In non-sparse situations, accurate estimation cannot be guaranteed. Section 4 summarizes our findings and presents an empirical example.

2. Geometry and properties of the GRR estimator when $p \geq n$

Here we present a novel geometric interpretation of the GRR estimator when $p \geq n$ and list some of its key properties. The following lemma plays a key role.

LEMMA 2.1. *Let $p \geq n$ and let $\mathbf{X} = \mathbf{U}\mathbf{D}\mathbf{V}^T$ be the singular value decomposition (SVD) of $\mathbf{X}$ where $\mathbf{U}(n \times n)$ and $\mathbf{V}(p \times n)$ are column orthonormal matrices ($\mathbf{U}\mathbf{U}^T = \mathbf{U}^T\mathbf{U} = \mathbf{V}^T\mathbf{V} = \mathbf{I}_n$) and $\mathbf{D}(n \times n)$ is a diagonal matrix with entries $d_1 \geq d_2 \geq \cdots \geq d_n \geq 0$ (the singular values of $\mathbf{X}$). Let $\mathbf{A} = \mathbf{V}^T(\mathbf{Q} + \lambda\mathbf{I}_p)$. Then for any $\lambda > 0$*

$$\mathbf{A}^+ = \mathbf{V}\mathbf{S}_\lambda^{-1}, \tag{2.1}$$

where $\mathbf{S}_\lambda = \operatorname{diag}\{d_i^2 + \lambda\}_{i=1}^n$ is an $n \times n$ diagonal matrix and $\mathbf{A}^+$ denotes the Moore–Penrose [**13**] *generalized inverse of $\mathbf{A}$. Furthermore, $\mathbf{A}\mathbf{A}^+ = \mathbf{I}_n$.*

Lemma 2.1 will enable us to re-express $\widehat{\boldsymbol{\beta}}_{\mathrm{G}}$ in a manner more conducive to studying its properties. First we observe that $\widehat{\boldsymbol{\beta}}_{\mathrm{G}}$ can be recast as a rescaled ridge estimator. Let $\mathbf{X}_* = \mathbf{X}\boldsymbol{\Lambda}^{-1/2}$ and $\mathbf{Q}_* = \mathbf{X}_*^T\mathbf{X}_*$ (hereafter "$*$" will be used to indicate a term mapped under the transformation $\mathbf{X}\boldsymbol{\Lambda}^{-1/2}$). Then

$$\widehat{\boldsymbol{\beta}}_{\mathrm{G}} = \boldsymbol{\Lambda}^{-1/2}(\mathbf{Q}_* + \mathbf{I}_p)^{-1}\mathbf{X}_*^T\mathbf{Y} = \boldsymbol{\Lambda}^{-1/2}\widehat{\boldsymbol{\beta}}_{\mathrm{R}}^*, \tag{2.2}$$

where $\widehat{\boldsymbol{\beta}}_{\mathrm{R}}^* = (\mathbf{Q}_* + \mathbf{I}_p)^{-1}\mathbf{X}_*^T\mathbf{Y}$ is the ridge estimator for the design matrix $\mathbf{X}_*$ with ridge parameter $\lambda = 1$. Let $\mathbf{X}_* = \mathbf{U}_*\mathbf{D}_*\mathbf{V}_*^T$ be the SVD for $\mathbf{X}_*$. Let $d_{1,*} \geq \cdots \geq d_{n,*} \geq 0$ denote the diagonal elements of $\mathbf{D}_*$. Lemma 2.1 implies the following result.

THEOREM 2.2. *If $p \geq n$ and $\lambda_k > 0$ for $k = 1, \ldots, p$, then*

$$\widehat{\boldsymbol{\beta}}_{\mathrm{G}} = \boldsymbol{\Lambda}^{-1/2}\mathbf{V}_*\mathbf{S}_{*1}^{-1}\mathbf{R}_*^T\mathbf{Y}, \tag{2.3}$$

where $S_{*1} = \operatorname{diag}\{d_{i,*}^2 + 1\}_{i=1}^n$ and $\mathbf{R}_* = \mathbf{U}_*\mathbf{D}_*$. Moreover, (2.3) *can be calculated in $O(pn^2)$ operations.*

2.1. Geometry.
We now describe a novel geometric interpretation for $\widehat{\boldsymbol{\beta}}_{\mathrm{G}}$. The MLS estimator will play a key role in this description and is formally defined below. For this, and all other results, we hereafter assume that $p \geq n$, $\lambda_k > 0$ for $k = 1, \ldots, p$, and $\lambda > 0$, unless otherwise stated.

DEFINITION 2.3. Call any vector $\boldsymbol{\beta} \in \mathbb{R}^p$ a least squares solution if $\|\mathbf{Y} - \mathbf{X}\boldsymbol{\beta}\|^2 \leq \|\mathbf{Y} - \mathbf{X}\mathbf{z}\|^2$ for all $\mathbf{z} \in \mathbb{R}^p$. A vector $\boldsymbol{\beta}$ is called a MLS solution if $\boldsymbol{\beta}$ is a least squares solution and $\|\boldsymbol{\beta}\|^2 < \|\mathbf{z}\|^2$ for all other least squares solutions $\mathbf{z}$.

A celebrated result, due to [**14**], is that the MLS estimator exists and is the unique estimator

$$\widehat{\boldsymbol{\beta}}_{\mathrm{MLS}} = \mathbf{X}^{+}\mathbf{Y} = \lim_{\lambda\to 0}\widehat{\boldsymbol{\beta}}_{\mathrm{R}} = \mathbf{V}\mathbf{S}_0^{+}\mathbf{R}^T\mathbf{Y},$$

where $\mathbf{R} = \mathbf{UD}$ and $\mathbf{S}_0^{+} = \mathrm{diag}\{s_{0i}^{+}\}_{i=1}^{n}$ is the Moore–Penrose generalized inverse of $\mathbf{S}_0$ defined by

$$s_{0i}^{+} = \begin{cases} 1/d_i^2 & \text{if } d_i > 0 \\ 0 & \text{otherwise.} \end{cases}$$

Our geometric result uses a slightly modified MLS estimator obtained using the transformed design matrix $\mathbf{X}_*$. The modified MLS estimator is defined as

$$\widehat{\boldsymbol{\beta}}^{*}_{\mathrm{MLS}} = \boldsymbol{\Lambda}^{-1/2}\mathbf{X}_*^{+}\mathbf{Y} = \boldsymbol{\Lambda}^{-1/2}\mathbf{V}_*\mathbf{S}_{*0}^{+}\mathbf{R}_*^T\mathbf{Y}. \tag{2.4}$$

Note that in the special case $\boldsymbol{\Lambda} = \lambda\mathbf{I}_p$ we obtain $\widehat{\boldsymbol{\beta}}^{*}_{\mathrm{MLS}} = \widehat{\boldsymbol{\beta}}_{\mathrm{MLS}}$. Consider the following geometric interpretation for GRR.

THEOREM 2.4. *$\widehat{\boldsymbol{\beta}}_{\mathrm{G}}$ is the solution to the following optimization problem:*

$$\underset{\boldsymbol{\beta}\in\mathbb{R}^p}{\text{minimize}}\ \mathbb{Q}(\boldsymbol{\beta}, \widehat{\boldsymbol{\beta}}^{*}_{\mathrm{MLS}}) \text{ subject to } \boldsymbol{\beta}^T\boldsymbol{\Lambda}\boldsymbol{\beta} \le L, \tag{2.5}$$

for some $L > 0$, where

$$\mathbb{Q}(\boldsymbol{\beta}, \widehat{\boldsymbol{\beta}}^{*}_{\mathrm{MLS}}) = \big(\boldsymbol{\beta} - \widehat{\boldsymbol{\beta}}^{*}_{\mathrm{MLS}}\big)^T\big(\boldsymbol{\Lambda}^{1/2}\mathbf{Q}_*\boldsymbol{\Lambda}^{1/2}\big)\big(\boldsymbol{\beta} - \widehat{\boldsymbol{\beta}}^{*}_{\mathrm{MLS}}\big)$$

defines an ellipsoid with contours $\boldsymbol{\Sigma}(c) = \{\boldsymbol{\beta} : \mathbb{Q}(\boldsymbol{\beta}, \widehat{\boldsymbol{\beta}}^{*}_{\mathrm{MLS}}) = c^2\}$ centered at $\widehat{\boldsymbol{\beta}}^{*}_{\mathrm{MLS}}$.

Theorem 2.4 shows that the GRR estimator is the solution to a constrained optimization problem involving the contours of an ellipsoid centered at the modified MLS estimator. This generalizes the classic setting $n > p$ from an optimization problem involving OLS to one involving MLS. The constraint region is generalized as well. For GRR, the constraint region is an ellipsoid that depends upon $\boldsymbol{\Lambda}$, whereas in the classic setting the constraint region is spherical. Another key difference is that the dimension of the subspace that $\widehat{\boldsymbol{\beta}}_G$ lies in depends upon n, and not p (see Theorem 2.5 below). Figure 1 provides an illustration of Theorem 2.4.

2.2. Properties. The following theorem summarizes key properties of GRR. It also gives an explicit representation for the linear predictor $\widehat{\boldsymbol{\mu}}_{\mathrm{G}} = \mathbf{X}\widehat{\boldsymbol{\beta}}_{\mathrm{G}}$. In the following, let $\mathbf{v}_{i,*}$ be the ith column vector of $\mathbf{V}_*$ and $\mathbf{u}_{i,*}$ be the ith column of $\mathbf{U}_*$.

THEOREM 2.5. $\widehat{\boldsymbol{\beta}}_{\mathrm{G}} = \boldsymbol{\Lambda}^{-1/2}\sum_{i=1}^{\mathfrak{d}} d_{i,*}\eta_{i,*}\mathbf{v}_{i,*}$, *where $\eta_{i,*} = (d_{i,*}^2 + 1)^{-1}\mathbf{u}_{i,*}^T\mathbf{Y}$. That is, $\widehat{\boldsymbol{\beta}}_{\mathrm{G}}$ lies in the $\mathfrak{d}$-dimensional subspace $\boldsymbol{\Lambda}^{-1/2}(\mathscr{V}_*) = \{\boldsymbol{\Lambda}^{-1/2}\mathbf{v} : \mathbf{v} \in \mathscr{V}_*\}$, where $\mathfrak{d} = \mathrm{rank}(\mathbf{X}) \le n$ and $\mathscr{V}_*$ is the span of $\{\mathbf{v}_{1,*}, \ldots, \mathbf{v}_{\mathfrak{d},*}\}$; i.e., $\mathscr{V}_*$ is the span of the eigenvectors of $\mathbf{Q}_*$. The linear predictor is expressible as $\widehat{\boldsymbol{\mu}}_{\mathrm{G}} = \sum_{i=1}^{\mathfrak{d}} d_{i,*}^2\eta_{i,*}\mathbf{u}_{i,*}$.*

REMARK 2.6. Using (2.4) and similar arguments as in the proof of Theorem 2.5 the modified MLS is expressible as $\widehat{\boldsymbol{\beta}}^{*}_{\mathrm{MLS}} = \boldsymbol{\Lambda}^{-1/2}\sum_{i=1}^{\mathfrak{d}} d_{i,*}^{-1}(\mathbf{u}_{i,*}^T\mathbf{Y})\mathbf{v}_{i,*}$ and its linear predictor can be written as $\widehat{\boldsymbol{\mu}}^{*}_{\mathrm{MLS}} = \sum_{i=1}^{\mathfrak{d}}(\mathbf{u}_{i,*}^T\mathbf{Y})\mathbf{u}_{i,*}$. These facts will become handy later in Section 3.

3. Implications for GRR when $p \ge n$

We now list several interesting facts that follow from our previous results.

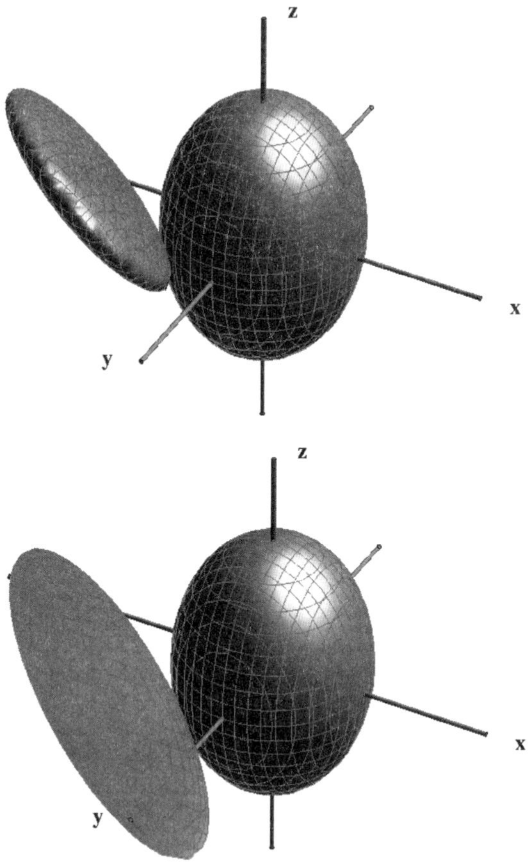

FIGURE 1. Illustration of GRR geometry. Top figure corresponds to a simulation where $p = 100$, $n = 25$, and $\beta_{0,k} = 0$ for $k \geq 3$ and $\lambda_k = \infty$ for $k \geq 4$. Only the first 3 coordinates of $\widehat{\boldsymbol{\beta}}_{\mathrm{G}}$ are nonzero, and these equal the point where the ellipsoid first touches the elliptical constraint region centered at zero. Bottom figure is $\lambda_k = \infty$ for $k \geq 3$. Now only the first two coordinates of $\widehat{\boldsymbol{\beta}}_{\mathrm{G}}$ are nonzero (the point touching the elliptical constraint region constrained to the x, y-plane and centered at zero).

3.1. Efficient calculation. In itself, Theorem 2.2 has immediate practical value as it permits efficient calculation of GRR in a linear number of operations in p; thus making GRR computationally feasible in $p \gg n$ settings. Note that as a special case of (2.3) the following representation for the ridge estimator holds

$$\widehat{\boldsymbol{\beta}}_{\mathrm{R}} = \mathbf{V}\mathbf{S}_{\lambda}^{-1}\mathbf{R}^T\mathbf{Y}. \tag{3.1}$$

Similar to (2.3), this shows that the ridge estimator can be computed in a linear number of operations. See [**4**] for related work.

3.2. Shrinkage. The expression for $\widehat{\boldsymbol{\mu}}_{\mathrm{G}}$ in Theorem 2.5 shows that the GRR predictor applies the greatest amount of shrinkage to those columns of $\mathbf{U}_*$ with smallest singular values. Thus, GRR is attempting to shrink the predictor in unfavorable directions relative to the column space of $\mathbf{X}_*$. This is a generalization of a well-known property of ridge regression (see, for example, [**5**, Chapter 3]). We can demonstrate the effect of this shrinkage by comparing the GRR predictor to the modified MLS predictor (we could compare GRR to MLS, but this is not as straightforward, and this extra effort may be unnecessary as the modified MLS estimator has been reported to have similar empirical behavior to the MLS; see [**12**]). By Remark 2.6 and Theorem 2.5, we have

$$\widehat{\boldsymbol{\mu}}^*_{\mathrm{MLS}} = \sum_{i=1}^{\mathfrak{d}} (\mathbf{u}_{i,*}^T \mathbf{Y})\,\mathbf{u}_{i,*} \tag{3.2}$$

$$\widehat{\boldsymbol{\mu}}_{\mathrm{G}} = \sum_{i=1}^{\mathfrak{d}} \frac{\delta_{i,*}}{\delta_{i,*}+1} (\mathbf{u}_{i,*}^T \mathbf{Y})\,\mathbf{u}_{i,*}, \tag{3.3}$$

where $\delta_{i,*} = d_{i,*}^2$. Notice how $\widehat{\boldsymbol{\mu}}_{\mathrm{G}}$ is shrunk in $\mathbf{u}_{i,*}$-directions corresponding to small singular values. Expressions (3.2) and (3.3) also show that the length of $\widehat{\boldsymbol{\mu}}^*_{\mathrm{MLS}}$ is always larger than $\widehat{\boldsymbol{\mu}}_{\mathrm{G}}$. Indeed, taking expectations, $\widehat{\boldsymbol{\mu}}^*_{\mathrm{MLS}}$ is always larger on average, because

$$\mathbb{E}\|\widehat{\boldsymbol{\mu}}^*_{\mathrm{MLS}}\|^2 = \mathbb{E}\|\widehat{\boldsymbol{\mu}}_{\mathrm{G}}\|^2 + \sum_{i=1}^{\mathfrak{d}} \frac{2\delta_{i,*}+1}{(\delta_{i,*}+1)^2}(\mu_{i,*}^2 + \sigma_0^2),$$

where $\mu_{i,*} = \mathbf{u}_{i,*}^T \boldsymbol{\mu}$.

Theorem 2.5 also generalizes the well known property of GRR as a shrinkage estimator. By Remark 2.6, the modified MLS estimator is expressible as

$$\widehat{\boldsymbol{\beta}}^*_{\mathrm{MLS}} = \boldsymbol{\Lambda}^{-1/2} \sum_{i=1}^{\mathfrak{d}} (d_{i,*} + d_{i,*}^{-1})\eta_{i,*}\mathbf{v}_{i,*}. \tag{3.4}$$

Comparing this to $\widehat{\boldsymbol{\beta}}_{\mathrm{G}}$, we see that each term in the summation of $\widehat{\boldsymbol{\beta}}^*_{\mathrm{MLS}}$ is larger than that of $\widehat{\boldsymbol{\beta}}_{\mathrm{G}}$ by an amount $d_{i,*}^{-1}$. Multiplying $\boldsymbol{\Lambda}^{1/2}$ throughout both expressions,

$$\|\boldsymbol{\Lambda}^{1/2}\widehat{\boldsymbol{\beta}}^*_{\mathrm{MLS}}\|^2 = \|\boldsymbol{\Lambda}^{1/2}\widehat{\boldsymbol{\beta}}_{\mathrm{G}}\|^2 + \sum_{i=1}^{\mathfrak{d}} (2 + \delta_{i,*}^{-1})\eta_{i,*}^2.$$

Thus, under a $\boldsymbol{\Lambda}^{1/2}$ rescaling, the squared-length of $\widehat{\boldsymbol{\beta}}_{\mathrm{G}}$ is always smaller than $\widehat{\boldsymbol{\beta}}^*_{\mathrm{MLS}}$. In particular, $\widehat{\boldsymbol{\beta}}^*_{\mathrm{MLS}}$ becomes elongated in the presence of small singular values. This is a generalization of the classic $n > p$ setting. There, it is well known that the ridge regression uniformly shrinks the OLS toward zero, so that its squared-length is always smaller than that of the OLS [**6**].

3.3. Sparsity. Theorem 2.5 shows that $\widehat{\boldsymbol{\beta}}_{\mathrm{G}}$ lies within the subspace $\boldsymbol{\Lambda}^{-1/2}(\mathscr{V}_*)$ containing the modified MLS estimator: this forces $\widehat{\boldsymbol{\beta}}_{\mathrm{G}}$ to lie in a low $\mathfrak{d}$-dimensional subspace which may degrade its performance in high dimensional non-sparse settings. To see why, consider the distance between the scaled GRR estimator, $\boldsymbol{\Lambda}^{1/2}\widehat{\boldsymbol{\beta}}_{\mathrm{G}}$, and the scaled regression parameter, $\boldsymbol{\beta}_0^* = \boldsymbol{\Lambda}^{1/2}\boldsymbol{\beta}_0$. We decompose $\boldsymbol{\beta}_0^*$ into its projection onto $\mathscr{V}_*$ and the orthogonal subspace $\mathscr{V}_*^{\perp}$. Because $\boldsymbol{\Lambda}^{1/2}\widehat{\boldsymbol{\beta}}_{\mathrm{G}} \in \mathscr{V}_*$, the distance between $\mathbf{\Lambda}^{1/2}\widehat{\boldsymbol{\beta}}_{\mathrm{G}}$ and $\boldsymbol{\beta}_0^*$ can be bounded below by the projection of $\boldsymbol{\beta}_0^*$ onto $\mathscr{V}_*^{\perp}$. Consequently

$$\left\|\mathbf{\Lambda}^{1/2}(\widehat{\boldsymbol{\beta}}_{\mathrm{G}} - \boldsymbol{\beta}_0)\right\|_2^2 \geq \left\|\left(\mathbf{I}_p - \mathbf{V}_*\mathbf{D}_*(\mathbf{V}_*\mathbf{D}_*)^+\right)\boldsymbol{\beta}_0^*\right\|^2.$$

If $\boldsymbol{\beta}_0$ lies in a high-dimensional subspace, then it may not be possible to find a $\mathbf{\Lambda}$ to make this distance zero. On the other hand, if $\boldsymbol{\beta}_0$ sits in a low-dimensional subspace of dimension no larger than $\mathfrak{d}$, then there always exists a ridge matrix making the right-hand side zero. The dimensionality of $\boldsymbol{\beta}_0$ is a sparsity condition. Because $\mathfrak{d} \leq n$, accurate estimation is guaranteed only when the sparsity condition $p_0 \leq n$ is met, where p_0 equals the number of nonzero coefficients of $\boldsymbol{\beta}_0$. But in non-sparse conditions, where $p_0 > n$, no such guarantee holds.

3.4. Prediction. It is reasonable to expect that GRR will outperform traditional least squares (i.e., MLS) in high dimensions. To formally investigate this we consider the difference in prediction performance of the GRR to the modified MLS (as we have remarked, working directly with the MLS is difficult, thus we instead work with the modified MLS which serves as a reasonable proxy). In the following, let $\boldsymbol{\mu} = \mathbb{E}(\mathbf{X}\boldsymbol{\beta}_0)$ be the true predicted value.

THEOREM 3.1.

$$\mathbb{E}\|\widehat{\boldsymbol{\mu}}_{\mathrm{MLS}}^* - \boldsymbol{\mu}\|^2 = \mathbb{E}\|\widehat{\boldsymbol{\mu}}_{\mathrm{G}} - \boldsymbol{\mu}\|^2 + \sum_{i=1}^{\mathfrak{d}} \frac{(2\delta_{i,*}+1)\sigma_0^2 - \mu_{i,*}^2}{(\delta_{i,*}+1)^2}. \tag{3.5}$$

Theorem 3.1 gives the mean-squared error for the modified MLS relative to GRR and hence provides insight into the risk behavior for the MLS relative to GRR. Interestingly, (3.5) identifies scenarios where the risk for $\widehat{\boldsymbol{\mu}}_{\mathrm{MLS}}^*$ may be smaller than $\widehat{\boldsymbol{\mu}}_{\mathrm{G}}$. One example is a noiseless system in which $\sigma_0^2 = 0$. Then the second term on the right of (3.5) is negative and $\widehat{\boldsymbol{\mu}}_{\mathrm{MLS}}^*$ will have smaller risk than $\widehat{\boldsymbol{\mu}}_{\mathrm{G}}$. In general, however, the risk for $\widehat{\boldsymbol{\mu}}_{\mathrm{MLS}}^*$ will be larger than $\widehat{\boldsymbol{\mu}}_{\mathrm{G}}$ if σ_0^2 is nonzero and $\delta_{i,*}$ is large — the latter occurs if the singular values are large. Thus, outside of low noise, high signal systems, the risk for $\widehat{\boldsymbol{\mu}}_{\mathrm{G}}$ is expected to be smaller.

3.5. Estimation. In a similar fashion we can compare the MSE performance of GRR to the modified MLS estimator. Consider the following MSE decomposition.

THEOREM 3.2. *Let* $\alpha_i = (\mathbf{\Lambda}^{1/2}\boldsymbol{\beta}_0)^T\mathbf{v}_{i,*}$. *Then*

$$\begin{aligned} &\mathbb{E}\|\mathbf{\Lambda}^{1/2}(\widehat{\boldsymbol{\beta}}_{\mathrm{MLS}}^* - \boldsymbol{\beta}_0)\|^2 \\ &= \mathbb{E}\|\mathbf{\Lambda}^{1/2}(\widehat{\boldsymbol{\beta}}_{\mathrm{G}} - \boldsymbol{\beta}_0)\|^2 + \sum_{i=1}^{\mathfrak{d}} \frac{(2\delta_{i,*}+1)(\mu_{i,*}^2+\sigma_0^2) - 2\alpha_i\delta_{i,*}^{1/2}(\delta_{i,*}+1)\mu_{i,*}}{\delta_{i,*}(\delta_{i,*}+1)^2}. \end{aligned} \tag{3.6}$$

Interpreting (3.6) is not as straightforward as (3.5). However, one interesting conclusion is that unlike the prediction scenario, a noiseless system with $\sigma_0^2 = 0$ does not necessarily confer a MSE advantage for $\widehat{\boldsymbol{\beta}}_{\mathrm{MLS}}^*$. Also, from the second term in (3.6), we see that small singular values will inflate the MSE for $\widehat{\boldsymbol{\beta}}_{\mathrm{MLS}}^*$ and that this inflation is further enhanced by the presence of $\mu_{i,*}^2$ (and the term σ_0^2). These results are consistent with our earlier comments that small singular values create instability in $\widehat{\boldsymbol{\beta}}_{\mathrm{MLS}}^*$. Thus we expect $\widehat{\boldsymbol{\beta}}_{\mathrm{MLS}}^*$ (and consequently $\widehat{\boldsymbol{\beta}}_{\mathrm{MLS}}$) to have poor MSE performance in high-dimensional correlated scenarios.

4. Discussion

The p bigger than n setting presents an ill-conditioned scenario where spurious correlations between variables can make estimation and prediction difficult. This paper studied how GRR estimators would perform in this setting given their stabilizing effects seen in $n > p$ situations.

Using geometric arguments, it was shown that the properties of GRR when $p > n$ shared similar features to the solution in the classic $n > p$ setting but also differed in several important ways. Like the classic setting, shrinkage plays a role which can lead to both improved estimation and prediction over MLS (least squares). However, an important difference in high dimensions is that the GRR solution is constrained to lie in a subspace containing the MLS estimator of dimension at most n (as opposed to a subspace of dimension p in the classic setting). This implies that for accurate estimation the true parameter vector should be sparse in the sense that $p_0 \leq n$. In non-sparse situations, accurate estimation cannot be guaranteed.

The high-dimensional sparse setting has attracted a considerable amount of research interest with a large focus on lasso and lasso-type regularization. Our results suggest that GRR can also have excellent performance in such settings if the ridge matrix is selected appropriately. One way to proceed would be to use a Bayesian approach which naturally lend themselves to ridge estimation. In particular, let $\mathbf{\Gamma} = \text{diag}\{\gamma_k\}_{k=1}^p$ be a $p \times p$ diagonal matrix with diagonal entries $\gamma_k > 0$. Consider the following Bayesian normal-normal hierarchy:

$$\begin{aligned} (\mathbf{Y} \mid \mathbf{X}, \boldsymbol{\beta}, \sigma^2) &\sim \text{N}(X\boldsymbol{\beta}, \sigma^2\mathbf{I}_n) \\ (\boldsymbol{\beta} \mid \mathbf{\Gamma}) &\sim \text{N}(\mathbf{0}, \mathbf{\Gamma}). \end{aligned} \tag{4.1}$$

Conjugacy ensures that the posterior distribution for $\boldsymbol{\beta}$ is multivariate normal. By standard calculations, the posterior distribution of $\boldsymbol{\beta}$ is

$$(\boldsymbol{\beta} \mid \mathbf{Y}, \mathbf{\Gamma}, \sigma^2) \sim \text{N}(\boldsymbol{\mu}_{\mathbf{\Gamma}}, \sigma^2\mathbf{\Sigma}_{\mathbf{\Gamma}})$$

where $\boldsymbol{\mu}_{\mathbf{\Gamma}} = \mathbf{\Sigma}_{\mathbf{\Gamma}}\mathbf{X}^T\mathbf{Y}$ is the posterior mean of $\boldsymbol{\beta}$ and $\mathbf{\Sigma}_{\mathbf{\Gamma}} = (\mathbf{Q}+\sigma^2\mathbf{\Gamma}^{-1})^{-1}$. Observe that $\boldsymbol{\mu}_{\mathbf{\Gamma}}$ is a GRR estimator.

Model (4.1) is a "plain vanilla" Bayesian hierarchy that assumes a fixed ridge matrix. However, in practice a more sophisticated hierarchy using a prior for $\mathbf{\Gamma}$ is a preferred way to estimate $\boldsymbol{\beta}$. This reduces the risk of poor estimation if $\mathbf{\Gamma}$ is misspecified. In particular, a non-generate prior for $\mathbf{\Gamma}$ results in a posterior mean that is no longer a GRR estimator but instead is a weighted (averaged) GRR estimator (WGRR). Such estimators are closely tied to mixed GRR estimators which are known to have a minimax property [**12**]. An example of such a hierarchy is the rescaled spike and slab model used in [**11**] which utilizes a continuous bimodal prior for $\mathbf{\Gamma}$ that allows γ_k to be adaptively determined.

We illustrate this method using the diabetes data of [**2**]; a popular benchmark dataset used in regression. The data consists of $n = 442$ patients in which the response is a quantitative measure of disease progression for a patient. The original data included 10 baseline measurements for each patient, age, sex, body mass index, average blood pressure and six blood serum measurements, in addition to 45 interactions formed by all pairwise interactions of the 10 baseline variables and 9 quadratic terms for the 9 baselines measurements that were continuous. To this we added 1000 "noise" variables, each sampled independently from a multivariate

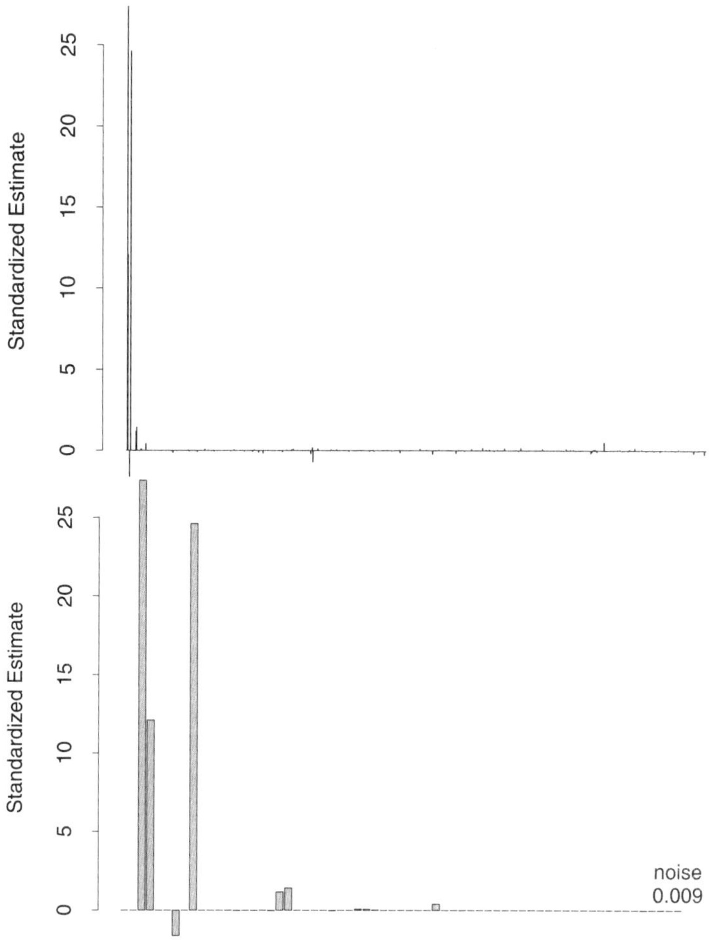

FIGURE 2. Standardized coefficient estimates for diabetes data with $q = 1000$ correlated noise variables ($n = 442$, $p = 1064$). Coefficient estimates obtained using a WGRR estimator. Top figure displays all $p = 1064$ coefficient estimates. Bottom figure displays the original 64 variables with right most estimate labeled "noise" equaling the mean absolute coefficient value of the $q = 1000$ noise variables.

normal distribution with mean zero and equicorrelation matrix with correlation $\rho = 0.5$. In total our modified data contained $p = 1064$ variables.

The standardized posterior coefficient values are displayed in Figure 2. The results were obtained using the R-package "spikeslab" [**9**] which fits a rescaled and slab model. All variables were standardized to have a sample mean of zero and sample variance of one. This yields standardized posterior coefficient values that can be compared against one another. The top figure displays all $p = 1064$ variables while the bottom figure displays the coefficient estimates for the original 64 variables. The right most estimate labeled "noise" is the averaged value of the absolute posterior coefficient estimates for the 1000 noise variables. Even in the presence of high correlation, the posterior estimates are shrunk towards zero for nearly all noise variables (average of.009) and only a subset of the original 64 variables appear informative. For technical details and further empirical illustrations the interested reader should consult [**11**].

Appendix A. Proofs

PROOF OF LEMMA 2.1. We first show that $\mathbf{A}\mathbf{A}^+ = \mathbf{I}_n$. Using $\mathbf{V}^T\mathbf{V} = \mathbf{I}_n$, $\mathbf{Q} = \mathbf{V}\mathbf{D}^2\mathbf{V}^T$, and $\mathbf{V}\mathbf{S}_\lambda^{-1} = \mathbf{V}(\mathbf{D}^2 + \lambda\mathbf{I}_n)^{-1}$, deduce that

$$\begin{aligned}\mathbf{A}\mathbf{A}^+ &= (\mathbf{D}^2\mathbf{V}^T + \lambda\mathbf{V}^T)\big[\mathbf{V}(\mathbf{D}^2 + \lambda\mathbf{I}_n)^{-1}\big]\\ &= (\mathbf{D}^2 + \lambda\mathbf{I}_n)(\mathbf{D}^2 + \lambda\mathbf{I}_n)^{-1}\\ &= \mathbf{I}_n.\end{aligned}$$

From this it immediately follows that: (i) $\mathbf{A}\mathbf{A}^+\mathbf{A} = \mathbf{A}$; (ii) $\mathbf{A}^+\mathbf{A}\mathbf{A}^+ = \mathbf{A}^+$; and (iii) $(\mathbf{A}\mathbf{A}^+)^T = \mathbf{A}\mathbf{A}^+$. Furthermore,

$$\mathbf{A}^+\mathbf{A} = \mathbf{V}(\mathbf{D}^2 + \lambda\mathbf{I}_n)^{-1}(\mathbf{D}^2\mathbf{V}^T + \lambda\mathbf{V}^T) = \mathbf{V}\mathbf{V}^T.$$

Therefore: (iv) $(\mathbf{A}^+\mathbf{A})^T = \mathbf{V}\mathbf{V}^T = \mathbf{A}^+\mathbf{A}$. Properties (i)–(iv) are the four criteria required for a Moore–Penrose generalized inverse. □

PROOF OF THEOREM 2.2. By (2.2), the GRR estimator can be expressed as a ridge estimator scaled by a diagonal matrix. Therefore, it suffices to prove that (3.1) holds and that the number of operations required to calculate (3.1) is of order $O(pn^2)$. Set $\mathbf{\Lambda} = \lambda\mathbf{I}_p$. Taking the derivative with respect to $\boldsymbol{\beta}$ in (1.2), setting this to zero, and multiplying right and left-hand sides by $\mathbf{V}^T$, it follows that $\widehat{\boldsymbol{\beta}}_{\mathrm{R}}$ must satisfy

$$\mathbf{A}\widehat{\boldsymbol{\beta}}_{\mathrm{R}} = \mathbf{V}^T\mathbf{X}^T\mathbf{Y}.$$

The solution must be $\widehat{\boldsymbol{\beta}}_{\mathrm{R}} = \mathbf{A}^+\mathbf{V}^T\mathbf{X}^T\mathbf{Y}$ because upon substituting this into the left-hand side we obtain

$$\mathbf{A}\widehat{\boldsymbol{\beta}}_{\mathrm{R}} = \mathbf{A}\mathbf{A}^+\mathbf{V}^T\mathbf{X}^T\mathbf{Y} = \mathbf{V}^T\mathbf{X}^T\mathbf{Y},$$

where we have used the fact that $\mathbf{A}\mathbf{A}^+ = \mathbf{I}_n$ from Lemma 2.1. Now substituting the right-hand side of (2.1) for $\mathbf{A}^+$, yields

$$\widehat{\boldsymbol{\beta}}_{\mathrm{R}} = \mathbf{V}\mathbf{S}_\lambda^{-1}\mathbf{V}^T\mathbf{X}^T\mathbf{Y} = \mathbf{V}\mathbf{S}_\lambda^{-1}\mathbf{R}^T\mathbf{Y}.$$

To determine the number of operations required to compute $\widehat{\boldsymbol{\beta}}_{\mathrm{R}}$, note that the SVD for $\mathbf{X}$ requires $O(pn^2)$ operations. Once the SVD is obtained, inverting $\mathbf{S}_\lambda^{-1}$ requires $O(n)$ operations. Multiplying this (of size $n \times n$) by $\mathbf{V}(p \times n)$ requires $O(pn^2)$ operations (note that because S_λ^{-1} is diagonal, this can be reduced further to $O(pn)$ operations, but this level of refinement is not essential). Multiplying by $\mathbf{R}^T\mathbf{Y}$ requires a total of $O(pn)$ operations. The total number of operations equals

$$O(pn^2) + O(n) + O(pn^2) + O(pn) = O(pn^2). \qquad \square$$

PROOF OF THEOREM 2.4. By (2.2), $\widehat{\boldsymbol{\beta}}_{\mathrm{G}} = \mathbf{\Lambda}^{-1/2}\widehat{\boldsymbol{\beta}}^*_{\mathrm{R}}$ where $\widehat{\boldsymbol{\beta}}^*_{\mathrm{R}}$ is the ridge estimator under X_* with ridge parameter $\lambda = 1$. We show that $\mathbf{\Lambda}^{-1/2}\widehat{\boldsymbol{\beta}}^*_{\mathrm{R}}$ is the solution to (2.5). As a Lagrangian problem, (2.5) can be written as

$$\underset{(\boldsymbol{\beta},\ell)\in\mathbb{R}^p\times\mathbb{R}_+}{\text{minimize}} \big\{\mathbb{Q}(\boldsymbol{\beta}, \widehat{\boldsymbol{\beta}}^*_{\mathrm{MLS}}) + \ell(\boldsymbol{\beta}^T\mathbf{\Lambda}\boldsymbol{\beta} - L)\big\},$$

where ℓ is the Lagrangian multiplier. Because L is arbitrary we can assume that $\ell = 1$ without loss of generality. Taking the derivative with respect to $\boldsymbol{\beta}$, the solution is

$$2\boldsymbol{\Lambda}\boldsymbol{\beta} + 2\boldsymbol{\Lambda}^{1/2}\mathbf{Q}_*\boldsymbol{\Lambda}^{1/2}(\boldsymbol{\beta} - \widehat{\boldsymbol{\beta}}^*_{\mathrm{MLS}}) = \mathbf{0}.$$

Multiplying throughout by $\boldsymbol{\Lambda}^{-1/2}$, $\boldsymbol{\beta}$ must satisfy

$$\begin{aligned}(\mathbf{Q}_* + \mathbf{I}_p)\boldsymbol{\Lambda}^{1/2}\boldsymbol{\beta} &= \mathbf{Q}_*\boldsymbol{\Lambda}^{1/2}\widehat{\boldsymbol{\beta}}^*_{\mathrm{MLS}}\\ &= \mathbf{V}_*\mathbf{D}_*^2\mathbf{V}_*^T\mathbf{V}_*\mathbf{S}_{*0}^+\mathbf{R}_*^T\mathbf{Y}\\ &= \mathbf{V}_*\mathbf{D}_*^2\mathbf{S}_{*0}^+\mathbf{D}_*\mathbf{U}_*^T\mathbf{Y}\\ &= \mathbf{V}_*\mathbf{D}_*\mathbf{U}_*^T\mathbf{Y}\\ &= \mathbf{X}_*^T\mathbf{Y}.\end{aligned}$$

Therefore, $\boldsymbol{\beta} = \boldsymbol{\Lambda}^{-1/2}\widehat{\boldsymbol{\beta}}^*_{\mathrm{R}}$; thus verifying that (2.5) is the optimization problem for $\widehat{\boldsymbol{\beta}}_{\mathrm{G}}$. □

Proof of Theorem 2.5. By (2.3), we can write $\widehat{\boldsymbol{\beta}}_{\mathrm{G}} = \boldsymbol{\Lambda}^{-1/2}\mathbf{V}_*\mathbf{S}_{*1}^{-1}\mathbf{D}_*\mathbf{U}_*^T\mathbf{Y}$. The stated representation for $\widehat{\boldsymbol{\beta}}_{\mathrm{G}}$ follows with some simple rearrangement. From this it is clear that $\widehat{\boldsymbol{\beta}}_{\mathrm{G}}$ lies in the subspace $\mathscr{V}_*$ (and hence $\boldsymbol{\Lambda}^{-1/2}(\mathscr{V}_*)$). To see that $\mathscr{V}_*$ can be interpreted as the span of the eigenvectors of $\mathbf{Q}_*$, note that $\mathbf{Q}_* = \mathbf{V}_*\mathbf{D}_*^2\mathbf{V}_*^T$. Thus, $\mathbf{Q}_*\mathbf{v}_{i,*} = d_{i,*}^2\mathbf{v}_{i,*}$, and hence $\mathbf{v}_{i,*}$ is an eigenvector of Q_* under the condition that $d_{i,*}^2 > 0$. Finally, to prove the claim for $\widehat{\boldsymbol{\mu}}_{\mathrm{G}} = \mathbf{X}\widehat{\boldsymbol{\beta}}_{\mathrm{G}}$, using $\mathbf{X}_* = \mathbf{X}\boldsymbol{\Lambda}^{-1/2}$ and the representation for $\widehat{\boldsymbol{\beta}}_{\mathrm{G}}$, we have

$$\begin{aligned}\widehat{\boldsymbol{\mu}}_{\mathrm{G}} &= \mathbf{X}^*\mathbf{V}_*\mathbf{S}_{*1}^{-1}\mathbf{D}_*\mathbf{U}_*^T\mathbf{Y}\\ &= \mathbf{U}_*\mathbf{D}_*\mathbf{V}_*^T\mathbf{V}_*\mathbf{S}_{*1}^{-1}\mathbf{D}_*\mathbf{U}_*^T\mathbf{Y}\\ &= \mathbf{U}_*(\mathbf{D}_*\mathbf{S}_{*1}^{-1}\mathbf{D}_*)\mathbf{U}_*^T\mathbf{Y}.\end{aligned}$$

It is easily checked that this corresponds to the stated expression for $\widehat{\boldsymbol{\mu}}_{\mathrm{G}}$. □

Proof of Theorem 3.1. We can write $\boldsymbol{\mu}$ in terms of the orthonormal basis $(\mathbf{u}_{i,*})_{i=1}^n$. We have $\boldsymbol{\mu} = \sum_{i=1}^n \mu_{i,*}\mathbf{u}_{i,*}$. Therefore, by (3.2) and (3.3),

$$\begin{aligned}\widehat{\boldsymbol{\mu}}^*_{\mathrm{MLS}} - \boldsymbol{\mu} &= \sum_{i=1}^{\partial}\left(\frac{\delta_{i,*}}{\delta_{i,*}+1} - \frac{\mu_{i,*}}{\mathbf{u}_{i,*}^T\mathbf{Y}}\right)(\mathbf{u}_{i,*}^T\mathbf{Y})\,\mathbf{u}_{i,*}\\ &\qquad + \sum_{i=1}^{\partial}\left(1 - \frac{\delta_{i,*}}{\delta_{i,*}+1}\right)(\mathbf{u}_{i,*}^T\mathbf{Y})\,\mathbf{u}_{i,*}\\ &= (\widehat{\boldsymbol{\mu}}_{\mathrm{G}} - \boldsymbol{\mu}) + \sum_{i=1}^{\partial}\left(\frac{1}{\delta_{i,*}+1}\right)(\mathbf{u}_{i,*}^T\mathbf{Y})\,\mathbf{u}_{i,*}.\end{aligned}$$

Squaring and collecting terms, deduce that

$$\begin{aligned}\text{(A.1)}\quad &\|\widehat{\boldsymbol{\mu}}^*_{\mathrm{MLS}} - \boldsymbol{\mu}\|^2\\ &= \|\widehat{\boldsymbol{\mu}}_{\mathrm{G}} - \boldsymbol{\mu}\|^2 + \sum_{i=1}^{\partial}\frac{(2\delta_{i,*}+1)(\mathbf{u}_{i,*}^T\mathbf{Y})^2 - 2\mu_{i,*}(\delta_{i,*}+1)(\mathbf{u}_{i,*}^T\mathbf{Y})}{(\delta_{i,*}+1)^2}.\end{aligned}$$

One can easily verify that

$$\text{(A.2)}\quad \begin{aligned}\mathbb{E}[(2\delta_{i,*}+1)(\mathbf{u}_{i,*}^T\mathbf{Y})^2 - 2\mu_{i,*}(\delta_{i,*}+1)(\mathbf{u}_{i,*}^T\mathbf{Y})] \\ = (2\delta_{i,*}+1)(\mu_{i,*}^2+\sigma_0^2) - 2\mu_{i,*}(\delta_{i,*}+1)\mu_{i,*} \\ = (2\delta_{i,*}+1)\sigma_0^2 - \mu_{i,*}^2.\end{aligned}$$

The theorem is proved by using this, and taking expecations in (A.1). □

PROOF OF THEOREM 3.2. The proof is similar to that of Theorem 3.1. Note that $(\mathbf{v}_{i,*})_{i=1}^{\eth}$ is an orthonormal basis for $\mathscr{V}_*$. Let $\boldsymbol{\beta}_0^* = \boldsymbol{\Lambda}^{1/2}\boldsymbol{\beta}_0$ and write $\mathscr{V}_*^\perp$ for the orthogonal subspace to $\mathscr{V}_*$. Then, $\boldsymbol{\beta}_0^* = \sum_{i=1}^n \alpha_i \mathbf{v}_{i,*} + \Delta^*$, where Δ^* is the projection of $\boldsymbol{\beta}_0^*$ onto $\mathscr{V}_*^\perp$. Using the representation (3.4) for $\widehat{\boldsymbol{\beta}}_{\text{MLS}}^*$ and the representation for $\widehat{\boldsymbol{\beta}}_{\text{G}}$ given in Theorem 2.5, we have

$$\begin{aligned}\boldsymbol{\Lambda}^{1/2}(\widehat{\boldsymbol{\beta}}_{\text{MLS}}^* - \boldsymbol{\beta}_0) &= \sum_{i=1}^{\eth}\left(\frac{d_{i,*}}{\delta_{i,*}+1} - \frac{\alpha_i}{\mathbf{u}_{i,*}^T\mathbf{Y}}\right)(\mathbf{u}_{i,*}^T\mathbf{Y})\,\mathbf{v}_{i,*} - \Delta^* \\ &\quad + \sum_{i=1}^{\eth}\left(\frac{d_{i,*}+d_{i,*}^{-1}}{\delta_{i,*}+1} - \frac{d_{i,*}}{\delta_{i,*}+1}\right)(\mathbf{u}_{i,*}^T\mathbf{Y})\,\mathbf{v}_{i,*} \\ &= \boldsymbol{\Lambda}^{1/2}(\widehat{\boldsymbol{\beta}}_{\text{G}} - \boldsymbol{\beta}_0) + \sum_{i=1}^{\eth}\left(\frac{d_{i,*}^{-1}}{\delta_{i,*}+1}\right)(\mathbf{u}_{i,*}^T\mathbf{Y})\,\mathbf{v}_{i,*}.\end{aligned}$$

Squaring, collecting terms, and taking expectations, deduce that

$$\begin{aligned}&\mathbb{E}\|\boldsymbol{\Lambda}^{1/2}(\widehat{\boldsymbol{\beta}}_{\text{MLS}}^* - \boldsymbol{\beta}_0)\|^2 \\ &\quad = \mathbb{E}\|\boldsymbol{\Lambda}^{1/2}(\widehat{\boldsymbol{\beta}}_{\text{G}} - \boldsymbol{\beta}_0)\|^2 + \mathbb{E}\left[\sum_{i=1}^{\eth}\frac{(2\delta_{i,*}+1)(\mathbf{u}_{i,*}^T\mathbf{Y})^2 - 2\alpha_i d_{i,*}(\delta_{i,*}+1)(\mathbf{u}_{i,*}^T\mathbf{Y})}{\delta_{i,*}(\delta_{i,*}+1)^2}\right].\end{aligned}$$

The result follows by taking the expectation inside the sum and using $\mathbb{E}(\mathbf{u}_{i,*}^T\mathbf{Y})^2 = \mu_{i,*}^2 + \sigma_0^2$ and $\mathbb{E}(\mathbf{u}_{i,*}^T\mathbf{Y}) = \mu_{i,*}$. □

References

[1] T. T. Cai and J. Lv, *Discussion: "The Dantzig selector: statistical estimation when p is much larger than n" [Ann. Statist.* **35** *(2007), no. 6, 2313–2351; MR2382644] by E. Candes and T. Tao*, Ann. Statist. **35** (2007), no. 6, 2365–2369, DOI 10.1214/009053607000000442. MR2382647 (2009b:62015)

[2] B. Efron, T. Hastie, I. Johnstone, and R. Tibshirani, *Least angle regression. With discussion, and a rejoinder by the authors*, Ann. Statist. **32** (2004), no. 2, 407–499, DOI 10.1214/009053604000000067. MR2060166 (2005d:62116)

[3] J. Fan and J. Lv, *Sure independence screening for ultrahigh dimensional feature space*, J. R. Stat. Soc. Ser. B Stat. Methodol. **70** (2008), no. 5, 849–911, DOI 10.1111/j.1467-9868.2008.00674.x. MR2530322

[4] T. Hastie and R. Tibshirani, *Efficient quadratic regularization for expression arrays*, Biostatistics **5** (2004), no. 3, 329–340, DOI 10.1093/biostatistics/kxh010.

[5] T. Hastie, R. Tibshirani, and J. Friedman, *The elements of statistical learning. Data mining, inference, and prediction*, Springer Series in Statistics, Springer-Verlag, New York, 2001. MR1851606 (2002k:62048)

[6] A. E. Hoerl and R. W. Kennard, *Ridge regression: Biased estimation for nonorthogonal problems*, Technometrics **12** (1970), no. 1, 55–67, DOI 10.1080/00401706.1970.10488634.

[7] A. E. Hoerl and R. W. Kennard, *Ridge regression: Applications to nonorthogonal problems.*, Technometrics **12** (1970), no. 1, 69–82; Erratum, Technometrics **12**, no. 3, 723.

[8] J. Huang, S. Ma, and C.-H. Zhang, *Adaptive Lasso for sparse high-dimensional regression models*, Statist. Sinica **18** (2008), no. 4, 1603–1618. MR2469326 (2010a:62214)
[9] H. Ishwaran, U. B. Kogalur, and S. J. Rao, *spikeslab: Prediction and variable selection using spike and slab regression*, R Journal **7** (2010), no. 2, 68–73.
[10] H. Ishwaran and J. Sunil Rao, *Consistency of spike and slab regression*, Statist. Probab. Lett. **81** (2011), no. 12, 1920–1928, DOI 10.1016/j.spl.2011.08.005. MR2845909 (2012h:62258)
[11] H. Ishwaran and J. S. Rao, *Generalized ridge regression: geometry and computational solutions when p is larger than n*, Technical Report 01/2011, Division of Biostatistics, University of Miami, 2011.
[12] H. Ishwaran and J. S. Rao, *Mixing generalized ridge regressions*, unpublished.
[13] R. Penrose, *A generalized inverse for matrices*, Proc. Cambridge Philos. Soc. **51** (1955), 406–413. MR0069793 (16,1082a)
[14] R. Penrose, *On best approximation solutions of linear matrix equations*, Proc. Cambridge Philos. Soc. **52** (1956), 17–19. MR0074092 (17,536d)
[15] R. Tibshirani, *Regression shrinkage and selection via the lasso*, J. Roy. Statist. Soc. Ser. B **58** (1996), no. 1, 267–288. MR1379242 (96j:62134)
[16] H. Zou, *The adaptive lasso and its oracle properties*, J. Amer. Statist. Assoc. **101** (2006), no. 476, 1418–1429, DOI 10.1198/016214506000000735. MR2279469 (2008d:62024)
[17] H. Zou and T. Hastie, *Regularization and variable selection via the elastic net*, J. R. Stat. Soc. Ser. B Stat. Methodol. **67** (2005), no. 2, 301–320, DOI 10.1111/j.1467-9868.2005.00503.x. MR2137327
[18] H. Zou and H. H. Zhang, *On the adaptive elastic-net with a diverging number of parameters*, Ann. Statist. **37** (2009), no. 4, 1733–1751, DOI 10.1214/08-AOS625. MR2533470 (2010j:62210)

Division of Biostatistics, Department of Public Health Sciences, University of Miami, Miami Florida 33136
E-mail address: hemant.ishwaran@gmail.com

Division of Biostatistics, Department of Public Health Sciences, University of Miami, Miami Florida 33136
E-mail address: rao.jsunil@gmail.com

Contemporary Mathematics
Volume **622**, 2014
http://dx.doi.org/10.1090/conm/622/12440

Multiple testing for high-dimensional data

Guoqing Diao, Bret Hanlon, and Anand N. Vidyashankar

ABSTRACT. This article is concerned with testing in ultra-high-dimensional regression problems allowing for data dimension p to exceed the sample size n substantially. Specifically, motivated by the large-sample considerations, we propose a new statistic for testing the regression coefficients. We provide heuristic evidence that the proposed method controls for the family-wise error rate. We use a multiplier bootstrap with Rademacher weights for computing the critical values of the test. Extensive simulations and real data analyses are used to demonstrate the validity and usefulness of our results.

1. Introduction

With the advent of modern technologies, "big data" is becoming an inherent part of the society. These data arise in multiple scientific areas including health care, bioinformatics, proteomics, and genetic epidemiology. A typical data set arising in applications may be represented by an $n \times p$ matrix, where n represents the number of observations and p represents the number of features. A common aspect of this problem is that $p \gg n$, i.e., p is much larger than n. Questions concerning joint effects of the features lead to hypothesis about multiple parameters. Standard statistical methods such as Hotelling's T^2 and Fisher's discriminant analysis are not immediately applicable since the sample covariance matrix is not invertible. These methods typically have low power to detect deviations from the null hypothesis due to the small sample size; see [**FUS10**] and the references therein for a discussion of these issues. A popular solution to this problem is to use regularization [**THNC02, BL04, FF08**], while more recently methods using random projections have been attempted [**LJW11**]. The regularization methods assume a particular sparsity structure, whereas theoretical properties of the methods based on random projections have not been well-investigated. The purpose of this paper is to study testing in ultra-high-dimensional problems.

The starting point of this paper is based on a recent work of [**KV10**] who established a central limit theorem (CLT) for the sample mean of infinite-dimensional random vectors under possibly missing data and other regularity assumptions. As a consequence of their CLT, they proposed a hypothesis testing methodology which

2010 *Mathematics Subject Classification.* Primary 62-07, 62H15, 62J12, 62J15, 62P10.
Key words and phrases. General class of regression models, high-dimensional models, multiple testing, multiplier bootstrap, Rademacher sequences.
The first author was supported in part by NSF DMS-1107108 and NIH CA150698.
The third author was supported by NSF DMS-1107108.

involves estimating the sample covariance matrix. Their test is referred to as the KV test. When the dimension p is substantially larger than n, this is not computationally feasible. One of the goals of this paper is to address this issue.

Additionally, motivated by the applications mentioned previously, it is natural to make inference about regression parameters across a large number of features. Furthermore, from a statistical perspective, one can use a test statistic based on efficient scores (which reduces to the student-type statistics for comparing the means, for example, see Example 2 in Section 2.3). In this paper, we provide an extension of KV methodology that encompasses the above considerations. Moreover, we also describe a multiplier bootstrap with Rademacher weights for calibrating the critical values of the test. Our methods are scalable to very high-dimensional problems compared to other existing methods; for instance, when $n = 20$ and $p = 10^4$.

Finally, we comment on the extension of the KV procedure to multiple testing. The original KV test was proposed as a multivariate test. It is well known that the KV test with the choice of the sup-norm can control for the family-wise error rate (FWER) and hence can be adopted in a multiple testing context. This is demonstrated in our simulation studies and data analyses.

The remainder of this paper is organized as follows. Section 2 describes the KV test and develops the high-dimensional efficient score test for regression problems. We derive the efficient scores for several examples encountered in practice. Section 3 is devoted to the implementation of the methodology. Extensive simulation results are presented in Section 4 while Section 5 is devoted to real data applications. Section 6 contains extensions and discussions.

2. High-dimensional efficient score test

2.1. Kuelbs and Vidyashankar (KV) test. We begin with the description of methodology for one-sample problem developed in [**KV10**]. Let $\mathbf{U}_i \equiv (U_{i1}, \ldots, U_{ip})^t$ be the observation for the ith sample, $i = 1, \ldots, n$; U_{ij} can be viewed as the jth feature for the ith sample. Let $\mathbf{U}_1, \ldots, \mathbf{U}_n$ be independent and identically distributed (i.i.d.) random vectors with mean $\boldsymbol{\mu} \in \mathbb{R}^p$ and covariance $\boldsymbol{\Sigma} \in \mathbb{R}^{p \times p}$. Furthermore, let $\overline{\mathbf{U}}^{(n)}$ denote the p-dimensional vector of componentwise averages; that is, $\overline{\mathbf{U}}^{(n)} = (\overline{U}_1^{(n)}, \ldots, \overline{U}_p^{(n)})^t$, where $\overline{U}_j^{(n)} = (1/n)\sum_{i=1}^n U_{ij}$, $j = 1, \ldots, p$, and t represents transpose. Let $\boldsymbol{\mu}_0 \equiv (\mu_{0,1}, \ldots, \mu_{0,p})^t \in \mathbb{R}^p$ be a p-dimensional mean vector. Consider the hypothesis testing problem $H_0\colon \boldsymbol{\mu} = \boldsymbol{\mu}_0$; [**KV10**] introduced the test statistic

$$T_d^{(n)} = \begin{cases} \|\sqrt{n}(\overline{\mathbf{U}}^{(n)} - \boldsymbol{\mu}_0)\|_d & \text{if } 0 < d < \infty \\ \|\sqrt{n}(\overline{\mathbf{U}}^{(n)} - \boldsymbol{\mu}_0)\|_\infty & \text{if } d = \infty, \end{cases}$$
$$= \begin{cases} \{\sum_{j=1}^p |\sqrt{n}(\overline{U}_j^{(n)} - \mu_{0,j})|^d\}^{1/d} & \text{if } 0 < d < \infty \\ \sup\{|\sqrt{n}(\overline{U}_j^{(n)} - \mu_{0,j})| : j = 1, \ldots, p\} & \text{if } d = \infty. \end{cases}$$

where $\|.\|_d$ is the L_d norm and $\|.\|_\infty$ denotes the sup-norm. To obtain the limit distribution of $T_d^{(n)}$, as a first step [**KV10**] embed the $\sqrt{n}(\overline{\mathbf{U}}^{(n)} - \boldsymbol{\mu}_0)$ into an appropriate infinite-dimensional space and derive the joint distribution in the infinite-dimensional space. This limit distribution is referred to as an infinite-dimensional normal distribution. They then apply the continuous mapping theorem to obtain the limit distribution of $T_d^{(n)}$. That is,

$$T_d^{(n)} \approx \|N_p(\mathbf{0}, \mathbf{\Sigma})\|_d. \tag{2.1}$$

Clearly for (2.1) to be utilized in constructing a test, the unknown covariance matrix must be estimated which, as explained in Section 1, is a difficult problem [**TC09**]. To address this issue, the KV test employs the shrinkage based estimator of Ledoit and Wolf [**LW04**]. The choice of this covariance estimator is dictated by good finite sample properties; in particular, the estimator is guaranteed to be positive definite. For recent development in covariance matrix estimation, see [**LW12**].

Returning to the KV test, the algorithm proceeds as follows: fix a significance level α;

(1) Compute the observed test statistic, $T_d^{(n)}$.
(2) Estimate the covariance matrix $\mathbf{S}^{(n)}$, using shrinkage.
(3) Generate B random vectors $\mathbf{L}_1, \ldots, \mathbf{L}_B \sim N_p(\mathbf{0}, \mathbf{S}^{(n)})$; compute the norm of these vectors, $T_i^\star \equiv \|\mathbf{L}_i\|_d, i = 1, \ldots, B$; finally compute the $1-\alpha$ sample quantile $q_{1-\alpha}^*$ from $T_1^\star, \ldots, T_B^\star$.
(4) Reject H_0 if $T_d^{(n)} > q_{1-\alpha}^*$.

Kuelbs and Vidyashankar used extensive simulations and real examples to illustrate the power of the approach when $n = 20$ and $p = 1000$. In many genetic studies, the number of features or genetic markers is often much larger than 1000. The shrinkage methods in those scenarios are not feasible to implement. Also typically, one encounters regression models at each marker and scientific questions of interest concern the regression parameters. To address these issues, we describe a high-dimensional efficient score test below.

2.2. Efficient score test. In the context of hypotheses testing in classical problems, it is well-known that t-tests yield more accurate results than the z-tests (based on normal distribution) in small samples. Hence it is natural to replace the standardized test statistic in the KV test by the Studentized test statistic yielding the t-test if the underlying distribution is normal. To address this issue and generalize the KV test to regression problems, we proceed as follows.

Suppose that there are n i.i.d. observations $\{Y_i, \mathbf{Z}_i, \mathbf{X}_i, i = 1, \ldots, n\}$, where Y_i is the outcome variable, $\mathbf{Z}_i$ is a set of explanatory variables and $\mathbf{X}_i = (X_{i1}, \ldots, X_{ip})^t$ are the features of interest. For example, in the context of genetic epidemiology studies, Y_i can represent a complex human disease trait, $\mathbf{Z}_i$ can include age, gender and other potential confounders, and $\mathbf{X}_i$ are the genotype scores at p single nucleotide polymorphisms (SNPs). We assume all the covariates are centered. Let the conditional distribution of Y_i given $\mathbf{Z}_i$ and X_{ij} be represented by $f(y|\mathbf{Z}_i, X_{ij}; \beta_j, \gamma_j)$, where β_j characterizes the the effect of X_{ij} on the outcome variable Y_i and γ_j contains nuisance parameters. Typical goal is to make inference on $\boldsymbol{\beta} \equiv (\beta_1, \ldots, \beta_p)$ and in particular test the null hypothesis $H_0\colon \boldsymbol{\beta} = \mathbf{0}$. It is worthwhile to observe that this setup encompasses both the one-sample and two-sample problems.

Many of the commonly-used test statistics for testing the jth component of $\boldsymbol{\beta}$, $\beta_j = 0$, can be represented in the form

$$W_j^{(n)} = U_j^{(n)} / \sqrt{V_j^{(n)}},$$

where $U_j^{(n)} = \sum_{i=1}^n U_{ij}$ and $V_j^{(n)} = \sum_{i=1}^n U_{ij}^2$. The quantity U_{ij} is frequently referred to as the efficient score function for β_j that involves only the data from the ith subject and the nuisance parameter γ_j. The unknown nuisance parameter γ_j in U_{ij} can be replaced by $\widetilde{\gamma}_j^{(n)}$, the estimator of γ_j under H_0.

We notice that in the context of statistical applications, under H_0,

$$E(U_j^{(n)}) = 0, \quad \forall\, 1 \le j \le p.$$

Also for every fixed j, $W_j^{(n)}$ converges to a standard normal distribution. Hence, as n increases, the variances of $W_j^{(n)}$ are close to 1, under appropriate moment conditions.

Thus, to study the joint asymptotic behavior of $(W_1^{(n)}, \ldots, W_p^{(n)})$, as in [**KV10**], we need to embed this vector in the space $\mathbb{R}^\infty$ and study the central limit theorem in appropriate subspaces of $\mathbb{R}^\infty$. A similar idea was adopted by [**KV11**], where central limit theorems were obtained for branching processes. Following the techniques from [**KV11**], we can study the central limit behavior of $\mathbf{W}^{(n)}$ embedded into $\mathbb{R}^\infty$. That is for

$$\mathbf{W}^{(n)} \equiv (W_1^{(n)}, W_2^{(n)}, \ldots, W_p^{(n)}, 0, \cdots).$$

To prove a theorem concerning the joint distribution of $\mathbf{W}^{(n)}$, we need conditions that guarantee convergence of the components. A key required condition is that the covariances of the components of $\mathbf{W}^{(n)}$ converge to $\boldsymbol{\Sigma}$, which is assumed to be a covariance matrix. Furthermore, we also need assumptions that are similar to the conditions needed to study the asymptotics of the score test of each component. Under all these conditions and further regularity conditions, one can show that

$$\lim_{n\to\infty} P(\|\mathbf{W}^{(n)}\|_d \le x) = P(\|\mathbf{G}^{(\infty)}\|_d \le x), \tag{2.2}$$

where $\mathbf{G}^{(\infty)} = (G_1, G_2, \ldots)$ and G_i's are standard normal random variables and $\mathrm{Cov}(G_i, G_j) = \sigma_{ij}$. It is worthwhile to point out that $\mathbf{G}^{(\infty)}$ is defined in an appropriate sub-space of $\mathbb{R}^\infty$.

Recently, Arlot et al. [**ABR10a, ABR10b**] studied alternative forms of the above result in the context of the sample means. Specifically, using concentration inequalities, they derive a test of the form

$$\left\| \frac{1}{n} \sum_{i=1}^{n} U_{ij} \right\|_d \ge c,$$

where c can be chosen according to certain bootstrap methods controlling for the type I error.

Motivated by these considerations, we adopt the following methodology for testing high-dimensional regression coefficients. We reject H_0 if

$$\|\mathbf{W}^{(n)}\|_d \ge c$$

where c is chosen to be the quantiles of $\|\mathbf{G}^{(\infty)}\|_d$, where $\mathbf{G}^{(\infty)}$ is defined in equation (2.2). In Section 3 below, we describe a multiplier bootstrap with Rademacher weights to implement this methodology.

While many choices for d are possible, we notice that $d = 2$ corresponds to the statistic in [**SD08**] when testing for the mean vectors in one sample and two-sample problems. We emphasize here that they do not consider the regression problems described in this paper. The case $d = \infty$ corresponds to the sup-norm of the test statistic. It is known that the choice $d = \infty$ leads to the control of FWER [**Lin05**].

2.3. Examples. Below we derive the efficient score for several models encountered in applications.

EXAMPLE 1 (Linear regression model). For each $j = 1, \ldots, p$, we consider the following simple linear regression model

$$Y_i = \gamma_j + \beta_j X_{ij} + \epsilon_{ij}$$

where the X_{ij}'s have been standardized such that $\sum_{i=1}^n X_{ij} = 0$ and $n^{-1} \sum_{i=1}^n X_{ij}^2 = 1$, and ϵ_{ij} is the residual error with mean 0 and variance σ_j^2. By the standard results of parametric likelihood, it can be shown that the efficient score function for β_j from the ith subject takes the form

$$U_{ij} = (Y_i - \gamma_j) X_{ij}.$$

Under H_0, γ_j can be estimated by $\overline{Y}^{(n)} = (1/n) \sum_{i=1}^n Y_i$.

EXAMPLE 2 (Two-sample problem). In the two sample problem, for the ith subject we observed the outcome data $(Y_{i1}, \ldots, Y_{ip})$ and the group indicator X_i which takes values 0 if the ith subject belongs to group 1 and takes value 1 otherwise. Number of subjects in groups 1 and 2 are n_1 and n_2, respectively. For $j = 1, \ldots, p$, we fit the following simple linear regression model

$$Y_{ij} = \gamma_j + \beta_j X_i + \epsilon_{ij},$$

where α_j is the intercept and β_j is the mean group difference for the jth outcome. Under H_0, γ_j can be estimated by $\overline{Y}_j^{(n)} = (1/n) \sum_{i=1}^n Y_{ij}$, yielding

$$U_{ij} = (Y_{ij} - \overline{Y}_j^{(n)})(X_i - \overline{X}^{(n)}), .$$

where $\overline{X}^{(n)} = n_2/n$.

EXAMPLE 3 (Logistic regression model). In this case, the outcome variable Y_i is binary. We consider the following logistic model

$$\log \frac{P(Y_i = 1 \mid X_{ij})}{P(Y_i = 0 \mid X_{ij})} = \alpha_j + \beta_j X_{ij},$$

where X_{ij}'s have been standardized. Denote

$$\pi_{ij} = P(Y_i = 1 \mid X_{ij}) = \frac{e^{\alpha_j + \beta_j X_{ij}}}{1 + e^{\alpha_j + \beta_j X_{ij}}}.$$

We can obtain

$$U_{ij} = (Y_i - \pi_{ij}) X_{ij}.$$

Under H_0, we can replace π_{ij} in U_{ij} by $\widehat{\pi}^{(n)} = (1/n) \sum_{i=1}^n Y_i$.

EXAMPLE 4 (Cox proportional hazards model). Suppose that we have n i.i.d. observations $\{Y_i = \min(T_i, C_i), \Delta_i = I(T_i \leq C_i), X_{i1}, \ldots, X_{ip}; i = 1, \ldots, n\}$, where T_i is the failure time for the ith subject, C_i is the censoring time, Δ_i is the censoring indicator, and $(X_{i1}, \ldots, X_{ip})$ are the standardized covariates of interest. For each covariate, we consider the Cox proportional hazards model

$$\lambda(t \mid X_{ij}) = \lambda(t) \exp(\beta_j X_{ij}),$$

where $\lambda(t|X_{ij})$ is the hazard function of the failure time given covariate X_{ij}, $\lambda(t)$ is the baseline hazard function and β_j is the regression coefficient for the jth covariate.

Follow the results in [**Kos08**, Chapter 3], we can show that the efficient score function for β_j under the null hypothesis such that $\beta_j = 0$ is given by

$$U_{ij} = \{\Delta_i - \Lambda(Y_i)\}X_{ij},$$

where $\Lambda(\cdot)$ is the cumulative baseline hazard function.

3. Implementation

3.1. Multiplier bootstrap procedure. Direct implementation of the above test will involve the estimation of the covariance matrix of $(W_1^{(n)}, W_2^{(n)}, \dots, W_p^{(n)})$, which is unstable for large p. To address this issue, borrowing the ideas from the multiplier bootstrap methodology, we propose the following Monte Carlo approach:

(1) Generate i.i.d. random variables $(L_1, \dots, L_n)$ such that $E(L_1) = 0$ and $\mathrm{Var}(L_1) = 1$, which are independent of the data.
(2) Calculate $\widetilde{U}_j^{(n)} = \sum_{i=1}^n U_{ij} L_i$ and $\widetilde{W}_j^{(n)} = \widetilde{U}_j^{(n)} / \sqrt{V_j^{(n)}}$. Define

$$\widetilde{\mathbf{W}}^{(n)} = (\widetilde{W}_1^{(n)}, \widetilde{W}_2^{(n)}, \dots, \widetilde{W}_p^{(n)}).$$

(3) Repeat steps 1 and 2 B times.
(4) For a given FWER α, compute the $1-\alpha$ sample quantile of the B values of $\|\widetilde{\mathbf{W}}^{(n)}\|_d$. Reject the null hypothesis if $\|\mathbf{W}^{(n)}\|_d$ exceeds this threshold.

We refer to the above procedure as the multiplier bootstrap procedure, following the terminology from [**Præ90, PW93**]. While several choices for the distributions of L are available, some yield improved performances compared to others. In particular, [**Lin05**] proposed Gaussian distribution for the choice of L. It is known in the learning theory literature [**BM02, Kol06**], among all possible distributions for L, two point distributions yield the most accurate results in terms of the FWER. Amongst these two point distributions, symmetric Rademacher sequence $(L_1, \dots, L_n)$, where L_i's are independent and take values -1 and 1 with equal probability, yields improved accuracy when the distributions of the data are symmetric. In this paper, we refer to the above procedure with the Rademacher sequence and Gaussian sequence as the multiplier bootstrap Rademacher weights (MBRW) test and Lin's test, respectively.

3.2. Classical bootstrap procedure. We also describe a Monte-Carlo procedure based on the classical bootstrap methodology. Specifically, we replace the first two steps in Section 3.1 with:

(1) Draw n random samples $(\widetilde{U}_{1j}, \dots, \widetilde{U}_{nj})$ from $(U_{1j}, \dots, U_{nj})$ with replacement, for each $j = 1, \dots, p$.
(2) Calculate $\widetilde{U}_j^{(n)} = \sum_{i=1}^n \widetilde{U}_{ij}$, $\widetilde{V}_j^{(n)} = \sum_{i=1}^n \widetilde{U}_{ij}^2$, and $\widetilde{W}_j^{(n)} = \widetilde{U}_j^{(n)} / \sqrt{\widetilde{V}_j^{(n)}} - W_j^{(n)}$. Define

$$\widetilde{\mathbf{W}}^{(n)} = (\widetilde{W}_1^{(n)}, \widetilde{W}_2^{(n)}, \dots, \widetilde{W}_p^{(n)}).$$

We then continue the steps (3) and (4) in Section 3.1 to perform the hypothesis test.

4. Simulation study

In this section, we describe simulation studies to evaluate the performance of the proposed methodology and compare it with other existing methods. Specifically, using the test statistic with sup-norm (i.e., $d = \infty$), we compare our proposed MBRW test to the classical bootstrap test, the KV test, and Lin's test. Note that the KV test in the simulation studies is an extension of the original KV test using the shrinkage estimator of the covariance matrix of the efficient scores. We assess the size and power as a function of the the sample size n, the total number of features p, the number of important features s, the pairwise correlation between the coefficients ρ, and the value of the signal μ. For each simulation setting, we generate 2000 simulated data sets and set $B = 2000$.

For each model we consider the following simulation settings: $n = 20, 100$; $p = 100, 1000$; $s = 0, 1, 2$; $\rho = 0, .5$; and μ are chosen such that we have reasonable power comparisons. The case of $s = 0$ corresponds to the evaluation of the size of the test, and $s = 1, 2$ to the evaluation for the power. We begin with the two-sample problem.

4.1. Two sample problem. We observe n observations from two groups, $\mathbf{X}_{i,0}$ $\mathbf{X}_{i,1}$, $i = 1, \ldots, n$. The two groups are independent normal random vectors, which have pairwise correlation given by ρ. Therefore,

$$\mathbf{X}_{i,0} \sim N_p(\boldsymbol{\mu}_0, \boldsymbol{\Sigma}_0), \quad \mathbf{X}_{i,1} \sim N_p(\boldsymbol{\mu}_1, \boldsymbol{\Sigma}_1),$$

with

$$\boldsymbol{\Sigma}_{0,ij} = \begin{cases} 1 & \text{if } i = j, \\ \rho & \text{if } i \neq j. \end{cases}, \quad \boldsymbol{\Sigma}_{1,ij} = \begin{cases} 1 & \text{if } i = j, \\ \rho & \text{if } i \neq j. \end{cases}$$

For ease of notation we set $\boldsymbol{\mu}_0 = \mathbf{0}$. The vector $\boldsymbol{\mu}_1$ has the first s components equal to μ, and the remaining $p - s$ equal to 0. In this model, we also consider the case when $n = 20$ and $p = 10^4$, i.e., $p \gg n$. Table 1 displays the type I error rates and powers of the four tests under comparison. Lin's test appears to be conservative for small sample size and has the lowest power among the four tests in all situations. The MBRW test has accurate control of type I error rates for all settings whereas the KV test has inflated type I error rates for $n = 20$. The KV test and the MBRW test have comparable powers and are more powerful than the other two tests. As expected, power increases with increasing of ρ, decreasing of p, and increasing of s. To see an intuitive reason why larger ρ tends to produce greater power, larger values of ρ are associated with fewer number of "independent" features. When $\rho = 1$, basically there is only one independent feature. Note that the results for the KV test were not available when $p = 10^4$ as it is computationally intensive for large p.

We compared the computation time of the four tests for different configurations of n and p. Under all situations, the computation time is not impacted by the choice of ρ or s. Therefore we fix $\rho = 0$ and $s = 0$. Table 2 presents the computation time with 1000 replicates for each test on a Dell PowerEdge 2900 server. As expected, the computation time increases as n or p increases. The computation time of Lin's test, the classical Bootstrap test, and the MBRW test were comparable whereas the KV test is computationally more intensive than other three tests especially when p is large.

TABLE 1. Type I error rates and powers (%) of the four tests for two sample problem at significance level of 0.05. — indicates that the results are not available.

n	p	ρ	μ	KV	Lin	Bootstrap	MBRW
				$H_0: s=0$			
20	100	0	0	9.70	2.65	4.15	6.80
20	100	0.5	0	8.95	3.85	6.25	7.05
20	1000	0	0	10.45	1.00	1.70	6.60
20	1000	0.5	0	8.45	2.45	4.70	6.70
20	10000	0	0	—	0.45	1.20	7.50
20	10000	0.5	0	—	1.65	4.45	6.75
100	100	0	0	5.15	3.65	4.00	4.95
100	100	0.5	0	5.60	4.50	5.55	5.20
100	1000	0	0	5.95	3.80	4.05	5.80
100	1000	0.5	0	5.90	4.80	5.25	5.65
100	10000	0	0	—	3.25	3.45	5.70
100	10000	0.5	0	—	4.45	5.05	5.80
				$H_1: s=1$			
20	100	0	1.6	61.60	45.05	50.00	58.05
20	100	0.5	1.6	67.30	57.30	62.90	66.00
20	1000	0	1.6	44.20	19.90	24.20	38.60
20	1000	0.5	1.6	50.95	36.85	42.25	50.70
20	10000	0	2.2	—	45.70	50.60	68.40
20	10000	0.5	2.2	—	70.10	75.45	81.20
100	100	0	0.7	52.65	48.85	49.95	52.40
100	100	0.5	0.7	61.00	58.45	59.70	60.80
100	1000	0	0.7	34.80	28.90	30.30	34.20
100	1000	0.5	0.7	43.50	40.25	40.75	43.10
100	10000	0	1.0	—	63.05	64.20	70.15
100	10000	0.5	1.0	—	79.75	80.70	82.45
				$H_1: s=2$			
20	100	0	1.6	84.15	69.45	74.30	80.60
20	100	0.5	1.6	81.70	72.45	78.00	80.30
20	1000	0	1.6	65.10	35.05	41.55	59.90
20	1000	0.5	1.6	66.50	51.75	57.80	65.95
20	10000	0	2.2	—	69.35	74.30	89.30
20	10000	0.5	2.2	—	83.55	87.90	91.85
100	100	0	0.7	77.50	74.45	74.95	76.80
100	100	0.5	0.7	76.75	74.25	75.85	76.85
100	1000	0	0.7	53.90	46.55	47.95	53.15
100	1000	0.5	0.7	58.80	55.20	56.10	58.70
100	10000	0	1.0	—	85.55	86.10	90.60
100	10000	0.5	1.0	—	90.90	91.60	92.95

TABLE 2. Computation time (seconds) of the four tests for two sample problem with 1000 replicates. — indicates that the results are not available.

n	p	KV	Lin	Bootstrap	MBRW
20	100	44	10	10	9
20	1000	3180	77	92	93
20	2000	18700	145	185	180
20	10000	—	760	940	940
100	100	66	59	55	55
100	1000	5000	340	470	450
100	2000	25000	660	940	880
100	10000	—	7200	8400	9800

4.2. Regression models. In this section, we consider linear regression and logistic regression models. We partition the covariates into covariates with zero coefficients $\mathbf{X}_{i,0}$ and covariates with non-zero coefficients $\mathbf{X}_{i,1}$. The two partitions are independent normal random vectors, which have pairwise correlation given by ρ.

Therefore,

$$\mathbf{X}_{i,0} \sim N_s(\mathbf{0}, \mathbf{\Sigma}_0), \quad \mathbf{X}_{i,1} \sim N_{p-s}(\mathbf{0}, \mathbf{\Sigma}_1),$$

with

$$\mathbf{\Sigma}_{0,ij} = \begin{cases} 1 & \text{if } i = j, \\ \rho & \text{if } i \neq j. \end{cases}, \quad \mathbf{\Sigma}_{1,ij} = \begin{cases} 1 & \text{if } i = j, \\ \rho & \text{if } i \neq j. \end{cases}$$

We generate the outcome variable Y_i using

(1) Normal: $Y_i = 0.5 + \boldsymbol{\beta}^t \mathbf{X}_i + \epsilon_i$, where $\epsilon_i \sim N(0,1)$;
(2) Logistic: $Y_i \sim Bern(p_i)$, where $\text{logit}(p_i) = 0.5 + \boldsymbol{\beta}^t \mathbf{X}_i$,

where the first s components of $\boldsymbol{\beta}$ are equal to $\mu \neq 0$, and the remaining $p - s$ are zero. We point out that the correlation in the design matrix $\mathbf{X}_{i,j}$ introduces the correlations amongst $W_j^{(n)}$'s defined in Section 2.

Tables 3 and 4 present the type I error rates and powers of the four tests for the simple linear regression model and the logistic regression model, respectively. The proposed MBRW test has accurate control of type I error rates under all settings whereas the other three tests are conservative especially when n is small and p is large. Furthermore, the MBRW test is substantially more powerful than other tests.

5. Applications

5.1. COGA study. We applied the four tests with sup-norm described in Sections 2 and 3 to the Collaborative Study on the Genetics of Alcoholism (COGA), a multi-center study designed to identify and characterize genes that influence susceptibility to alcohol dependence and related phenotypes [**BRH+95**]. The COGA study contains 143 multi-generation pedigrees with a total of 1,614 individuals, of which 1,353 individuals were selected for genotyping of SNPs conducted by Affymetrix and Illumina. The family sizes range from 5 to 32. We considered the quantitative electrophysiological phenotype TTTH1 (electric potential FP1, far frontal left side channel). Of the 1,353 genotyped individuals, 901 had the TTTH1

TABLE 3. Type I error rates and powers (%) of the four tests for simple linear regression model at significance level of 0.05.

n	p	ρ	μ	KV	Lin	Bootstrap	MBRW
				$H_0 : s = 0$			
20	100	0	0	1.40	0.00	0.20	6.55
20	100	0.5	0	3.05	0.70	1.75	6.95
20	1000	0	0	0.05	0.00	0.00	7.45
20	1000	0.5	0	1.00	0.00	0.25	6.95
100	100	0	0	3.80	1.90	1.85	4.95
100	100	0.5	0	4.30	3.50	3.25	4.70
100	1000	0	0	2.85	0.90	0.75	4.70
100	1000	0.5	0	3.90	2.55	2.70	5.20
				$H_1 : s = 1$			
20	100	0	1.25	64.50	22.05	33.45	83.00
20	100	0.5	1.25	65.95	46.05	58.25	85.70
20	1000	0	1.25	11.80	0.15	0.55	63.00
20	1000	0.5	1.25	29.70	9.55	16.55	71.75
100	100	0	0.3	30.45	23.15	21.65	32.20
100	100	0.5	0.3	34.90	32.15	32.05	38.05
100	1000	0	0.3	13.10	7.00	6.50	17.00
100	1000	0.5	0.3	19.65	17.35	16.85	23.65
				$H_1 : s = 2$			
20	100	0	1.25	29.05	4.00	8.60	54.10
20	100	0.5	1.25	81.90	59.60	74.8	96.15
20	1000	0	1.25	1.25	0.00	0.00	28.40
20	1000	0.5	1.25	41.80	16.70	29.30	88.65
100	100	0	0.3	43.10	34.95	33.90	45.75
100	100	0.5	0.3	91.25	89.80	89.90	92.65
100	1000	0	0.3	19.40	11.45	10.20	24.20
100	1000	0.5	0.3	78.45	75.80	74.45	82.15

measurements. In the case of missing genotype data, we treat them as missing at random as described in [**KV10**].

[**DL06**] used a semiparametric variance-component (VC) model and performed a chromosome-wide association analysis on Chromosome 7 using the genotype data provided by Affymetrix for 577 SNPs. Age, age^2, gender, and maximum number of drinks consumed in a 24-hr period were included as covariates in the semiparametric VC model. The maximum of the likelihood ratio test statistics was 15.06 at the location of 130.405cM for SNP tsc0022400. The same SNP was also detected to be associated with TTTH1 by [**HDR$^+$05**]. To account for missing genotype data, we set U_{ij} to be 0 if no individual in the ith family has genotype data at marker j. To compare the performance of different tests, we obtain the adjusted p-value by comparing the test-statistic for each feature/at each marker location to the null-distribution of the sup-norm statistic. Incidentally, this controls the FWER. [**DL06**] obtained the adjusted p-value of 0.050 using Lin's test [**Lin05**]. The adjusted p-values are 0.059, 0.066, and 0.019 from the KV, classical Bootstrap, and MBRW tests, respectively. As expected, the MBRW test is more powerful than

TABLE 4. Type I error rates and powers (%) of the four tests for logistic regression model at significance level of 0.05

n	p	ρ	μ	KV	Lin	Bootstrap	MBRW
				$H_0 : s = 0$			
20	100	0	0	1.15	0.40	1.00	6.65
20	100	0.5	0	1.90	1.55	3.40	7.05
20	1000	0	0	0.05	0.00	0.00	8.25
20	1000	0.5	0	0.75	0.35	1.30	6.15
100	100	0	0	4.05	3.65	3.85	5.40
100	100	0.5	0	4.50	4.40	4.55	5.00
100	1000	0	0	2.85	2.50	2.40	5.45
100	1000	0.5	0	3.85	3.55	4.00	4.80
				$H_1 : s = 1$			
20	100	0	3	20.00	20.40	27.45	56.85
20	100	0.5	3	33.30	38.25	46.15	65.05
20	1000	0	3	0.10	0.55	1.85	29.35
20	1000	0.5	3	4.90	8.05	13.30	38.65
100	100	0	0.6	22.80	21.65	22.65	26.00
100	100	0.5	0.6	27.35	27.80	28.75	30.50
100	1000	0	0.6	9.00	8.65	8.60	12.95
100	1000	0.5	0.6	14.40	14.75	15.30	17.25
				$H_1 : s = 2$			
20	100	0	3	6.60	7.45	11.05	30.75
20	100	0.5	3	40.25	47.05	57.05	74.75
20	1000	0	3	0.05	0.15	0.60	13.15
20	1000	0.5	3	5.15	10.00	17.10	46.20
100	100	0	0.6	32.70	32.35	33.40	37.05
100	100	0.5	0.6	82.00	82.55	83.30	84.40
100	1000	0	0.6	11.25	10.90	11.20	16.40
100	1000	0.5	0.6	61.50	62.40	63.20	66.30

other tests and detected more significant association between SNP tsc0022400 and the quantitative trait TTTH1.

5.2. Breast Cancer Study. We also applied the proposed methodology to a breast cancer study described in [**ZMY+09**]. One objective of the study was to identify copy number alterations that were correlated with the time in developing distant metastasis in breast cancer patients. The data set submitted to the Gene Expression Omnibus data base [**BTW+11**] with accession number 10099 by the authors contains the clinical and SNP data for 313 breast cancer patients.

Median age of the 313 patients at the time of surgery was 54 years (range 26–83 years). One hundred fourteen tumors were classified as estrogen receptor (ER) negative and others as ER positive. One hundred fifty two patients were before menopause and 161 were after menopause. Among the 313 patients, , 114 patients developed distant metastasis with a median failure time of 28 months (range 2–80 months). Copy number data at 115,353 SNPs across the whole genome were available for each patient. We notice here that both applications involve the survival data. However, the ratio of p/n in the Breast Cancer Study is about 100 times that in the COGA study.

We performed genomewide association analysis to evaluate the relationship between copy number alternations and time to distant metastasis by fitting a Cox proportional hazards model accounting for the effects of age, ER status, and menopause status. To save computational burden, we calculated the score test statistic at each SNP, which is asymptotically equivalent to the likelihood ratio test statistic. We obtained the adjusted p-values using the classical Bootstrap, MBRW, and Lin's tests with $B = 10^4$. Note that the KV test is computationally prohibitive for large p therefore it is not included in the data analysis. The maximum of the score test statistic was 24.17 at SNP rs10486488. The adjusted p-value using the classical Bootstrap, MBRW, and Lin't tests were 0.0214, 0.0017, and 0.0056, respectively. As expected, the MBRW test obtained the most significant result among three tests. Figure 1 presents the negative logarithm (base 10) of the adjusted p-values at 115,353 SNPs using three multivariate tests. The MBRW test detected 79 SNPs at genomewide significance level of 0.01, compared to 40 SNPs and 0 SNP based on the Lin's test and the classical Bootstrap test. The proposed methods are computationally efficient; it took less than 75 minutes to complete the analysis of these data on a Dell PowerEdge 2900 server.

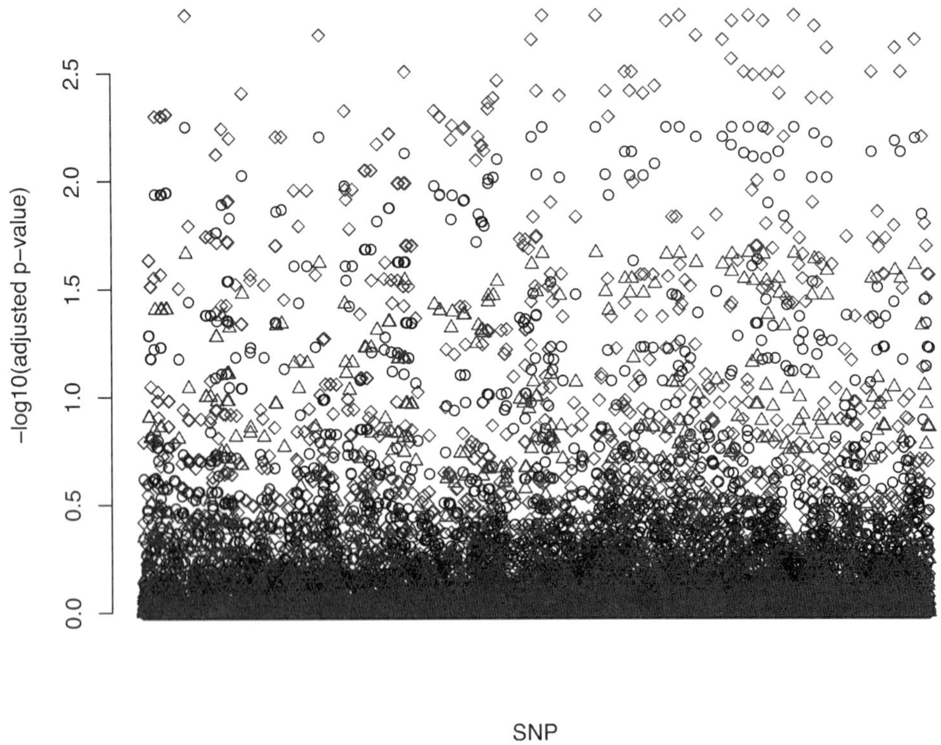

FIGURE 1. Logarithm (base 10) of the adjusted p-values at 115,353 SNPs from the Breast Cancer Study using the Lin's test (∘), the MBRW test (⋄), and the classical Bootstrap test (△).

6. Discussion

The primary objective of the paper is to describe a methodology to test for regression parameters in ultra-high-dimensional regression problems. We studied a test statistic namely the maximum of efficient scores, taken across the features, obtained by constructing an appropriate regression model for each component of the data vector. The key point of the approach that we presented here is the generality.

We provided heuristic justification for the validity of the described procedure. A rigorous justification will necessarily involve specific models and the assumptions will need to be adapted to the specific problems under consideration. However, roughly speaking, the test statistics described in the paper are t-type statistics whose joint asymptotics can be described along the lines of the work described in [**ABR10a**, **ABR10b**, **KV10**, **KV11**]. The details are available in our joint working paper [**VDH13**].

Our simulation results show that the proposed methodology yields nominal FWER and improved power over the other competing methods. The methods work even in situations when $n = 20$ and $p = 10^4$. Importantly, as shown in Table 2, our methods are scalable. In the specific example concerning copy number variation, the proposed method identified more genes than the other existing methods.

There are several questions that need to be further resolved. A rigorous theoretical study of the power analysis is lacking and methods to identify the correct norm for data analysis is still lacking. It is also feasible to extend our methods to random effects models and other multivariate settings. We are pursuing some of these extensions.

Acknowledgement. The authors thank the editor and the referees for a careful reading of the manuscript and several useful suggestions.

References

[ABR10a] S. Arlot, G. Blanchard, and E. Roquain, *Some nonasymptotic results on resampling in high dimension. I. Confidence regions*, Ann. Statist. **38** (2010), no. 1, 51–82, DOI 10.1214/08-AOS667. MR2589316 (2011c:62155)

[ABR10b] S. Arlot, G. Blanchard, and E. Roquain, *Some nonasymptotic results on resampling in high dimension. II. Multiple tests*, Ann. Statist. **38** (2010), no. 1, 83–99, DOI 10.1214/08-AOS668. MR2589317 (2011c:62150)

[BL04] P. J. Bickel and E. Levina, *Some theory of Fisher's linear discriminant function, 'naive Bayes', and some alternatives when there are many more variables than observations*, Bernoulli **10** (2004), no. 6, 989–1010, DOI 10.3150/bj/1106314847. MR2108040 (2006a:62081)

[BM02] P. L. Bartlett and S. Mendelson, *Rademacher and Gaussian complexities: risk bounds and structural results*, J. Mach. Learn. Res. **3** (2002), no. Spec. Issue Comput. Learn. Theory, 463–482, DOI 10.1162/153244303321897690. MR1984026 (2004d:68134)

[BRH+95] H. Begleiter, T. Reich, V. Hesselbrock, B. Porjesz, T. K. Li, M. A. Schuckit, H. J. Edenberg, and J. P. Rice, *The collaborative study on the genetics of alcoholism*, Alcohol, Health & Research World **19** (1995), no. 3, 228–236.

[BTW+11] T. Barrett, D. B. Troup, S. E. Wilhite, P. Ledoux, C. Evangelista, I. F. Kim, M. Tomashevsky, K. A. Marshall, K. H. Phillippy, P. M. Sherman, R. N. Muertter, M. Holko, O. Ayanbule, A. Yefanov, and A. Soboleva, *NCBI GEO: archive for functional genomics data sets—10 years on*, Nucleic Acids Research **39** (2011), 1005–1010, DOI 10.1093/nar/gkq1184.

[DL06] G. Diao and D. Y. Lin, *Improving the power of association tests for quantitative traits in family studies*, Genetic Epidemiology **30** (2006), no. 4, 301–313, DOI 10.1002/gepi.20145.

[FF08] J. Fan and Y. Fan, *High-dimensional classification using features annealed independence rules*, Ann. Statist. **36** (2008), no. 6, 2605–2637, DOI 10.1214/07-AOS504. MR2485009 (2010f:62109)

[FUS10] Y. Fujikoshi, V. V. Ulyanov, and R. Shimizu, *Multivariate statistics*, Wiley Series in Probability and Statistics, John Wiley & Sons Inc., Hoboken, NJ, 2010. High-dimensional and large-sample approximations. MR2640807 (2011h:62206)

[HDR+05] L. M. Havill, T. D. Dyer, D. K. Richardson, M. C. Mahaney, and J. Blangero, *The quantitative trait linkage disequilibrium test: a more powerful alternative to the quantitative transmission disequilibrium test for use in the absence of population stratification*, BMC Genetics **6** (2005), S91, DOI 10.1186/1471-2156-6-S1-S91.

[Kol06] V. Koltchinskii, *Local Rademacher complexities and oracle inequalities in risk minimization*, Ann. Statist. **34** (2006), no. 6, 2593–2656, DOI 10.1214/009053606000001019. MR2329442 (2009h:62060)

[Kos08] M. R. Kosorok, *Introduction to empirical processes and semiparametric inference*, Springer Series in Statistics, Springer, New York, 2008. MR2724368 (2012b:62005)

[KV10] J. Kuelbs and A. N. Vidyashankar, *Asymptotic inference for high-dimensional data*, Ann. Statist. **38** (2010), no. 2, 836–869, DOI 10.1214/09-AOS718. MR2604698 (2011c:60009)

[KV11] J. Kuelbs and A. N. Vidyashankar, *Weak convergence results for multiple generations of a branching process*, J. Theoret. Probab. **24** (2011), no. 2, 376–396, DOI 10.1007/s10959-009-0266-y. MR2795044 (2012h:60260)

[Lin05] D. Y. Lin, *An efficient Monte Carlo approach to assessing statistical significance in genomic studies*, Bioinformatics **21** (2005), no. 6, 781–787, DOI 10.1093/bioinformatics/bti053.

[LJW11] M. E. Lopes, L. J. Jacob, and M. J. Wainwright, *A more powerful two-sample test in high dimensions using random projection* (2011), preprint, available at `arXiv:1108.2401`.

[LW04] O. Ledoit and M. Wolf, *A well-conditioned estimator for large-dimensional covariance matrices*, J. Multivariate Anal. **88** (2004), no. 2, 365–411, DOI 10.1016/S0047-259X(03)00096-4. MR2026339 (2004m:62130)

[LW12] O. Ledoit and M. Wolf, *Nonlinear shrinkage estimation of large-dimensional covariance matrices*, Ann. Statist. **40** (2012), no. 2, 1024–1060, DOI 10.1214/12-AOS989. MR2985942

[Præ90] J. Præstgaard, *Bootstrap with general weights and multiplier central limit theorems*, Technical Report 195, Department of Statistics, University of Washington, 1990.

[PW93] J. Præstgaard and J. A. Wellner, *Exchangeably weighted bootstraps of the general empirical process*, Ann. Probab. **21** (1993), no. 4, 2053–2086. MR1245301 (94k:60054)

[SD08] M. S. Srivastava and M. Du, *A test for the mean vector with fewer observations than the dimension*, J. Multivariate Anal. **99** (2008), no. 3, 386–402, DOI 10.1016/j.jmva.2006.11.002. MR2396970 (2009e:62261)

[TC09] C.-A. Tsai and J. J. Chen, *Multivariate analysis of variance test for gene set analysis*, Bioinformatics **25** (2009), no. 7, 897–903, DOI 10.1093/bioinformatics/btp098.

[THNC02] R. Tibshirani, T. Hastie, B. Narasimhan, and G. Chu, *Diagnosis of multiple cancer types by shrunken centroids of gene expression*, Proc. Natl. Acad. Sci. USA **99** (2002), no. 10, 6567–6572, DOI 10.1073/pnas.082099299.

[VDH13] A. N. Vidyashankar, G. Diao, and B. Hanlon, *Asymptotic and non-asymptotic results for high-dimensional regression problems* (2013), in preparation.

[ZMY+09] Y. Zhang, J. W. M. Martens, J. X. Yu, J. Jiang, A. M. Sieuwerts, M. Smid, J. G. M. Klijn, Y. Wang, and J. A. Foekens, *Copy number alterations that predict metastatic capability of human breast cancer*, Cancer Research **69** (2009), no. 9, 3795–3801, DOI 10.1158/0008-5472.CAN-08-4596.

Department of Statistics, George Mason University, Fairfax, VA 22030, USA
E-mail address: `gdiao@gmu.edu`

Department of Statistics, University of Wisconsin—Madison, Madison, WI 53706, USA
E-mail address: `hanlon@stat.wisc.edu`

Department of Statistics, George Mason University, Fairfax, VA 22030, USA
E-mail address: `avidyash@gmu.edu`

Contemporary Mathematics
Volume **622**, 2014
http://dx.doi.org/10.1090/conm/622/12431

On multiple contrast tests and simultaneous confidence intervals in high-dimensional repeated measures designs

Frank Konietschke, Yulia R. Gel, and Edgar Brunner

ABSTRACT. In this paper we study multiple contrast tests and simultaneous confidence intervals for arbitrary contrasts of means in high-dimensional repeated measures designs with normal data. Such designs involving more variables than subjects are widely met in a variety of disciplines but are of particular interest in modern biostatistical applications, e.g., various "omics" studies and medical imaging. In order to derive a robust and reliable statistical inference for both high-dimensional global hypothesis and multiple comparison procedures, we propose to combine modern regularization techniques such as banding and thresholding for estimation of an empirical covariance matrix, with the parametric bootstrap. We establish efficiency and consistency properties of the new regularized tests and investigate their finite sample behavior by simulations. Our studies indicate that the new regularized statistics yield competitive results both in terms of the delivered Type I error rates and power performance, even for a small number of subjects and large number of variables. We also illustrate the proposed methodology in application to the sleep disorder study.

1. Introduction

Repeated measures designs, i.e., designs in which the same subject is repeatedly observed under several conditions such as time points or treatments, constitute one of the most widely employed classes of designs in applications, ranging from medicine and biology to social sciences. For instance, a simple case of a paired design refers to a situation when the same subject is observed twice under two different conditions. However, in most studies more than two conditions (variables) are considered. Moreover, many modern applications and especially various studies related to high-throughput technology in the "omics" field, e.g., genomics, transcriptomics, metabolomics, proteomics, and others, as as to functional magnetic resonance imaging (fMRI), often involve substantially more variables than subjects in the trial, which leads to a high-dimensional hypothesis testing framework and

2010 *Mathematics Subject Classification.* Primary 62F05, 62F40; Secondary 62K15, 62P10 .
Key words and phrases. Multiple testing, repeated measures designs, matrix regularization, familywise error rate, resampling and bootstrap.
The first author was supported by the German Science Foundation projects BR-655/16-1 and HO 1687/9-1.
The research of the second author was supported by grant from the Natural Science and Engineering Research Council of Canada and by the Waterloo–Göttingen Collaboration grant.

the so-called *large p-small n problem*. Furthermore, we often observe different variances for the response variables, e.g., in the case of time curves, for the response variables, and generally the covariance matrix has to be assumed unstructured, which constitutes an additional challenge in deriving reliable conclusions from repeated measures designs when the number of variables is substantially larger than the number of subjects.

Inference methods for testing global hypotheses in high-dimensional repeated measures and multivariate data received substantial attention in the recent years (see, for example, **Bai and Saranadasa (1996)**, **Rauf Ahmad et al. (2008)**, **Kuelbs and Vidyashankar (2010)**, **Chi et al. (2012)**, **Srivastava et al. (2013)**, **Srivastava and Kubokawa (2013)**, and references therein). However, in most repeated measures designs the experimenter is not only interested in testing the overall hypotheses about the treatment effects but also in multiple comparisons in terms of multiple contrasts and related confidence intervals. Indeed, regulatory authorities routinely require simultaneous confidence intervals (SCI) in randomized clinical trials: "*Estimates of treatment effects should be accompanied by confidence intervals, whenever possible* [...]" **ICH (1998**, E9 Guideline, Chapter 5.5, p. 25). However, the frequently applied testing strategy of (1) testing the overall null hypothesis by an appropriate procedure (e.g., Repeated Measures ANOVA), then (2) performing multiple comparisons to identify the significant factor levels (e.g., paired t-tests with the Bonferroni adjustment), and finally (3) computing SCI for the effects of interest, may result in neither consonant nor coherent conclusions and should be avoided (**Konietschke et al., 2012**). The related multiple comparisons for means of high-dimensional independent observations are considered by Cao and Kosorok (**2011**), whereas Fan et al. (**2007**) discuss the question "How many t-tests can be simultaneously performed?" In the context of microarray analysis with false (FDR) control, Kosorok and Ma (**2007**) consider uniform convergence for a range of univariate statistics constructed for each data dimension. In this article, we propose a new single-step multiple comparison procedure and derive the related simultaneous confidence intervals for arbitrary contrasts of means in high-dimensional repeated measures designs that are, by construction, consonant and coherent. Our key idea is to combine parametric bootstrap with the modern regularization techniques such as banding and thresholding for estimation of covariance matrices (**Bickel and Levina, 2008a**; **2008b**), in a high-dimensional multiple hypothesis testing framework. Our studies indicate that the new regularized multiple contrast procedures deliver competitive performance both in terms of size and power of the test, even for a small number of subjects and a substantial number of variables.

The paper is organized as follows. In Section 2 the statistical model, hypotheses and test statistics are introduced. We present the new regularization based multiple contrast tests in Section 3. Section 4 is devoted to assessment of finite sample properties of the proposed methodology for high-dimensional repeated measures designs using simulations. In Section 5, we illustrate application of the regularized multiple contrast tests and simultaneous confidence intervals to the sleep disorder study. The paper concludes with a discussion and future work in Section 6. All technical details can be found in the Appendix.

2. Statistical model, hypotheses and test statistics

Let us consider a general model for repeated measures designs driven by independent and identically distributed random vectors

$$\mathbf{Y}_k = (Y_{1k}, \ldots, Y_{dk})' \sim N(\boldsymbol{\mu}, \mathbf{V}), \quad k = 1, \ldots, n, \tag{2.1}$$

with an expectation $\boldsymbol{\mu} = (\mu_1, \ldots, \mu_d)'$ and a positive definite covariance matrix $\mathbf{V} = [v_{ij}]$. We assume that $\mathbf{V}$ belongs to the uniformity class of matrices that satisfy both banding and thresholding conditions (**Bickel and Levina, 2008a**; **2008b**):

$$\begin{aligned}\big\{\mathbf{V} : |v_{ij}| \leq C|i-j|^{-(\alpha+1)}, \forall i,j : |i-j| \geq 1,\\ 0 < \epsilon_0 \leq \lambda_{\min}(\mathbf{V}) \leq \lambda_{\max}(\mathbf{V}) \leq 1/\epsilon_0\big\},\end{aligned} \tag{2.2}$$

where C and α are constants, and $\lambda_{\min}(\mathbf{V})$ and $\lambda_{\max}(\mathbf{V})$ denote the smallest and largest eigenvalues of $\mathbf{V}$, respectively.

REMARK 2.1. Note that the assumption (2.2) is general and can be relaxed if a particular regularization procedure, i.e., banding or thresholding, is selected, as well as if we a-priori know what kind of contrast matrix is to be employed in the design, and, hence, the respective sparsity of the contrast matrix is accounted for.

Throughout the paper, we are mainly interested in the situation when $d > n$, and d might be a function of the sample size n, such that as $d \to \infty$, also $n \to \infty$ and $\log(d)/n \to 0$. Furthermore, the repeated measures may have a factorial structure, i.e., in this setup the factorial design can be described by incorporating a "factor pattern" into d in a similar way as in the theory of linear mixed models (**Rauf Ahmad et al., 2008**).

REMARK 2.2. Note that the assumption of multivariate normality in (2.1) can be relaxed, and all the results in this paper still hold under the assumption of bounded random variables (see the discussion by Van Der Laan and Bryan (**2001**)). However, the proposed statistics employ critical values from the extreme tail portion of multivariate t-distributions and, hence, as noted by Gao et al. (**2008**) and Konietschke et al. (**2012**), are particularly sensitive to departures from normality. Moreover, simulation studies indicate that even for low-dimensional skewed data of small sample sizes, satisfactory results for mean based multiple contrast tests with arbitrary contrasts cannot be expected.

Let

$$\mathbf{C} = \begin{pmatrix} \mathbf{c}_1' \\ \vdots \\ \mathbf{c}_q' \end{pmatrix} = \begin{pmatrix} c_{11} & \cdots & c_{1d} \\ \vdots & \ddots & \vdots \\ c_{q1} & \cdots & c_{qd} \end{pmatrix}$$

denote a user-defined $q \times d$-contrast matrix, i.e., $\mathbf{C}\mathbf{1}_q = \mathbf{0}$, where $\mathbf{1}_q = (1, \ldots, 1)'$, and each row vector $\mathbf{c}_\ell' = (c_{\ell 1}, \ldots, c_{\ell d})$ represents one contrast. For example, multiple comparisons to a control group are defined by using the contrast matrix

$$\mathbf{C} = \begin{pmatrix} -1 & 1 & 0 & \cdots & 0 \\ -1 & 0 & 1 & \cdots & 0 \\ \vdots & \vdots & \ddots & \ddots & \vdots \\ -1 & 0 & \cdots & \cdots & 1 \end{pmatrix}.$$

Here we are interested in simultaneous testing of q linear hypotheses $H_0^\ell\colon \mathbf{c}_\ell'\boldsymbol{\mu} = 0$, $\ell = 1, \ldots, q$, and the respective *familywise error rate* (FWER) should be strongly controlled at α level of significance (**Hochberg and Tamhane, 1987**, p. 2ff). Examples of such multiple comparisons include Dunnett's many-to-one comparisons (**Dunnett, 1955**), Tukey's all-pairwise comparisons (**Tukey, 1994**), multiple comparisons to the overall means, change point analysis, etc. For a detailed overview of different contrasts, we refer to **Bretz et al. (2001)**.

Now let $\mathbf{X}_k = \mathbf{C}\mathbf{Y}_k$, $k = 1, \ldots, n$ be random vectors with expectation $E(\mathbf{X}_1) = \mathbf{C}\boldsymbol{\mu} = (\delta_1, \ldots, \delta_q)'$ and covariance matrix $\mathrm{Cov}(\mathbf{X}_1) = \mathbf{C}\mathbf{V}\mathbf{C}' = \boldsymbol{\Sigma} = [\sigma_{\ell m}]$, $\ell, m = 1, \ldots, q$. Let $\overline{\mathbf{X}}_\cdot = (\overline{X}_{1\cdot}, \ldots, \overline{X}_{q\cdot})'$ denote the vector of the means $\overline{X}_{i\cdot} = n^{-1}\sum_{k=1}^n X_{ik}$, and let

$$\widehat{\boldsymbol{\Sigma}}_n = \frac{1}{n-1}\sum_{k=1}^n (\mathbf{X}_k - \overline{\mathbf{X}}_\cdot)(\mathbf{X}_k - \overline{\mathbf{X}}_\cdot)' \tag{2.3}$$

be a $q \times q$-empirical covariance matrix, where $\widehat{\boldsymbol{\Sigma}}_n = [\widehat{\sigma}_{\ell m}]$, $\ell, m = 1, \ldots, q$. It is well known that the empirical covariance matrix is singular when $d > n$ or when the contrast matrix is not of full rank. Hence, procedures being based on the inverse of $\widehat{\boldsymbol{\Sigma}}_n$, such as Hotelling's T^2- test, are not applicable in this case (see **Rauf Ahmad et al. (2008)** and references therein), and thus we need to employ test statistics which do not require the inverse of $\widehat{\boldsymbol{\Sigma}}_n$.

In particular, we start from testing the individual null hypothesis $H_0^\ell\colon \delta_\ell = 0$, and computing a simultaneous confidence interval (SCI) for the effect δ_ℓ, using the paired t-test type pivotal quantity

$$T_\ell^\mu = \sqrt{n}\frac{\overline{X}_{\ell\cdot} - \delta_\ell}{\sqrt{\widehat{\sigma}_{\ell\ell}}}, \quad \ell = 1, \ldots, q.$$

It is well known that T_ℓ^μ follows a central t-distribution with $n-1$ degrees of freedom and that the classical Bonferroni adjustment can be used to perform multiple comparisons as well as for the computation of SCI (see, e.g., **Hochberg and Tamhane (1987**, pp. 3, 363)). However, this approach typically delivers a low power, particularly when the test statistics are not independent (**Konietschke et al., 2012**). Hence, in order to account for correlation in simultaneous inference, it is necessary to consider the multivariate distribution $\mathbf{F}_n$ of the vector

$$\mathbf{T} = (T_1^\mu, \ldots, T_q^\mu)' \sim \mathbf{F}_n. \tag{2.4}$$

Note that $\mathbf{T}$ is not computed by using an inverse of $\widehat{\boldsymbol{\Sigma}}_n$, but only by studentizing each mean with the corresponding diagonal element from $\widehat{\boldsymbol{\Sigma}}_n$. Hence, the statistic is invariant under scalar transformations of each component (**Srivastava and Du, 2008**). Although in case of normality each component of $\mathbf{T}$ follows a central t-distribution, $\mathbf{T}$ does not follow a multivariate t-distribution. This is due to the fact that each component of $\overline{\mathbf{X}}_\cdot$ is studentized with "its own" variance estimator. Only if each component of $\overline{\mathbf{X}}_\cdot$ had been studentized by an overall variance estimate, $\mathbf{T}$ would be multivariate t-distributed (**Cornish, 1954**; **Bretz et al., 2001**; **Hothorn et al., 2008**). Hence, the finite sample distribution of $\mathbf{T}$ is yet unknown, while its asymptotic distribution under $n \to \infty$, $d \to \infty$ such that $\log(d)/n \to 0$, is a multivariate normal distribution with an unknown correlation matrix $\mathbf{R} = \mathbf{D}^{-1/2}\boldsymbol{\Sigma}\mathbf{D}^{-1/2}$ (Theorem 3.1), where $\mathbf{D} = \mathrm{diag}(\sigma_{11}, \ldots, \sigma_{qq})$ is a diagonal matrix, and σ_{jj}, $j = 1, \ldots, q$ are the diagonal elements of the population covariance matrix $\boldsymbol{\Sigma}$. For large sample sizes, the individual null hypotheses

$H_0\colon \delta_\ell = 0$ are then rejected at a two-sided multiple level α, if

$$|T_\ell^0| \geq z_{1-\alpha}(\mathbf{R}), \quad \ell = 1, \ldots, q, \tag{2.5}$$

where $z_{1-\alpha}(\mathbf{R})$ denotes the two-sided equicoordinate quantile from $N(\mathbf{0}, \mathbf{R})$ (see **Bretz et al. (2001)**, **Hothorn et al. (2008)**, and references therein). Asymptotic $(1-\alpha)$- SCI for the treatment effects δ_ℓ are then given by

$$\mathrm{CI}_\ell = [\overline{X}_{\ell\cdot} \pm z_{1-\alpha}(\mathbf{R})\sqrt{\widehat{\sigma}_{\ell\ell}/n}]. \tag{2.6}$$

Note that these SCI are compatible with the individual test decisions in (2.5), i.e., it cannot occur that an individual null hypothesis is rejected, but the SCI contains the value null coming from the null hypothesis. For large sample sizes, the global null hypothesis $H_0\colon \mathbf{C}\boldsymbol{\mu} = \mathbf{0}$ is then rejected, if

$$T_0 = \max\{|T_1^0|, \ldots, |T_q^0|\} \geq z_{1-\alpha}(\mathbf{R}). \tag{2.7}$$

Hence, the multiple contrast test procedure (MCTP), T_0, can be equivalently defined as follows:

$$\psi_n = \mathcal{I}_{(z_{1-\alpha}(\mathbf{R}),\infty)}(T_0). \tag{2.8}$$

Note that the joint distribution of $\mathbf{T}$ as defined in its pivotal version (2.4) does not depend on the quantities δ_ℓ, $\ell = 1, \ldots, q$, and hence the subset pivotality holds (**Westfall and Troendle, 2008**). Therefore, the MCTP ψ_n controls the FWER in the strong sense (see, for example, **Bretz et al. (2001)**, **Hothorn et al. (2008)**, and references therein). From a conservative point of view, the quantiles $z_{1-\alpha}(\mathbf{R})$ can be replaced by the $t_{1-\alpha}(n-1, \mathbf{R})$-quantiles from the multivariate $T(n-1, \mathbf{R})$-distribution with $n-1$ degrees of freedom and the correlation matrix $\mathbf{R}$. However, since $\mathbf{R}$ is unknown, the SCI in (2.6) are not computable in data evaluations. For low-dimensional data, i.e., when $d < n$, Hothorn et al. (**2008**) propose to replace $\mathbf{R}$ by a consistent (i.e., $d < n$, $n \to \infty$) plug-in estimator $\widehat{\mathbf{R}}_n$ in (2.5), (2.6), and (2.7), and prove that such a procedure is asymptotic valid. However, simulation studies show that the rate of convergence of the finite-sample distribution of $\mathbf{T}$ to its limiting multivariate normal (or t-) distribution is rather slow, even in the case of normality. Moreover, the quality of the approximation deteriorates even further for a high-dimensional case when $d > n$, especially for small samples of observations.

In the next section, we develop new regularized resampling-based versions of $\mathbf{T}$ that aim to improve finite-sample performance of the multiple contrast tests in a high-dimensional setting, and investigate their asymptotic properties. In particular, our key idea is to combine modern regularization procedures such as banding and thresholding, in application to $\widehat{\boldsymbol{\Sigma}}_n$, with the parametric bootstrap in a hypothesis testing framework.

3. Inference for T in a high-dimensional case

In this section we discuss three different strategies to improve the finite-sample performance of $\mathbf{T}$ in the high-dimensional setting and mimick its limiting distribution, with the aim to develop valid multiple testing procedures and simultaneous confidence intervals. As pointed out in Section 2, the approximation via a multivariate $T(n-1, \widehat{\mathbf{R}}_n)$-distribution is not satisfactory due to the singularity of $\widehat{\mathbf{R}}_n$, when $d > n$. Hence, our idea is to develop a method allowing us to avoid the direct estimation of $\mathbf{R}$, which can be achieved by regularizing $\widehat{\boldsymbol{\Sigma}}_n$ and then employing the parametric bootstrap. Van Der Laan and Bryan (**2000**; **2001**) investigate the parametric bootstrap for gene expression analysis in high-dimensional settings and discuss the range of its theoretical and practical applicability. In particular, as indicated by Van Der Laan and Bryan (**2001**) and Giné and Zinn (**1990**), the parametric bootstrap might be preferred over the nonparametric bootstrap due to the fact that the parametric bootstrap is asymptotically valid under relatively milder conditions compared to those required by the nonparametric bootstrap.

Now, note that the distribution of $\mathbf{T}$ under the null hypothesis is completely specified by the covariance matrix $\mathbf{\Sigma}$. Hence, our idea is first to construct a consistent regularized empirical estimator of $\widehat{\mathbf{\Sigma}}_n$, denoted by $\widetilde{\mathbf{\Sigma}}_n$, and then to embed $\widetilde{\mathbf{\Sigma}}_n$ into the parametric bootstrap scheme. To generalize this approach, we start from describing this procedure without specifying a particular type of regularization, which is then discussed in details the next subsection.

Given $\mathbf{X} = (\mathbf{X}_1', \ldots, \mathbf{X}_n')'$, let

$$\mathbf{X}_k^* = (X_{1k}^*, \ldots, X_{qk}^*)' \sim N(\mathbf{0}, \widetilde{\mathbf{\Sigma}}_n), \quad k = 1, \ldots, n, \tag{3.1}$$

denote n independent and identically distributed multivariate normal vectors with expectation $\mathbf{0}$ and covariance matrix $\widetilde{\mathbf{\Sigma}}_n$. Now consider the multivariate distribution of the vector of t-test type statistics

$$\mathbf{T}^* = (T_1^*, \ldots, T_q^*)', \quad \text{where } T_\ell^* = \sqrt{n}\frac{\overline{X}_{\ell\cdot}^*}{\sqrt{\widetilde{\sigma}_{\ell\ell}^*}}, \tag{3.2}$$

where $\overline{X}_{\ell\cdot}^* = n^{-1}\sum_{k=1}^n X_{\ell k}^*$ and $\widetilde{\sigma}_{\ell\ell}^* = (n-1)^{-1}\sum_{k=1}^n (X_{\ell k}^* - \overline{X}_{\ell\cdot}^*)^2$ denote the sample mean and variance of the ℓ-th sample component $X_{\ell 1}^*, \ldots, X_{\ell n}^*$, respectively. Then, according to the definition of the unconditional multiple contrast test ψ_n in (2.8), the conditional resampling tests can be defined as follows

$$\psi_n^* = \mathcal{I}_{(z_{1-\alpha}^*(\mathbf{R}), \infty)}(T_0^*), \tag{3.3}$$

where $T_0^* = \max\{|T_1^*|, \ldots, |T_q^*|\}$, and $z_{1-\alpha}^*(\mathbf{R})$ is the $(1-\alpha)$-equicoordinate quantile from the distribution of $\mathbf{T}^*$. Note that the denominator in (3.2) is a part of the resampling procedure, which is in accordance with the guidelines for bootstrap testing (see **Hall and Wilson (1991)**, **Beran (1997)**, **Bickel and Freedman (1981)**, and **Janssen (2005)**). Delaigle et al. (**2011**) suggest that such studentized resampling t-statistics are more robust and accurate than their nonstudentized counterparts.

The next theorem shows that the distribution of $\mathbf{T}^*$ is asymptotically equivalent to the distribution of $\mathbf{T}$.

THEOREM 3.1. *Let $n \to \infty$ and $d \to \infty$ such that $\log(d)/n \to 0$ and let $\widetilde{\mathbf{\Sigma}}_n$ denote a consistent estimator for $\mathbf{\Sigma}$. Also, let ψ_n and ψ_n^* be the conditional and unconditional multiple contrast tests defined in* (2.8) *and* (3.3), *respectively. Then, the parametric bootstrap test ψ_n^* is*

(i) *asymptotically effective under H_0 with respect to ψ_n, i.e.,*

$$E(|\psi_n^* - \psi_n|) \to 0, \quad n, d \to \infty,$$

and

(ii) *consistent, i.e.,*

$$E(\psi_n^*) \to \mathcal{I}(\exists\, \mu_i > \mu_j, i \neq j)$$

for $\mathbf{C}\boldsymbol{\mu} \neq \mathbf{0}$, as $n, d \to \infty$.

The proof is given in the Appendix.

We can then construct the simultaneous confidence intervals, according to the following algorithm:

(1) Given $\mathbf{X}$, compute $\overline{\mathbf{X}}_{\cdot}$, $\widetilde{\mathbf{\Sigma}}_n$, $\mathbf{T}$ and T_0.
(2) Generate $\mathbf{X}_1^*, \ldots, \mathbf{X}_n^*$ and compute T_0^*.
(3) Repeat the previous two steps B times, and in each step compute $T_{0,j}^*$ for $j = 1, \ldots, B$.
(4) Estimate the $(1-\alpha)$-quantile $z_{1-\alpha}^*(\mathbf{R})$ from the resampling distribution of $\{T_{0,1}^*, \ldots, T_{0,B}^*\}$.

Finally, the $(1-\alpha)$-simultaneous confidence intervals are the given by

$$\mathrm{CI}_\ell^* = [\overline{X}_{\ell\cdot} \pm z_{1-\alpha}^*(\mathbf{R})\sqrt{\widehat{\sigma}_{\ell\ell}/n}]. \tag{3.4}$$

Now, let us discuss three different estimation methods for the empirical covariance matrix $\widehat{\mathbf{\Sigma}}_n$.

3.1. Regularization of $\widehat{\mathbf{\Sigma}}_n$ with banding and thresholding. While there exists a variety of regularization techniques aiming to deliver consistent estimators of covariance matrices in a high-dimensional setting (see the overview **Bickel and Li (2006)** and references therein), here we focus on two regularization procedures, namely banding and thresholding (see **Bickel and Levina (2008b; 2006)**, **Bickel and Gel (2011)**, and references therein). In particular, the banded estimator $\widetilde{\mathbf{\Sigma}}_n^B$ of $\mathbf{\Sigma}$ is defined as

$$\widetilde{\mathbf{\Sigma}}_n^{B(s)} = [\widetilde{\sigma}_{\ell m}] = [\widehat{\sigma}_{\ell m}\mathcal{I}(|\ell - m| \le s)], \tag{3.5}$$

for $0 \le s < q$. In turn, the thresholded estimator $\widetilde{\mathbf{\Sigma}}_n^T$ of $\widehat{\mathbf{\Sigma}}_n$ is given by

$$\widetilde{\mathbf{\Sigma}}_n^{T(t)} = [\widetilde{\sigma}_{\ell m}] = [\widehat{\sigma}_{\ell m}\mathcal{I}(|\widehat{\sigma}_{\ell m}| \ge t)]. \tag{3.6}$$

Note that under the conditions of Theorem 3.1, $\mathbf{X} = (\mathbf{X}_1', \ldots, \mathbf{X}_n')'$ satisfies the conditions of Theorems 1 in **Bickel and Levina (2008a; 2008b)**, which implies consistency of $\widetilde{\mathbf{\Sigma}}_n^{B(s)}$ and $\widetilde{\mathbf{\Sigma}}_n^{T(t)}$.

THEOREM 3.2. *Under the conditions of Theorem* 3.1, *the banded and thresholded estimators,* $\widetilde{\mathbf{\Sigma}}_n^{B(s)}$ *and* $\widetilde{\mathbf{\Sigma}}_n^{T(t)}$, *respectively, are consistent estimators of* $\mathbf{\Sigma}$ *under an operator norm.*

Hence, Theorem 3.1 coupled with Theorem 3.2 implies that the regularized resampling-based multiple contrast tests are consistent in a high-dimensional setting when $n \to \infty$, $d \to \infty$ and $\log(d) = o(n)$.

Following Bickel and Levina (**2008a**; **2008b**), we estimate the "optimal" regularizing parameters, i.e., band s_{opt} and threshold t_{opt}, using crossvalidation. In particular, we split the data $\mathbf{X}_1, \ldots, \mathbf{X}_n$ randomly into two parts, $\mathbf{X}^{(1,j)}$ and $\mathbf{X}^{(2,j)}$, with sample sizes n_1 and n_2, respectively. We find little sensitivity in terms of the choice of n_1 and n_2, and use $n_1 = n/3$ in our studies. Let $\widetilde{\mathbf{\Sigma}}_{n_1}^R(\mathbf{X}^{(1,j)})$ denote the regularized covariance estimator being obtained from $\mathbf{X}^{(1,j)}$, where R is a banding or thresholding operator, and let $\widehat{\mathbf{\Sigma}}_{n_2}(\mathbf{X}^{(2,j)})$ be the empirical covariance matrix being computed with $\mathbf{X}^{(2,j)}$. Then, the optimal regularizer is an argument that minimizes the following loss function:

$$\min_m \frac{1}{N}\sum_{j=1}^{N} \|\widetilde{\mathbf{\Sigma}}_{n_1}^{R(m)}(\mathbf{X}^{(1,j)}) - \widehat{\mathbf{\Sigma}}_{n_2}(\mathbf{X}^{(2,j)})\|_F, \tag{3.7}$$

where $\|\cdot\|_F$ denotes the Frobenius norm, N is the number of random splits and m are possible regularizers, i.e., bands or thresholds.

As shown by Bickel and Levina (**2008a**; **2008b**) and Bickel and Gel (**2011**), the regularized estimators, with the regularizing parameter being selected from the crossvalidation procedure, satisfy the oracle properties under the Frobenius norm, and, hence, $\widetilde{\mathbf{\Sigma}}_n^{B(s_{\text{opt}})}$ and $\widetilde{\mathbf{\Sigma}}_n^{T(t_{\text{opt}})}$ are also consistent estimators of $\mathbf{\Sigma}$ with the same rates of convergence under the Frobenius norm as the banded and thresholded estimators, with the band and thresholding parameter being selected by the oracle.

Finally, following the algorithm on page 115, we generate $\mathbf{X}_k^*$ from the parametric bootstrap:

$$\mathbf{X}_k^* = \mathbf{Z}_k\big(\widetilde{\mathbf{\Sigma}}_n^{R_{\text{opt}}}\big)^{1/2}, \tag{3.8}$$

where $\mathbf{Z}_k \sim N(\mathbf{0}, \mathbf{I})$, $\big(\widetilde{\mathbf{\Sigma}}_n^{R_{\text{opt}}}\big)^{1/2}$ is obtained via the Cholesky decomposition, i.e., $\big(\widetilde{\mathbf{\Sigma}}_n^{R_{\text{opt}}}\big)^{1/2} = \mathbf{L}_n\mathbf{D}_n^{1/2}\mathbf{L}_n'$, and $\widetilde{\mathbf{\Sigma}}_n^{R_{\text{opt}}}$ corresponds to $\widetilde{\mathbf{\Sigma}}_n^{B(s_{\text{opt}})}$ or $\widetilde{\mathbf{\Sigma}}_n^{T(t_{\text{opt}})}$. Routine calculations show that $\text{Cov}(\mathbf{X}_1^*) = \widetilde{\mathbf{\Sigma}}_n^{R_{\text{opt}}}$.

REMARK 3.3. In our studies we frequently observe a situation when the cross-validation procedure yields no positive definite regularized matrix $\widetilde{\mathbf{\Sigma}}_n^{R(s)}$, except of the diagonal matrix. In these cases, we taper $\widehat{\mathbf{\Sigma}}_n$ with an exponential filter, i.e., $\ddot{\mathbf{\Sigma}}_n = [\ddot{\sigma}_{\ell m}] = (1+r)^{-(|\ell - m|)}\widehat{\sigma}_{\ell m}$, and then perform banding or thresholding. We use $r = 0.1$ throughout our studies.

3.2. Singular value decomposition of $\widehat{\mathbf{\Sigma}}_n$. As an alternative to more computationally intensive regularized procedures, we can also employ a singular value decomposition (SVD) of $\widehat{\mathbf{\Sigma}}_n$, i.e.,

$$\widehat{\mathbf{\Sigma}}_n = \mathbf{A}_n\mathbf{B}_n\mathbf{E}_n, \tag{3.9}$$

where $\mathbf{A}_n$ and $\mathbf{E}_n$ are unitary matrices, and $\mathbf{B}_n$ is a diagonal matrix embedding the singular values of $\widehat{\mathbf{\Sigma}}_n$. The generation of $\mathbf{X}_k^*$ is then given by $\mathbf{X}_k^* = \mathbf{Z}_k\mathbf{A}_n\mathbf{B}_n^{1/2}\mathbf{E}_n$, where $\mathbf{Z}_k \sim N(\mathbf{0}, \mathbf{I})$, $k = 1, \ldots, n$. Routine calculations show that $\text{Cov}(\mathbf{X}_1) = \widehat{\mathbf{\Sigma}}_n$. Although $\widehat{\mathbf{\Sigma}}_n$ is also a consistent estimate of $\mathbf{\Sigma}$ in model (2.1) (**Van Der Laan and Bryan, 2001**), the main disadvantage of this method is that the informative part of $\widehat{\mathbf{\Sigma}}_n$ is truncated too roughly via SVD, which results in more liberal test decisions (see Section 4 for details).

4. Simulations

In this section we evaluate the finite-sample performance of four versions of the multiple contrast tests, i.e., when the critical values are approximated by a multivariate $T(n-1, \widehat{\mathbf{R}}_n)$ distribution (TAPP) and the three resampling-based methods with the banding-type regularized covariance estimator (3.5) (BAND), thresholding-type regularized estimator (3.6) (THRES) and the singular value decomposition (3.9) (SVD). We investigate the behavior of these test statistics with regard to (i) maintaining the pre-assigned Type I error level α under the hypothesis $H_0 : \mathbf{C}\boldsymbol{\mu} = \mathbf{0}$, and (ii) power of the test under the alternative in a high-dimensional framework.

The observations are generated from d-variate multivariate heteroscedastic normal distributions with two different types of covariance matrices:

$$\mathbf{V}_1 = \mathbf{I} + 0.5(\mathbf{J} - \mathbf{I}) \quad \text{and} \quad \mathbf{V}_2 = [v_{ij}] = 0.6^{(|i-j|-1)/4}, \tag{4.1}$$

where $\mathbf{J} = \mathbf{1}\mathbf{1}'$ is a $d \times d$-matrix of 1's. The matrices $\mathbf{V}_1$ and $\mathbf{V}_2$ are also known as compound-symmetric and auto-regressive covariance structures, respectively. Furthermore, we will investigate two-different kinds of contrast matrices, namely

$$(4.2) \qquad \mathbf{C}_1 = \mathbf{I} - \frac{1}{d}\mathbf{J} \quad \text{and} \quad \mathbf{C}_2 = \begin{pmatrix} -1 & 1 & 0 & \dots & \dots & 0 & 0 \\ -1 & 0 & 1 & 0 & \dots & \dots & 0 \\ \vdots & \vdots & \vdots & \vdots & \vdots & \vdots & \vdots \\ -1 & 0 & 0 & 0 & \dots & \dots & 1 \\ 0 & -1 & 1 & 0 & \dots & 0 & 0 \\ 0 & -1 & 0 & 1 & 0 & \dots & 0 \\ \vdots & \vdots & \vdots & \vdots & \vdots & \vdots & \vdots \\ 0 & \dots & \dots & \dots & \dots & -1 & 1 \end{pmatrix},$$

which are known as the *Average*-type ($\mathbf{C}_1$) and Tukey-type ($\mathbf{C}_2$) contrasts, respectively. We consider low, medium and high-dimensional settings by varying the dimension $d \in \{5, 10, 20, 30, 40, 50\}$ for the Average-type and $d \in \{4, 6, 8, 10, 12, 14\}$ for the Tukey-type comparisons. The number of comparisons is $q = d - 1$ for the Average-type and $q = d(d-1)/2$ for the Tukey-type, respectively. We only consider small sample sizes of $n = 10$ and $n = 15$ observations, since this is a typical case in many biostatistical applications. The results reported here constitute a representative set from a much larger simulation study using R (`www.r-project.org`). All simulation results were obtained from 10,000 simulation and 10,000 bootstrap runs. The equicoordinate quantiles were computed with the R-package *mvtnorm* (**Genz and Bretz, 2009**). Throughout this section, the parameters s and t for banding and thresholding, respectively, are chosen using the fully data-driven crossvalidation procedure for each simulation run. Hence, the selected values for s and t differ for each realization, and for example, in the case of banding range anywhere from 1 to $d - 1$.

The simulation results for average-type contrasts are displayed in Figure 1. Figure 1 indicates that the approximation by a multivariate $T(n-1, \widehat{\mathbf{R}}_n)$ distribution is an inadequate method for statistical inferences in a high-dimensional setting, as this procedure tends to be highly liberal with an increasing dimension d or respectively increasing numbers of comparisons q. Resampling its limit distribution via SVD tends to be more accurate, but the results are still liberal, especially for small samples of $n = 10$. In contrast, both the banding and thresholding methods accurately control Type I error rates across both covariance structures, even for n of 10 and number of comparisons of 49.

Next we investigate the simulation results for the Tukey-type comparisons (see Figure 2). As expected, the results for the Tukey-type contrasts are generally worse than the results for the Average-type contrasts (see Figure 2), which is due to a higher dimensionality of the Tukey-type contrasts, i.e., under the same fixed number of observations n, the maximum number of comparisons are $q = 49$ and $q = 91$ for the Average- and Tukey-types, respectively. As for the Average-type, we find that both the approximation by a multivariate $T(n-1, \widehat{\mathbf{R}}_n)$ distribution and the SVD resampling-based approximations yield liberal sizes of the tests and substantially deviate from the declared nominal level of 0.05, especially for an increasing number of comparisons and $n = 10$. In contrast, the banding and thresholding methods deliver a competitive performance across all covariance structures and sample sizes.

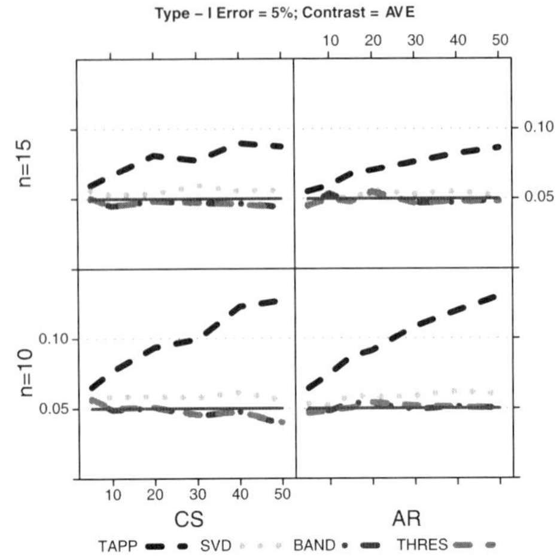

FIGURE 1. The empirical Type I error rates under the nominal level α of 0.05 and the Average contrasts (4.2), with the sample sizes of $n = 10$ (bottom) and $n = 15$ (top), different covariance matrices $\mathbf{V}_1$ (the left panel) and $\mathbf{V}_2$ (the right panel), varying dimensions d (x-axis) and by using 10,000 Monte Carlo replications and 10,000 bootstrap runs. The numbers of contrasts is given by $q = d - 1$.

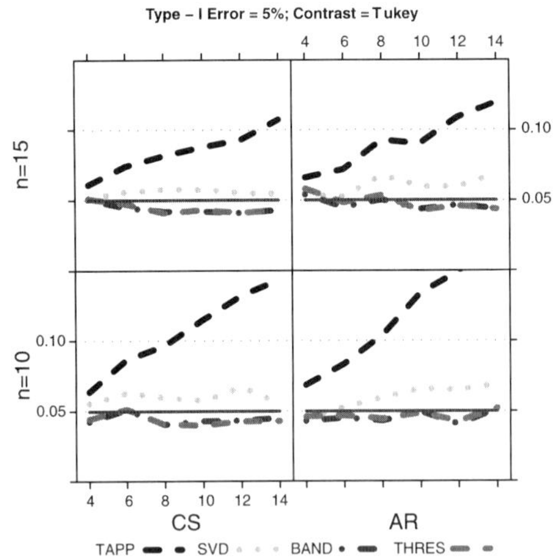

FIGURE 2. The empirical Type I error rates under the nominal level α of 0.05 and the Tukey contrasts (4.2), with the sample sizes of $n = 10$ (bottom) and $n = 15$ (top), different covariance matrices $\mathbf{V}_1$ (the left panel) and $\mathbf{V}_2$ (the right panel), varying dimensions d (x-axis) and by using 10,000 Monte Carlo replications and 10,000 bootstrap runs. The numbers of contrasts is given by $q = d(d-1)/2$.

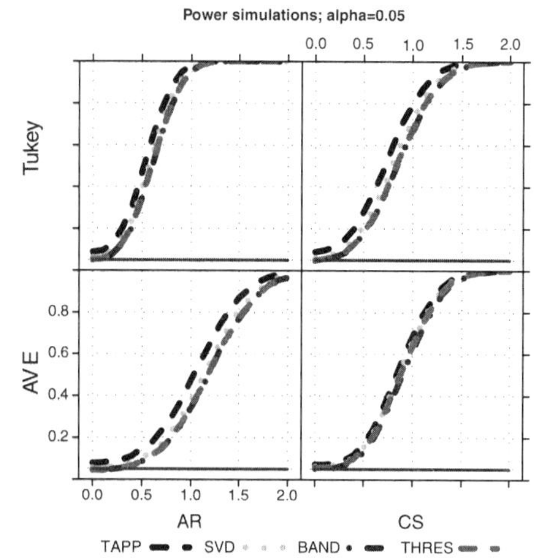

FIGURE 3. Power performance under the nominal level α of 0.05, the Average-type (bottom) and Tukey-type (top) contrasts, with the sample sizes of $n = 10$, different covariance matrices $\mathbf{V}_1$ (the left panel) and $\mathbf{V}_2$ (the right panel), dimension $d = 20$ for shift effects δ (x-axis)by using 10,000 Monte Carlo replications and 10,000 bootstrap runs.

Our simulation studies for other kinds of contrasts and covariance structures yield very similar findings and, hence, are not shown here.

To investigate the power performance among the four types of the multiple contrast tests, we generate a sample of 10 observations from the $d = 20$-variate normal distribution with covariance matrices $\mathbf{V}_1$ and $\mathbf{V}_2$ as defined in (4.1). Due to abundance of possible alternatives, we concentrate on the least favorable alternative, that is, $\boldsymbol{\mu} = (0, \ldots, 0, \delta)$, with varying shift parameter $0 \leq \delta \leq 2$. The simulation results are displayed in Figure 3. Remarkably, although the $T(n-1, \widehat{\mathbf{R}}_n)$-approximation is very liberal, no significant power increase can be detected, and the difference in power performance among the four procedures is negligible. The results for other kinds of alternatives, e.g., $\boldsymbol{\mu} = (0, \delta, 2\delta, \ldots, (d-1)\delta)$, are similar and, hence, are not shown here.

In final, we can conclude that while delivering the comparable power performance, the parametric bootstrap using the regularized covariance estimators yields the most accurate results in terms of the size of the test, across all considered types of contrasts, covariance structures and sample sizes, even when the number of comparisons q substantially exceeds the number of observations n and n is small. Hence, the new regularized resampling-based versions of the multiple contrast tests can be viewed as a preferred testing procedure in a high-dimensional setting.

5. Example

Now let us illustrate the developed methodology for high-dimensional multiple contrast tests in application to a sleep disorder study. In particular, we re-consider the sleep disorder experiment being analyzed by Rauf Ahmad et al. (**2008**), where the objective was to investigate the activity of prostaglandin-D-synthase (β-trace) in relation to human sleep. The variable of interest, serum concentration of lipocalintype prostaglandin D synthase, was measured on 10 young, healthy women. The selected part of the data consists of serum concentrations measured for first three nights at time 00:00, 04:00, 08:00, 12:00, 16:00, 20:00. Thus, there is a total of 18 repeated measurements taken on each of 10 subjects at 4 h time intervals for three consecutive nights, classified as *Normal Sleep*, *Total Sleep Deprivation* and *Recovery Night*. The design of this trial is a structured high-dimensional repeated measures design, involving the factors *Day*, *Time* and the interaction effect *Day* $\times$ *Time*. Boxplots and mean-error plots of the serum concentrations for each *Day* $\times$ *Time* combination are displayed in Figures 4 and 5, respectively. We illustrate the practical advantage of the new regularized resampling-based multiple contrast tests by comparing the results with the global testing procedures (**Rauf Ahmad et al., 2008**). First, we evaluate the data by using the modified ANOVA-type statistic for high-dimensional repeated measures (**Rauf Ahmad et al., 2008**) and test the global null hypotheses of "no day effect," "no time effect," and "no interaction effect." We obtain a non-significant difference between the serum concentrations between the three days (with p-value of 0.535), a statistically significant day effect with p-value of 0.014 and an interaction effect between day and time on the border of statistical significance (with p-value of 0.051). The questionable result on the interaction effect needs further medical and statistical investigations. However, the testing procedure of **Rauf Ahmad et al. (2008)** can only be employed to test the global null hypotheses and cannot shed light on the local information for the individual effects.

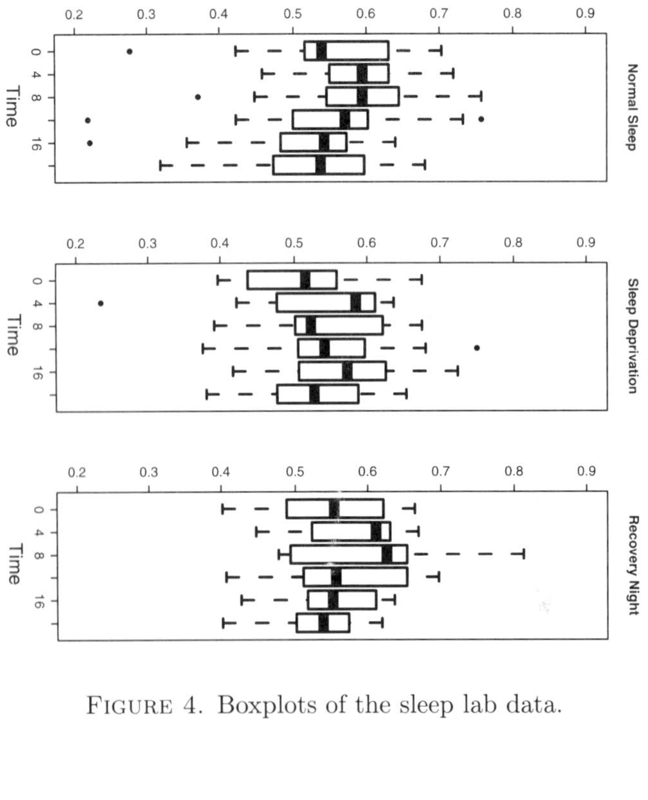

FIGURE 4. Boxplots of the sleep lab data.

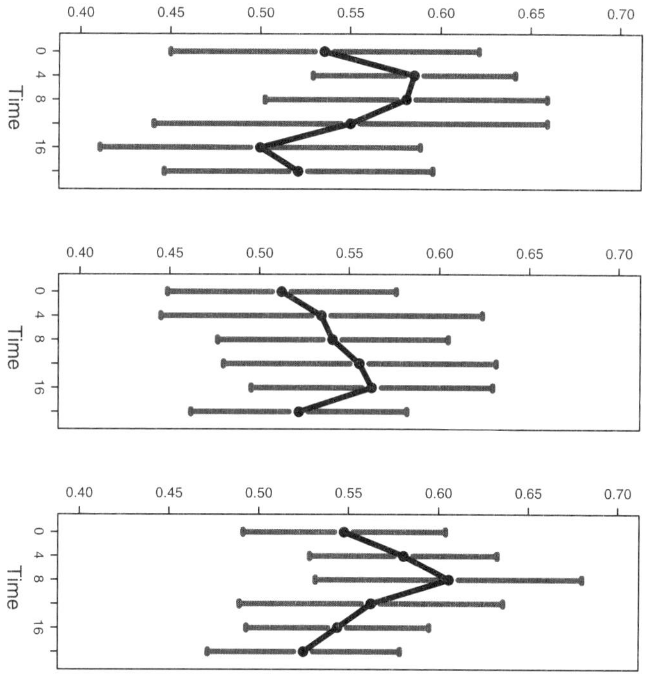

FIGURE 5. Mean profiles of the sleep lab data (normal sleep (left), sleep deprivation (middle), recovery night (right)).

We incorporate this local information by re-evaluating these data with a multiple contrast test and simultaneous confidence intervals for the interaction effects, i.e., by investigating the contrast matrix

$$\mathbf{C} = \mathbf{P}_3 \otimes \mathbf{P}_6, \tag{5.1}$$

where $P_d = \mathbf{I} - 1/d\mathbf{J}_d$ and $\mathbf{A} \otimes \mathbf{B}$ is the Kronecker product of matrices $\mathbf{A}$ and $\mathbf{B}$ (**Rauf Ahmad et al., 2008**). By utilizing the classical approximation by a multivariate $T(n-1, \widehat{\mathbf{R}}_n)$-distribution, we obtain a p-value of 0.047, which could be interpreted as a statistically significant interaction. The respective simultaneous confidence intervals are displayed in the upper panel of Figure 6, which immediately indicates that the time point 8am under recovery sleeps leads to the statistically significant interaction. However, as our simulation study indicates, the t-approximation in high-dimensional settings exhibit very liberal sizes of the test and hence is unreliable. We therefore apply the parametric bootstrap approach based on banding-type regularized estimates, with 10,000 bootstrap replications and band parameter being selected from crossvalidation. (The results using the parametric bootstrap with thresholded covariance estimators are very similar and, hence, are omitted.)

We find that the parametric bootstrap approach based on the banded covariance estimator results in a p-value of 0.052, which is in accordance with the global testing procedure by **Rauf Ahmad et al. (2008)** and the respective simultaneous confidence intervals are displayed in the lower panel of Figure 6.

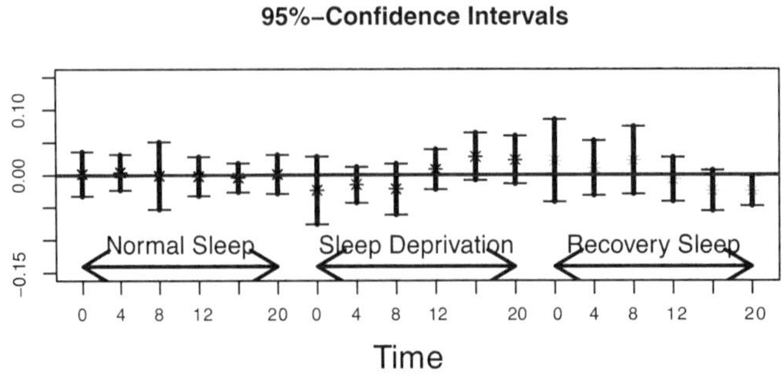

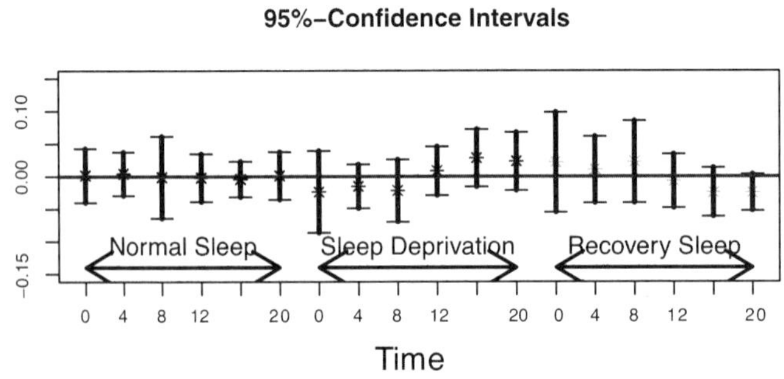

FIGURE 6. The 95%-simultaneous confidence intervals for the interaction effects using a $T(n-1, \widehat{\mathbf{R}}_n)$ approximation (the upper panel) and the banded resampling-based approximation (the lower panel) for the sleep lab data. Number of bootstrap replications is 10,000.

6. Discussion

In this paper we propose the new regularized resampling-based versions of multiple contrast tests in high-dimensional repeated measures designs when the number of observations (subjects) is substantially smaller than the number of hypotheses (variables). We employ two types of matrix regularization procedures, i.e., banding and thresholding, and show that the proposed regularized resampling-based multiple contrast tests deliver remarkably better performance than other currently available testing procedures.

In this paper, we restrict our analysis to normally distributed data, which is a limitation for many practical studies. As a future extension, we plan to relax this assumption and elaborate the proposed methodology for purely nonparametric rank-based multiple contrast tests and related simultaneous confidence intervals in a high-dimensional setting.

Appendix A. Proof of Theorem 3.1

By the multivariate normality of $\mathbf{Y}_k = (Y_{1k}, \ldots, Y_{dk})' \sim N(\boldsymbol{\mu}, \mathbf{V})$, we have $\mathbf{X}_k = \mathbf{C}\mathbf{Y}_k \sim N(\mathbf{C}\boldsymbol{\mu}, \Sigma)$, $k = 1, \ldots, n$. Hence, the vector of standardized means $\widetilde{\mathbf{T}} = (\widetilde{T}_1, \ldots, \widetilde{T}_q)'$, where $\widetilde{T}_\ell = \sqrt{n}(\overline{\mathbf{X}}_{\ell\cdot} - \delta_\ell)/\sqrt{\sigma_{\ell\ell}}$, asymptotically follows a multivariate normal distribution with expectation $\mathbf{0}$ and correlation matrix $\mathbf{R}$. Hence, if $\widetilde{\mathbf{\Sigma}}_n$ is consistent under the assumption $\log(d)/n \to 0$, $\mathbf{T}$ asymptotically follows a multivariate normal distribution with expectation $\mathbf{0}$ and correlation matrix $\mathbf{R}$. To prove (i) it remains to show that $\mathbf{T}^*$ follows that same limiting distribution as $\mathbf{T}$. Since $\mathbf{X}_k^*$ are multivariate normal, we need to check that the empirical variances of the resampling variables are consistent if $\log(d)/n \to 0$. This result follows from Theorem 3.1 by Van Der Laan and Bryan (**2001**) using the Bernstein inequality. For the proof of (ii) note that (3.3) also holds under the alternative $\mathbf{C}\boldsymbol{\mu} \neq \mathbf{0}$. Hence, the result follows from

$$T_0 = \max_{\ell=1,\ldots,q} \{|T_1^0|, \ldots, |T_q^0|\} + \max_{\ell=1,\ldots,q} \{\sqrt{n}\mathbf{c}_\ell'\boldsymbol{\mu}/\sqrt{\sigma_{\ell\ell}}\} \xrightarrow{P} \max_{\ell=1,\ldots,q} \{\text{sign}(\mathbf{c}_\ell'\boldsymbol{\mu})\} \cdot \infty.$$

References

Bai, Z. and H. Saranadasa. 1996. *Effect of high dimension: by an example of a two sample problem*, Statist. Sinica **6**, no. 2, 311–329. MR1399305 (97i:62062)

Beran, R. 1997. *Diagnosing bootstrap success*, Ann. Inst. Statist. Math. **49**, no. 1, 1–24, DOI 10.1023/A:1003114420352. MR1450689 (98f:62152)

Bretz, F., A. Genz, and L. A. Hothorn. 2001. *On the numerical availability of multiple comparison procedures*, Biom. J. **43**, no. 5, 645–656, DOI 10.1002/1521-4036(200109)43:5⟨645::AID-BIMJ645⟩3.0.CO;2-F. MR1863491

Bickel, P. J. and D. A. Freedman. 1981. *Some asymptotic theory for the bootstrap*, Ann. Statist. **9**, no. 6, 1196–1217. MR630103 (83a:62051)

Bickel, P. J. and Y. R. Gel. 2011. *Banded regularization of autocovariance matrices in application to parameter estimation and forecasting of time series*, J. R. Stat. Soc. Ser. B Stat. Methodol. **73**, no. 5, 711–728, DOI 10.1111/j.1467-9868.2011.00779.x. MR2867455 (2012k:62261)

Bickel, P. J. and E. Levina. 2008a. *Regularized estimation of large covariance matrices*, Ann. Statist. **36**, no. 1, 199–227, DOI 10.1214/009053607000000758. MR2387969 (2009a:62255)

Bickel, P. J. and E. Levina. 2008b. *Covariance regularization by thresholding*, Ann. Statist. **36**, no. 6, 2577–2604, DOI 10.1214/08-AOS600. MR2485008 (2010b:62197)

Bickel, P. J. and B. Li. 2006. *Regularization in statistics*, Test **15**, no. 2, 271–344, DOI 10.1007/BF02607055. With comments and a rejoinder by the authors. MR2273731

Cao, H. and M. R. Kosorok. 2011. *Simultaneous critical values for t-tests in very high dimensions*, Bernoulli **17**, no. 1, 347–394, DOI 10.3150/10-BEJ272. MR2797995 (2012c:62243)

Chi, Y.-Y., M. Gribbin, Y. Lamers, J. F. Gregory III, and K. E. Muller. 2012. *Global hypothesis testing for high-dimensional repeated measures outcomes*, Stat. Med. **31**, no. 8, 724–742, DOI 10.1002/sim.4435. MR2901796

Cornish, E. A. 1954. *The multivariate t-distribution associated with a set of normal sample deviates*, Austral. J. Physics **7**, 531–542. MR0066598 (16,602g)

Delaigle, A., P. Hall, and J. Jin. 2011. *Robustness and accuracy of methods for high dimensional data analysis based on Student's t-statistic*, J. R. Stat. Soc. Ser. B Stat. Methodol. **73**, no. 3, 283–301, DOI 10.1111/j.1467-9868.2010.00761.x. MR2815777 (2012e:62254)

Dunnett, C. W. 1955. *A multiple comparison procedure for comparing several treatments with a control*, J. Amer. Statist. Assoc. **50**, no. 272, 1096–1121, DOI 10.1080/01621459.1955.10501294.

Fan, J., P. Hall, and Q. Yao. 2007. *To how many simultaneous hypothesis tests can normal, Student's t or bootstrap calibration be applied?*, J. Amer. Statist. Assoc. **102**, no. 480, 1282–1288, DOI 10.1198/016214507000000969. MR2372536 (2009a:62080)

Gao, X., M. Alvo, J. Chen, and G. Li. 2008. *Nonparametric multiple comparison procedures for unbalanced one-way factorial designs*, J. Statist. Plann. Inference **138**, no. 8, 2574–2591, DOI 10.1016/j.jspi.2007.10.015. MR2432382

Genz, A. and F. Bretz. 2009. *Computation of multivariate normal and t probabilities*, Lecture Notes in Statistics, vol. 195, Springer, Dordrecht. MR2840595 (2012f:60002)

Giné, E. and J. Zinn. 1990. *Bootstrapping general empirical measures*, Ann. Probab. **18**, no. 2, 851–869. MR1055437 (91g:60042)

Hall, P. and S. R. Wilson. 1991. *Two guidelines for bootstrap hypothesis testing* (English, with French summary), Biometrics **47**, no. 2, 757–762, DOI 10.2307/2532163. MR1132543 (92f:62061)

Hochberg, Y. and A. C. Tamhane. 1987. *Multiple comparison procedures*, Wiley Series in Probability and Mathematical Statistics: Applied Probability and Statistics, John Wiley & Sons Inc., New York. MR914493 (89c:62125)

Hothorn, T., F. Bretz, and P. Westfall. 2008. *Simultaneous inference in general parametric models*, Biom. J. **50**, no. 3, 346–363, DOI 10.1002/bimj.200810425. MR2521547

ICH. 1998. *Statistical principles for clinical trials*, http://private.ich.org.

Janssen, A. 2005. *Resampling Student's t-type statistics*, Ann. Inst. Statist. Math. **57**, no. 3, 507–529, DOI 10.1007/BF02509237. MR2206536 (2008a:62036)

Konietschke, F., L. A. Hothorn, and E. Brunner. 2012. *Rank-based multiple test procedures and simultaneous confidence intervals*, Electron. J. Stat. **6**, 738–759, DOI 10.1214/12-EJS691. MR2988427

Kosorok, M. R. and S. Ma. 2007. *Marginal asymptotics for the "large p, small n" paradigm: with applications to microarray data*, Ann. Statist. **35**, no. 4, 1456–1486, DOI 10.1214/009053606000001433. MR2351093 (2009j:62010)

Kuelbs, J. and A. N. Vidyashankar. 2010. *Asymptotic inference for high-dimensional data*, Ann. Statist. **38**, no. 2, 836–869, DOI 10.1214/09-AOS718. MR2604698 (2011c:60009)

Rauf Ahmad, M., C. Werner, and E. Brunner. 2008. *Analysis of high-dimensional repeated measures designs: the one sample case*, Comput. Statist. Data Anal. **53**, no. 2, 416–427, DOI 10.1016/j.csda.2008.08.013. MR2649096

Srivastava, M. S. and M. Du. 2008. *A test for the mean vector with fewer observations than the dimension*, J. Multivariate Anal. **99**, no. 3, 386–402, DOI 10.1016/j.jmva.2006.11.002. MR2396970 (2009e:62261)

Srivastava, M. S., S. Katayama, and Y. Kano. 2013. *A two sample test in high dimensional data*, J. Multivariate Anal. **114**, 349–358, DOI 10.1016/j.jmva.2012.08.014. MR2993891

Srivastava, M. S. and T. Kubokawa. 2013. *Tests for multivariate analysis of variance in high dimension under non-normality*, J. Multivariate Anal. **115**, 204–216, DOI 10.1016/j.jmva.2012.10.011. MR3004555

Tukey, J. W. 1994. *The collected works of John W. Tukey. Vol. VIII*, Chapman & Hall, New York. Multiple comparisons: 1948–1983; With a preface by William S. Cleveland; With a biography by Frederick Mosteller; Edited and with an introduction and comments by Henry I. Braun. MR1263027 (95d:01025)

Van Der Laan, M. J. and J. Bryan. 2001. *Gene expression analysis with the parametric bootstrap*, Biostatistics **2**, 445–461.

Van Der Laan, M. J. and J. Bryan. 2000. *Gene expression analysis with the parametric bootstrap*, technical report.

Westfall, P. H. and J. F. Troendle. 2008. *Multiple testing with minimal assumptions*, Biom. J. **50**, no. 5, 745–755, DOI 10.1002/bimj.200710456. MR2542340

Department of Medical Statistics, University Medical Center Göttingen, Humboldtallee 32, 37073 Göttingen, Germany

E-mail address: fkoniet@gwdg.de

Department of Statistics and Actuarial Science, University of Waterloo, Waterloo, ON N2L 3G1, Canada

E-mail address: ygl@math.uwaterloo.ca

Department of Medical Statistics, University Medical Center Göttingen, Humboldtallee 32, 37073 Göttingen, Germany

E-mail address: brunner@ams.med.uni-goettingen.de

Contemporary Mathematics
Volume **622**, 2014
http://dx.doi.org/10.1090/conm/622/12433

Data-driven smoothing can preserve good asymptotic properties

Zhouwang Yang, Huizhi Xie, and Xiaoming Huo

ABSTRACT. The asymptotic properties of cubic smoothing splines is well known. The derivation of the cubic smoothing splines depends on special property of a penalty that is the integral of the squared second order derivative. We show that an alternative approach, whose derivation is completely based on data, can preserve nearly all the nice asymptotic properties that the cubic smoothing splines enjoy when the predictor variable follows an equally spaced design. The key idea is to derive the decaying rate of eigenvalues in a new smoothing matrix. Numerical experiments demonstrate superb performance of the proposed method.

1. Introduction

When (x_i, y_i), $i = 1, \dots, n$ ($x_i \in \mathbb{R}$, $y_i \in \mathbb{R}$) is a sample, to find the functional relationship f depicted by $y_i = f(x_i) + \varepsilon_i$, a standard approach is to assume $f \in \mathscr{F}$, where $\mathscr{F}$ includes all functions with squared integrable second order derivative, i.e., $\mathscr{F} = \{f : \int_{\mathbb{R}} |f^{(2)}(x)|^2 < \text{Constant}\}$, and solve

$$\min_{f \in \mathscr{F}} \sum_{i=1}^{n} \big(y_i - f(x_i)\big)^2 + \lambda \int_{\mathbb{R}} |f^{(2)}(x)|^2 \, dx. \tag{1.1}$$

It is well known that the solution to the above is the cubic smoothing spline [**Wah90**]. Its asymptotic properties have been explicitly derived: (a) It achieves the optimal convergence rate [**Spe85**] that is known in the general framework of nonparametric regression [**Sto82**]. (b) The hidden constant in the minimax risk is *approximately* optimal [**Nus85**] (up to an optimal filtering that will be described later).

There is an alternative way to derive an estimator of f. The derivation does not include an integration of the squared second order derivative. The basic idea is to replace the penalty term in (1.1) by its local least squares estimator, as will be shown in Section 2. It is interesting to find out whether the results of [**Spe85**] and [**Nus85**] still hold for the new estimator. This paper establishes affirmative answers when the predictor variable follows equally spaced design.

2010 *Mathematics Subject Classification.* Primary 62G05, 62G20.

Key words and phrases. Smoothing, penalization, density estimation.

The first author was supported in part by the NSF of China (No. 11171322) and the Program NCET-11-0881.

The third author was supported in part by USA NSF grants 1106940 and 0831300.

Other papers have considered adaptive smoothing splines, some representative ones are [**PSH06**, **LG10**]. In [**LG10**], the penalty function is assumed to be a step function, which is data driven. In [**PSH06**], a reproducing kernel Hilbert space representation of the penalty function is considered. In fact, the authors essentially considered a piecewise constant penalty function as they analyzed the numerical approach in Section 2.2 of their paper. Our method does not require the penalty function to be piecewise constant.

The remainder of the paper is organized as follows. In Section 2, a completely data-driven method is presented. The asymptotic properties of our data driven method, which include optimal rate of convergence and optimal minimax risk constant, are presented in Section 3.

2. The data-driven method

We derive our method in this section. Formulation and regularization are visited with needed additional details in Section 2.1. In Section 2.2, we derive a least squares estimator of the integration of the second order derivative. We then give the formula of our estimator in Section 2.3. Some basic properties of this estimator are presented as well. The tuning parameter λ is chosen by GCV as in smoothing splines [**Wah90**].

2.1. Formulation. Denote Ω as a bounded subset of the one-dimensional Euclidean space ($\mathbb{R}$). Assume the formulation in Introduction: (1.1). We avoid the integration of squared second order derivative by replacing the integration with its approximate:

$$\widetilde{\mathcal{J}}(f) = \sum_{i=1}^{n} \frac{x_{i+1} - x_{i-1}}{2} |f^{(2)}(x_i)|^2,$$

with sorted $\{x_1 < x_2 < \cdots < x_n\} \subset \Omega = [a, b] = [x_0, x_{n+1}]$. In this paper, we will consider equally spaced x_i's; furthermore, we denote $\delta = x_{i+1} - x_i$, $\forall i$. Therefore we come to optimizing the following:

$$\min_{f \in \mathscr{F}} \mathcal{Q}(f) = \frac{1}{n} \sum_{i=1}^{n} |y_i - f(x_i)|^2 + \lambda \left(\frac{\delta}{n} \sum_{i=1}^{n} h_i^2 \right), \tag{2.1}$$

where $h_i = f^{(2)}(x_i)$. In the following subsections, we will find the solution to (2.1).

2.2. Local estimate of the second-order derivative. The key step in our strategy is utilizing the function values at x_i and its neighbors to estimate h_i, i.e., the second order derivative $f^{(2)}(x_i)$. Let $\mathscr{B}_k(x_i) = \{x_{j_1}, \ldots, x_{j_k}\}$ be the k-nearest neighbors of x_i where k is a fixed positive integer. Then a local quadratic approximation of f is provided by the Taylor expansion:

$$f(x_j) - f(x_i) \approx (x_j - x_i) f^{(1)}(x_i) + \tfrac{1}{2}(x_j - x_i)^2 f^{(2)}(x_i), \quad x_j \in \mathscr{B}_k(x_i), \tag{2.2}$$

where $f^{(1)}(x_i)$ and $f^{(2)}(x_i)$ are the first-order and second-order derivatives of f at x_i, respectively. For each x_i with its k-nearest neighbors $\mathscr{B}_k(x_i) = \{x_{j_1}, \ldots, x_{j_k}\}$, we denote

$$\mathbf{L} = \begin{pmatrix} -1 & 1 & \ldots & \mathbf{0} \\ \vdots & \vdots & \ddots & \vdots \\ -1 & \mathbf{0} & \ldots & 1 \end{pmatrix} \in \mathbb{R}^{k \times (k+1)},$$

$\mathbf{f}_i = \big(f(x_i), f(x_{j_1}), \ldots, f(x_{j_k})\big)^T \in \mathbb{R}^{k+1}$, $p_\ell = x_{j_\ell} - x_i$, $v_\ell = \frac{1}{2}p_\ell^2$, $\ell = 1, \ldots, k$. A matrix version of the local approximation is given by

$$\mathbf{L}\mathbf{f}_i \approx \mathbf{P}g_i + \mathbf{V}h_i = (\mathbf{P}, \mathbf{V})\begin{pmatrix} g_i \\ h_i \end{pmatrix}, \tag{2.3}$$

where $\mathbf{P} = (p_1, \ldots, p_k)^T \in \mathbb{R}^k$, $\mathbf{V} = (v_1, \ldots, v_k)^T \in \mathbb{R}^k$ are constant vectors, and $g_i = f^{(1)}(x_i)$, $h_i = f^{(2)}(x_i)$ are unknowns. We can then easily derive the least squares estimator of h_i in a closed form

$$\hat{h}_i = \mathbf{H}_i\mathbf{f}_i, \tag{2.4}$$

where $\mathbf{H}_i = (\mathbf{V}^T\mathbf{V} - \mathbf{V}^T\mathbf{P}(\mathbf{P}^T\mathbf{P})^{-1}\mathbf{P}^T\mathbf{V})^{-1}(\mathbf{V}^T - \mathbf{V}^T\mathbf{P}(\mathbf{P}^T\mathbf{P})^{-1}\mathbf{P}^T)\mathbf{L} \in \mathbb{R}^{1\times(k+1)}$. Then the local estimator of squared second order derivative is given by

$$|f^{(2)}(x_i)|^2 \approx |\hat{h}_i|^2 = \mathbf{f}_i^T\mathbf{H}_i^T\mathbf{H}_i\mathbf{f}_i. \tag{2.5}$$

Since our method is based on local Taylor expansion and using nearest neighbors, we anticipate that it will be used when the sample size is large. Otherwise, the nearest neighbor based approximation can be problematic.

2.3. Global solution to the approximation model. For each $i \in \{1, \ldots, n\}$, we have $\mathbf{S}_i \in \mathbb{R}^{(k+1)\times n}$ with 0 and 1 being its components and satisfying

$$(x_i, x_{j_1}, \ldots, x_{j_k}) = (x_1, x_2, \ldots, x_n)\mathbf{S}_i^T;$$

i.e., $\mathbf{S}_i$ 'selects' the ith observation and its k-nearest neighbors. It is evident to have $\mathbf{f}_i = \mathbf{S}_i\mathbf{f}$ with $\mathbf{f} = \big(f(x_1), \ldots, f(x_n)\big)^T \in \mathbb{R}^n$. Then the objective function in (2.1) results in a quadratic form

$$\sum_{i=1}^n |y_i - f(x_i)|^2 + \lambda\sum_{i=1}^n \|\mathbf{h}_i\|^2 = \|\mathbf{y} - \mathbf{f}\|^2 + \lambda(\mathbf{f}^T\mathbf{M}\mathbf{f}), \tag{2.6}$$

where $\mathbf{y} = (y_1, \ldots, y_n)^T \in \mathbb{R}^n$, $\mathbf{M} = \sum_{i=1}^n \mathbf{Q}_i^T\mathbf{Q}_i \in \mathbb{R}^{n\times n}$ and $\mathbf{Q}_i = \mathbf{H}_i\mathbf{S}_i \in \mathbb{R}^{1\times n}$. The Hessian regularized model eventually leads to solving a linear system:

$$(\mathbf{I}_n + \lambda\mathbf{M})\widehat{\mathbf{f}} = \mathbf{y}, \tag{2.7}$$

where $\widehat{\mathbf{f}}$ is the global estimate of $\{f(x_i)\}_{i=1}^n$. We denote $\mathbf{A}_n(\lambda) = (\mathbf{I}_n + \lambda\mathbf{M})^{-1}$. Accordingly, the solution operator $\mathcal{A}_{n,\lambda}\colon \mathbb{R}^n \to \mathbb{R}^n$, is given for $\mathbf{y} \in \mathbb{R}^n$ by $\mathcal{A}_{n,\lambda}(\mathbf{y}) = \mathbf{A}_n(\lambda)\mathbf{y} = (\mathbf{I}_n + \lambda\mathbf{M})^{-1}\mathbf{y}$, which minimizes $\mathcal{Q}(f)$ in (2.1).

3. Theoretical properties

We discuss the asymptotic properties of the new estimator in this section. Specifically, we show that the new estimator enjoys all the nice properties presented for smoothing splines [**Spe85**, **Nus85**] *when the predictor variable follows equally spaced design.* In Section 3.1, we study the behavior of the eigenvalues of the matrix associated with our estimator. The optimal rate of convergence and sharp bound for the minimax risk are discussed in Sections 3.2 and 3.3, respectively.

3.1. Eigenvalues associated with equally spaced design. The sample knots $\{x_i\}_{i=1}^n$ are assumed to be sorted and equally spaced, i.e., (recall) $\delta = x_{i+1} - x_i$ for $1 \leq i \leq n-1$. Let $k = 2q$ be the fixed even number used in our algorithm for getting the k-nearest neighbors of each x_i. Denote $\mathbf{f} = \big(f(x_1), \ldots, f(x_n)\big)^T$. From the local least squares estimate of second order derivative $|h_i|^2 = \mathbf{f}^T \mathbf{Q}_i^T \mathbf{Q}_i \mathbf{f}$, we get

$$\mathbf{M} = \sum_{i=1}^n \mathbf{Q}_i^T \mathbf{Q}_i,$$

and every $\mathbf{Q}_i$ for $q+1 \leq i \leq n-q$ has the form

$$\frac{1}{\gamma_q \delta^2}(0, \ldots, 0, q^2, (q-1)^2, \ldots, 1, -2\beta_q, 1, \ldots, (q-1)^2, q^2, 0, \ldots, 0) \in \mathbb{R}^n, \tag{3.1}$$

where $-2\beta_q$ is the i-th component, $\beta_q = \sum_{\ell=1}^q \ell^2$ and $\gamma_q = \sum_{\ell=1}^q \ell^4$. The derivation detail of (3.1) is given in Appendix. It can be shown that $\mathbf{M}$ is a symmetric and banded matrix of bandwidth $2k+1$. After eliminating the first and last k rows and columns of $\mathbf{M}$, the remaining submatrix is Toeplitz. Let $\widetilde{\mathbf{M}}$ be a cyclic matrix that has the same dimension as matrix $\mathbf{M}$. In addition, after the aforementioned elimination, matrix $\mathbf{M}$ and $\widetilde{\mathbf{M}}$ share the same central submatrix, denoted by $\mathbf{M}^{(1)} \in \mathbb{R}^{(n-2k)\times(n-2k)}$. The advantage is that the eigenvalues of $\widetilde{\mathbf{M}}$ can be computed in closed form. Moreover, the Cauchy interlacing theorem will be used to give bounds on the eigenvalues of $\mathbf{M}$.

Denote

$$\begin{aligned}
\tau_j &= \frac{(j-\frac{1}{2})\pi}{n}, \\
\xi_j &= \frac{4}{\gamma_q^2 \delta^4}\left[\sum_{\ell=1}^q \ell^2(\cos \ell\tau_j - 1)\right]^2, \\
\widetilde{\mathbf{u}}_j &= \big(\sin(\tau_j), \sin(2\tau_j), \ldots, \sin(n\tau_j)\big)^T,
\end{aligned}$$

$j = 1, \ldots, n$. It is not difficult to verify the following eigen-equations

$$\widetilde{\mathbf{M}}\widetilde{\mathbf{u}}_j = \xi_j \widetilde{\mathbf{u}}_j, \quad j = 1, \ldots, n, \tag{3.2}$$

which indicate that ξ_j is the jth eigenvalue of $\widetilde{\mathbf{M}}$ in ascending order. Let $\mu_1 \leq \cdots \leq \mu_n$ and $\mu_1^{(1)} \leq \cdots \leq \mu_{n-2k}^{(1)}$ be the eigenvalues of $\mathbf{M}$ and $\mathbf{M}^{(1)}$, respectively. By twice applying the Cauchy interlacing theorem, for all $2k+1 \leq j \leq n-2k$, we have

$$\xi_{j-2k} \leq \mu_{j-2k}^{(1)} \leq \mu_j \leq \mu_j^{(1)} \leq \xi_{j+2k}. \tag{3.3}$$

Now we can bound the eigenvalues of $\mathbf{M}$ by using Jordan's inequality, which states that $\frac{2}{\pi}x \leq \sin(x) \leq x$ for any $x \in [0, \pi/2]$.

LEMMA 3.1. *For any $\ell \in \mathbb{N}$ and $0 \leq \tau \leq \pi$, the cosine function holds the following expansion*

$$\cos(\ell\tau) - 1 = (\cos\tau - 1)\left[\ell^2 - O_+\left(\left(\sin\frac{\tau}{2}\right)^2\right)\right], \tag{3.4}$$

where $O_+(\cdot)$ means that $0 \leq \lim_{x\to 0} O_+(x)/x < +\infty$.

The proof just needs basic calculus and thus the details are omitted.

LEMMA 3.2. *For $2k+1 \le j \le n$, there exist constants $B_1, B_2 > 0$ such that*

$$B_1 j^4 \le \mu_j \le B_2 j^4. \tag{3.5}$$

PROOF. Since $(\cos\tau - 1)^2 = \sin^4(\tau/2)$, by Jordan's inequality, the conclusion in (3.5) immediately follows from (3.2), (3.3) and Lemma 3.1. □

3.2. The optimal rate of convergence. Let $\widehat{\mathbf{f}}_{n,\lambda} = \mathcal{A}_{n,\lambda}(\mathbf{y}) = \mathbf{A}_n(\lambda)\mathbf{y}$ be the estimator from the approximation model, for an appropriate $\lambda > 0$ and given $\mathbf{y} = (y_1, \ldots, y_n)^T$. It is clear that $\mathcal{A}_{n,\lambda}$ is a linear operator, so $\mathcal{A}_{n,\lambda}(\mathbf{y}) = \mathcal{A}_{n,\lambda}(\mathbf{f}) + \mathcal{A}_{n,\lambda}(\varepsilon)$ where $\varepsilon = (\varepsilon_1, \ldots, \varepsilon_n)^T$ are uncorrelated random errors with zero-mean and finite variance $\sigma^2 > 0$. Thus the average mean squared error (AMSE) is given as

$$\begin{aligned}\text{AMSE}(\widehat{\mathbf{f}}_{n,\lambda}) &= E\left[\frac{1}{n}(\widehat{\mathbf{f}}_{n,\lambda} - \mathbf{f})^T(\widehat{\mathbf{f}}_{n,\lambda} - \mathbf{f})\right] \\ &= \frac{1}{n}\|\mathcal{A}_{n,\lambda}(\mathbf{f}) - \mathbf{f}\|^2 + E\left[\frac{1}{n}\|\mathcal{A}_{n,\lambda}(\varepsilon)\|^2\right],\end{aligned} \tag{3.6}$$

since $E(\varepsilon) = \mathbf{0}$. The rate of decay of AMSE can be observed by using the eigenvalue decomposition

$$\mathbf{M} = \mathbf{U}\Lambda\mathbf{U}^T, \tag{3.7}$$

where $\Lambda = \operatorname{diag}(\mu_1, \ldots, \mu_n)$ and $\mathbf{U} = (\mathbf{u}_1, \ldots, \mathbf{u}_n)$. So the distribution of eigenvalues of $\mathbf{A}_n(\lambda) = (\mathbf{I}_n + \lambda\mathbf{M})^{-1}$ will be $a_j(\lambda) = 1/(1+\lambda\mu_j)$, $j = 1, \ldots, n$. Then we get a further expression

$$\begin{aligned}\text{AMSE}(\widehat{\mathbf{f}}_{n,\lambda}) &= \frac{1}{n}\mathbf{f}^T(\mathbf{A}_n(\lambda) - \mathbf{I})^2\mathbf{f} + \frac{\sigma^2}{n}\operatorname{tr}[\mathbf{A}_n(\lambda)^2] \\ &= \frac{1}{n}\sum_{j=1}^{n}\frac{\lambda^2\mu_j^2 b_j^2}{(1+\lambda\mu_j)^2} + \frac{\sigma^2}{n}\sum_{j=1}^{n}\frac{1}{(1+\lambda\mu_j)^2},\end{aligned} \tag{3.8}$$

where $\mathbf{f} = \big(f(x_1), \ldots, f(x_n)\big)^T$ and $\mathbf{b} = \mathbf{U}^T\mathbf{f} = (b_1, \ldots, b_n)^T$. The first term of $\text{AMSE}(\widehat{\mathbf{f}}_{n,\lambda})$ in (3.8) is the average squared shrinkage bias and the second term is the average variance. We are now ready to show the convergence results of the new estimator.

We first present a result that is useful to prove the main theorem of the convergence rate.

PROPOSITION 3.3. *If for $B_1, B_2 > 0$, we have $B_1 j^m \le \mu_j \le B_2 j^m$ for a constant $m > 0$ and $j = 1, 2, \ldots$, then we have for $n > 0$, $\lambda > 0$,*

$$\sum_{j=1}^{n}\frac{1}{(1+\lambda\mu_j)^2} = O(\lambda^{-1/m});$$

i.e., there exist constant $0 < C_1 < C_2$, such that for large enough n, we have

$$C_1\lambda^{-1/m} < \sum_{j=1}^{n}\frac{1}{(1+\lambda\mu_j)^2} < C_2\lambda^{-1/m}.$$

The proof is similar to that of Theorem 1.1 in Section 6 of [**Utr88**] and the details are omitted.

THEOREM 3.4. *In the 2nd-order Sobolev space* W_2^2, *the estimator* $\widehat{\mathbf{f}}_{n,\lambda}$ *converges to the true function* $f(\cdot)$ *with an optimal rate. Concretely, by choosing the smoothing parameter* $\lambda = O(n^{-4/5})$,[1] *we have*

$$\mathrm{AMSE}(\widehat{\mathbf{f}}_{n,\lambda}) = O(n^{-4/5}).$$

PROOF. For the first term of $\mathrm{AMSE}(\widehat{\mathbf{f}}_{n,\lambda})$, we have

$$\begin{aligned}\frac{1}{n}\sum_{j=1}^{n}\frac{\lambda^2\mu_j^2 b_j^2}{(1+\lambda\mu_j)^2} &= \frac{\lambda}{n}\sum_{j=1}^{n}\frac{\lambda\mu_j}{(1+\lambda\mu_j)^2}\mu_j b_j^2 \\ &\le \frac{\lambda}{4n}\sum_{j=1}^{n}\mu_j b_j^2 = \frac{\lambda}{4n}\mathbf{f}^T\mathbf{M}\mathbf{f} = O(\lambda),\end{aligned} \tag{3.9}$$

since $(1/n)\mathbf{f}^T\mathbf{M}\mathbf{f} = O(\int_\Omega |f^{(2)}(t)|^2\,\mathrm{d}t)$ is finite as $f \in W_2^2(\Omega)$.

For the second term, by the bounds of eigenvalues $\mu_j = O(j^4)$ obtained in the Lemma 3.2, we have

$$\frac{\sigma^2}{n}\sum_{j=1}^{n}\frac{1}{(1+\lambda\mu_j)^2} = \frac{\sigma^2}{n}\sum_{j=1}^{n}\frac{1}{(1+\lambda O(j^4))^2} = O(\lambda^{-1/4}n^{-1}), \tag{3.10}$$

where the last equation is based on Proposition 3.3.

Furthermore, we achieve the optimal rate of decay of AMSE, i.e., $\mathrm{AMSE}(\widehat{\mathbf{f}}_{n,\lambda}) = O(n^{-4/5})$, as $\lambda = O(n^{-4/5})$ is chosen. □

3.3. Nearly sharp bound for the minimax risk. Let $L_2([0,1])$ be the Hilbert space of square-integrable functions on the unit interval, and $\|\cdot\|$ the usual norm therein. We denote $W_2^m = \{f \in L_2 \mid D^m f \in L_2\}$ the corresponding Sobolev space where $D^m f$ is the derivative of order m for $f \in L_2$, and let, for given $K > 0$,

$$W_2^{(m,K)} = \{f \in L_2 \mid \|D^m f\| \le K\}$$

be the nonparametric class of functions to which f is assumed to belong. $\mathscr{F}_n$ is defined to be the class of all estimators of f for given sample size n, i.e., measurable mappings $\hat{f}\colon \mathbb{R}^n \times [0,1] \to \mathbb{R}$. By applying the optimal filtering method of [**Pin**] to our Hessian regularized model, we can give an estimator $\tilde{f}$ which attains a sharp minimax risk bound presented in [**Nus85**].

For any real number x, denote by $\lfloor x \rfloor$ the largest integer not greater than x, and set $\tilde{n} = \lfloor n^{1/5}\log(n^{1/5})\rfloor$. Define a function $g(x) = (1-(\pi x)^m)_+(\mathrm{sgn}(x))_+$ and a number $\omega = (1/K)\int g(x)(1-g(x))\,\mathrm{d}x$. Further denote $\mathbf{G} = \sum_{j=1}^n g_j\mathbf{u}_j\mathbf{u}_j^T$, where $g_j = 1$, $j = 1,\dots,\tilde{n}$, $g_j = g(j\omega n^{-1/(2m+1)})$, $j = \tilde{n}+1,\dots,n$, and $\{\mathbf{u}_j\}_{j=1}^n$ is the orthonormal eigensystem of $\mathbf{M}$ so that $\mathbf{M} = \sum_{j=1}^n \mu_j\mathbf{u}_j\mathbf{u}_j^T$. Let $\widetilde{\mathbf{y}} = \mathbf{G}\mathbf{y}$, and define $\widetilde{\mathbf{f}} = \mathcal{A}_{n,\lambda}(\widetilde{\mathbf{y}})$ be the function values of estimator $\tilde{f}$ at the design points $\{x_j\}$. We first present two lemmas that are useful for later derivation.

LEMMA 3.5. *There is a sequence* $\{\eta_j\}$, *not depending on* n, $\lim_j \eta_j = 1$, *such that*

$$\sup_{1\le j\le\tilde{n}} \eta_j\mu_j(\pi j)^{-4} \le 1 + o(1), \quad n \to \infty. \tag{3.11}$$

[1]This $O(\cdot)$ notation means that we treat λ as a function of n (i.e., we have $\lambda(n)$) and for n large enough, we have $C_1 n^{-4/5} < \lambda(n) < C_2 n^{-4/5}$, where C_1 and C_2 are prescribed positive constants.

Proof of the above lemma is in Appendix.

LEMMA 3.6. *There is a sequence* $\{\eta_j\}$, *not depending on* n, $\lim_j \eta_j = 1$, *such that*

$$\sup_{1\le j\le \tilde{n}} \eta_j \mu_j (\pi j)^{-4} \ge 1 + o(1), \quad n \to \infty. \tag{3.12}$$

The proof is similar to that for the previous lemma and is thus omitted.

Based on the properties of eigenvalues $\{\mu_j\}$ (Lemmas 3.5 and 3.6), we are now ready to prove the result of optimal rate in a stronger sense than the usual rate-optimal ones.

THEOREM 3.7. *Let* $\mathscr{W} = W_2^{(2,K)}$ *and*

$$\Gamma(m, K) = (2m+1)^{1/(2m+1)} K^{1/(2m+1)} \left[\frac{m}{\pi(m+1)}\right]^{2m/(2m+1)}.$$

The normality of noise is assumed, i.e., $\{\varepsilon_i\}$ *are independent random variables with normal distribution* $N(0, \sigma^2)$. *Then*

$$\lim_{n\to\infty} \inf_{\hat{f}\in\mathscr{F}_n} \sup_{f\in\mathscr{W}} n^{4/5} E_f \|\hat{f} - f\|_T^2 = \sigma^2 \Gamma(2, K),$$

and the estimator $\tilde{f}$ *attains this sharp bound.*

PROOF. It is obvious from Theorems 2.1 and 2.2 in [**Nus85**], since the condition of Theorem 2.1 is fulfilled by Lemma 3.5 and the condition of Theorem 2.2 is implied by Lemma 3.6. □

Note that in the above theorem, we prove that the optimal constant in the minimax risk is achieved after the optimal filtering (that is described in [**Pin**]) is applied. To show that the optimal constant is achieved by our method, we must investigate the values of μ_j and g_j for $j \ge \tilde{n}$. This is still a challenging and open question. Unfortunately, we do not resolve it in this paper. This is why we used the phrase "nearly sharp" in the title of this section.

4. Numerical experiments

Our complete-data-driven smoothing (CDS in short) model has been implemented in Matlab R2011b, running on a laptop with an Intel Core i7-2640M @2.8GHz processor and 4.0 Gb of memory. In this section, we will present the performance on several simulated examples and compare our CDS estimator against the cubic smoothing spline method.

In our simulation, the CDS estimator and the cubic smoothing spline were both computed in Matlab R2011b, and their smoothing parameters were selected by GCV. Typically we set $k = 4$ for equally spaced sampling and $k = 24$ for uniformly distributed sampling, where k is the number of nearest neighbors specified for the local estimate of Hessian in the CDS model. We use a smooth sine curve and four challenging examples presented in [**DJ95**] for comparison. The test functions are equally-spaced sampled, followed by adding Gaussian noises, to form the noisy observation y_i's. In each case, we show the reconstructions of two methods, as well as the box-plots that are based on the root mean squared errors (RMSE) over 100 replications per method.

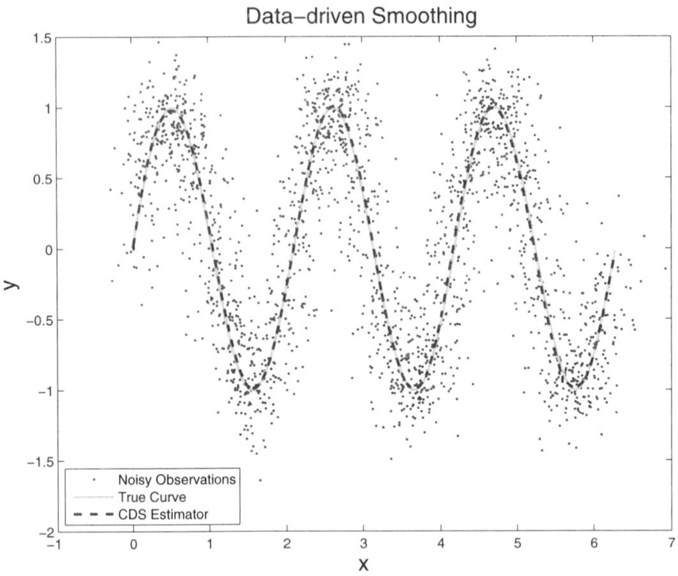

FIGURE 1. The CDS reconstruction from noisy samples on a sine curve (Sample size: $n = 2000$; Noise: $\sigma = 0.2$).

A sine curve is adopted in the first example to illustrate the effects of the CDS estimator and the cubic smoothing spline on smooth function. Figure 1 depicts the CDS reconstruction from $n = 2000$ data points that are uniformly distributed samples on a sine curve corrupted by Gaussian noise with a standard deviation of 0.2. From Figure 2 and 3, we can see that the CDS estimator and the cubic smoothing spline have comparable performance in terms of both reconstruction error and RMSE.

Figures 4, 6, 8 and 10 display the outcome of applying the CDS estimator and the cubic smoothing spline to four other challenging functions. The results of the CDS estimator are qualitatively appealing: the reconstructions of CDS catches the transient features of the underlying functions much better than the cubic smoothing splines do, while still preserving the smooth segments. The boxplots of RMSE's are displayed in Figures 5, 7, 9 and 11, respectively, where 100 replications were used in each case. This simulation study shows that with sufficiently large sample size, the CDS estimator is automatically smoothness adaptive and can perform significantly better than the cubic smoothing splines, especially in the cases of significant spatial variability.

5. Conclusion

A data-driven functional estimation (i.e., smoothing) method is introduced. The main idea of the new method is to use a numerical approximation of the second derivative, to replace the analytical penalty function that is adopted in the derivation of the cubic smoothing splines. We later call this method the complete data-driven smoothing (CDS). We show that the CDS estimator can preserve nearly all the nice asymptotic properties that the cubic smoothing splines hold. In simulations, we show that for spatially nonhomogeneous functions, the CDS has

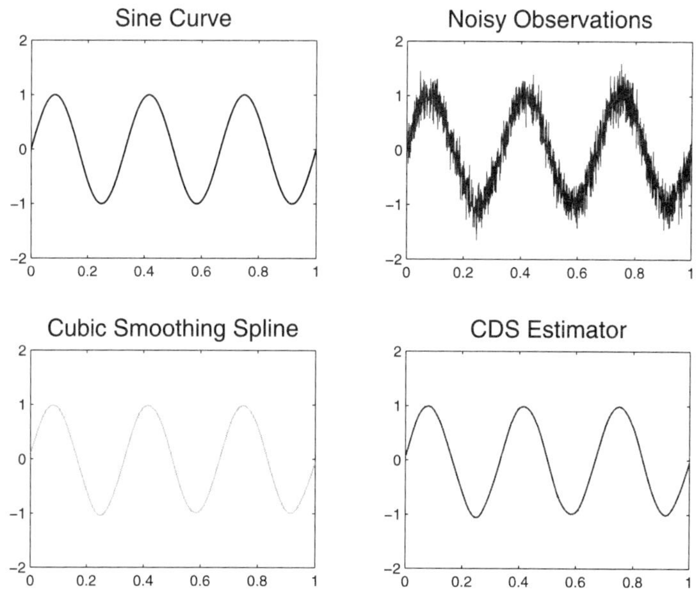

FIGURE 2. Example: SmoothSine (Sample size: $n = 2000$; Noise: $\sigma = 0.2$).

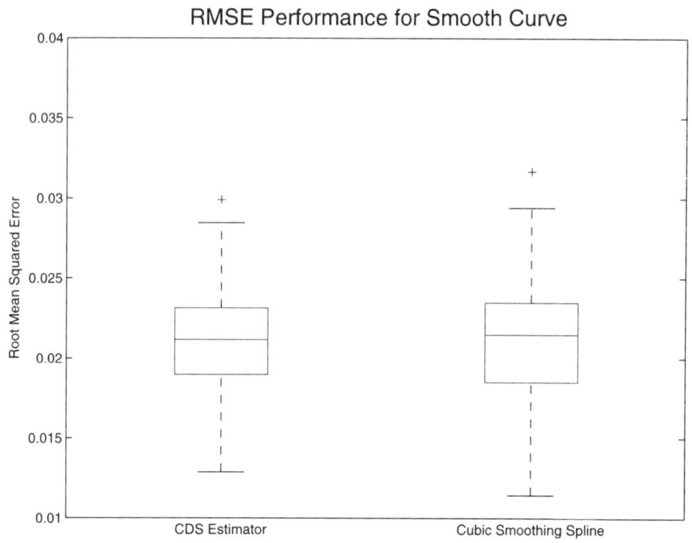

FIGURE 3. Root MSE's for simulation on Example: SmoothSine (Sample size: $n = 2000$; Noise: $\sigma = 0.2$) over 100 replications.

an advantage over the classical cubic smoothing splines: CDS better captures the transient feature in the underlying function. In the future, it will be of interests to compare CDS with wavelets smoothing. Because of time, we did not pursue in this direction in the current paper. We suspect that the advantage of CDS is more

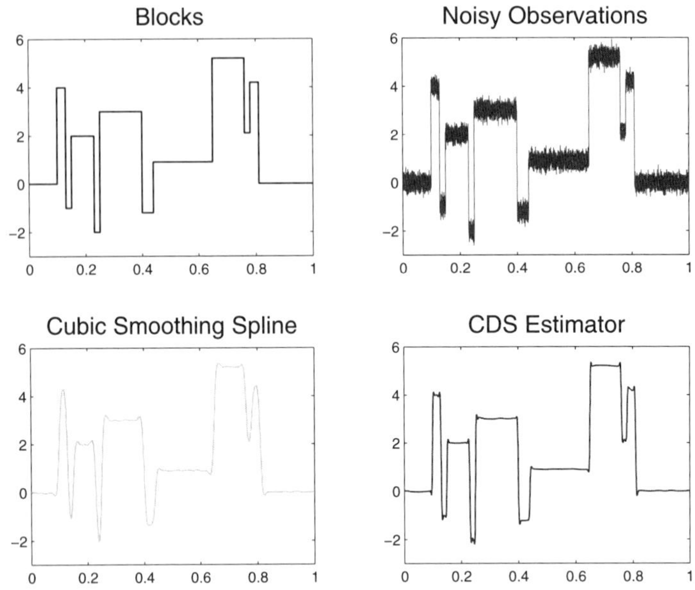

FIGURE 4. Example: Blocks (Sample size: $n = 8192$; Noise: $\sigma = 0.2$).

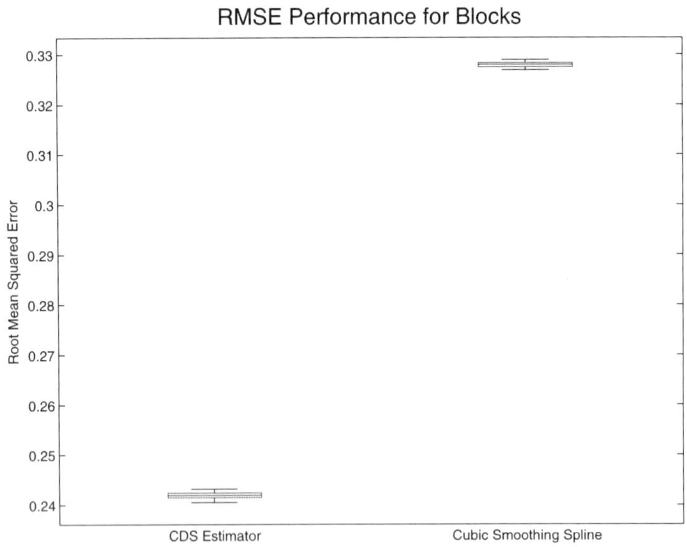

FIGURE 5. Root MSE's for simulation on Example: Blocks (Sample size: $n = 8192$; Noise: $\sigma = 0.2$) over 100 replications.

significant as the sample size increases – given the asymptotic properties that we've derived. In addition, with bigger sample size, one can choose bigger k (recall one needs to use k-nearest neighbors to estimate the local Hessian of the underlying function), hence the CDS estimator performs closer to what these asymptotic optimality theorems predict. At the end of this project, we were drawn attention to

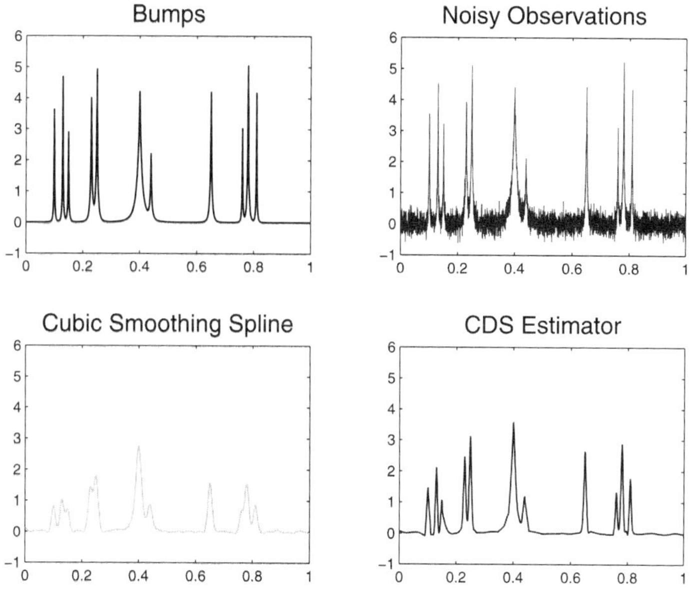

FIGURE 6. Example: Bumps (Sample size: $n = 4096$; Noise: $\sigma = 0.2$).

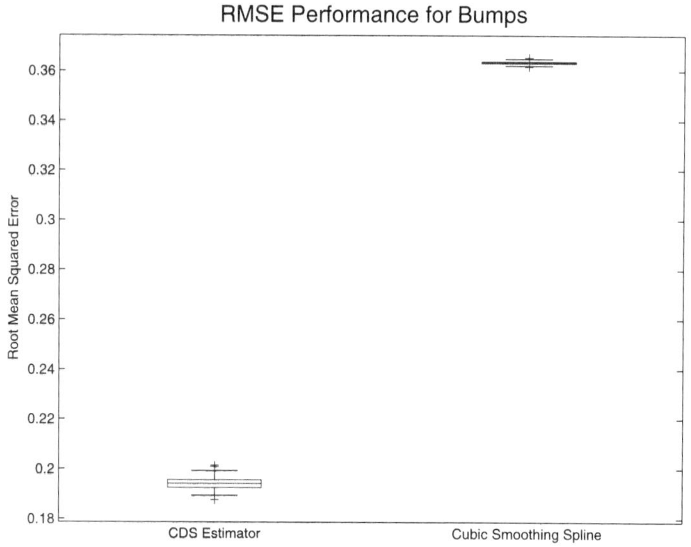

FIGURE 7. Root MSE's for simulation on Example: Bumps (Sample size: $n = 4096$; Noise: $\sigma = 0.2$) over 100 replications.

a recent work [**WDS13**], which explores adaptive penalty function. It is certainly another interesting direction in achieving spatial adaptivity.

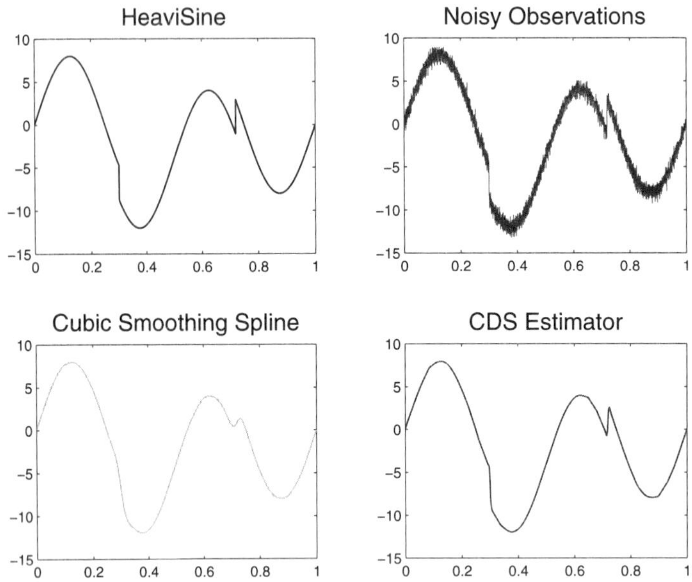

FIGURE 8. Example: HeaviSine (Sample size: $n = 4096$; Noise: $\sigma = 0.5$).

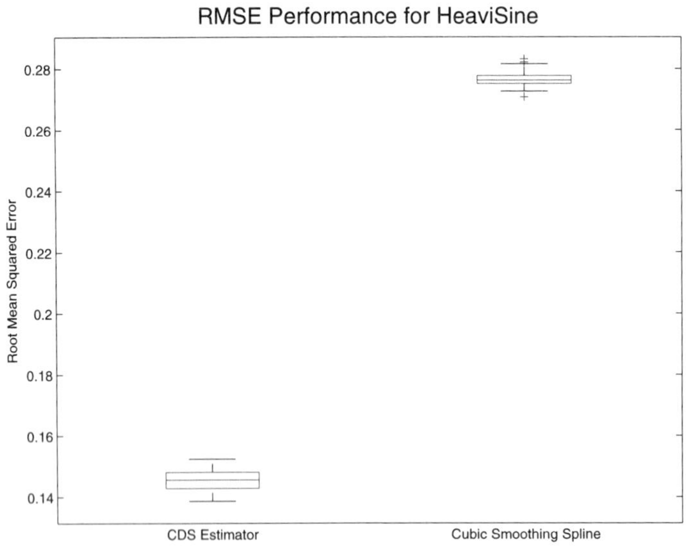

FIGURE 9. Root MSE's for simulation on Example: HeaviSine (Sample size: $n = 4096$; Noise: $\sigma = 0.5$) over 100 replications.

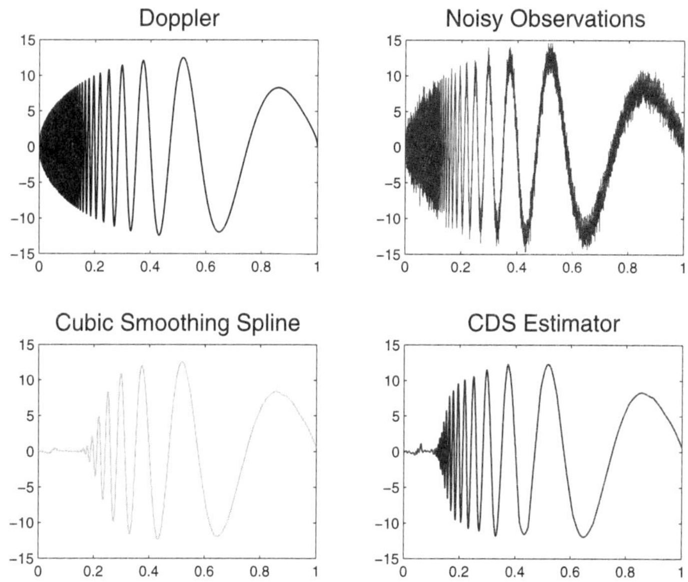

FIGURE 10. Example: Doppler (Sample size: $n = 8192$; Noise: $\sigma = 1.0$).

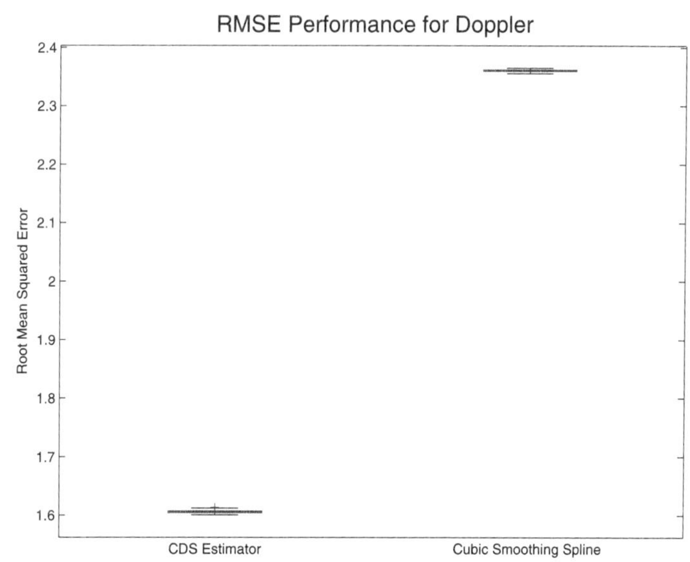

FIGURE 11. Root MSE's for simulation on Example: Doppler (Sample size: $n = 8192$; Noise: $\sigma = 1.0$) over 100 replications.

Appendix A

A.1. Derivation detail of (3.1). Recall we assumed that the sample knots $\{X_i\}_{i=1}^n$ are sorted and equally spaced, i.e., $\delta = X_{i+1} - X_i$ for $i = 1, \ldots, n-1$ and let $k = 2q$. At a sample point $X_i, q+1 \le i \le n-q$, we have

$$\begin{aligned}\mathscr{B}_k(X_i) &= \{X_{i-q}, \dots, X_{i-1}, X_{i+1}, \dots, X_{i+q}\},\\ \mathbf{P} &= (-q\delta, \dots, -\delta, \delta, \dots, q\delta)^T \in \mathbb{R}^{k\times 1},\\ \mathbf{V} &= \tfrac{1}{2}(q^2\delta^2, \dots, \delta^2, \delta^2, \dots, q^2\delta^2)^T \in \mathbb{R}^{k\times 1}.\end{aligned}$$

It is obvious that $\mathbf{P}^T\mathbf{V} = \mathbf{V}^T\mathbf{P} = 0$, $\mathbf{V}^T\mathbf{V} = \frac{1}{2}\gamma_q\delta^4$ and

$$\mathbf{V}^T\mathbf{L} = \tfrac{1}{2}\delta^2(-2\beta_q, q^2, \dots, 1, 1, \dots, q^2),$$

with $\beta_q = \sum_{\ell=1}^q \ell^2$ and $\gamma_q = \sum_{\ell=1}^q \ell^4$. From the local least squares estimator of Hessian (2.4), then we have

$$\widehat{\mathbf{h}}_i = \mathbf{H}_i\mathbf{f}_i = (\mathbf{V}^T\mathbf{V})^{-1}\mathbf{V}^T\mathbf{L}\mathbf{f}_i = \frac{1}{\gamma_q\delta^2}(-2\beta_q, q^2, \dots, 1, 1, \dots, q^2)\mathbf{f}_i$$

where $\mathbf{f}_i = \big(f(X_i), f(X_{i-q}), \dots, f(X_{i-1}), f(X_{i+1}), \dots, f(X_{i+q})\big)^T \in \mathbb{R}^{(k+1)\times 1}$. Finally, we get

$$\mathbf{Q}_i = \mathbf{H}_i\mathbf{S}_i = \frac{1}{\gamma_q\delta^2}(\cdots, 0, q^2, (q-1)^2, \dots, 1, -2\beta_q, 1, \dots, (q-1)^2, q^2, 0, \dots).$$

A.2. Proof of Lemma 3.5.

PROOF. Since

$$\mu_j \overset{(3.3)}{\le} \xi_{j+2k} = \frac{4}{\gamma_q^2\delta^4}\left[\sum_{\ell=1}^q \ell^2(\cos\ell\tau_{j+2k} - 1)\right]^2,$$

we can apply Lemma 3.1 to obtain

$$\begin{aligned}\mu_j(\pi j)^{-4} &\le \frac{4}{\gamma_q^2\delta^4(\pi j)^4}\left[\sum_{\ell=1}^q \ell^2(\cos\ell\tau_{j+2k} - 1)\right]^2\\ &\overset{\text{Lemma 3.1}}{=} \frac{4(\cos\tau_{j+2k} - 1)^2}{\gamma_q^2\delta^4(\pi j)^4}\left[\sum_{\ell=1}^q \ell^2\left(\ell^2 - O_+\left(\left(\sin\frac{\tau_{j+2k}}{2}\right)^2\right)\right)\right]^2\\ &= \frac{16\sin^4(\tau_{j+2k}/2)}{\delta^4(\pi j)^4}\cdot\frac{\left[\sum_{\ell=1}^q \ell^4 - O_+\left((\sin(\tau_{j+2k}/2))^2\right)\right]^2}{\gamma_q^2}\\ &\le \frac{16(\tau_{j+2k}/2)^4}{\delta^4(\pi j)^4} = \left(1 + \frac{2k - 1/2}{j}\right)^4,\end{aligned} \tag{A.1}$$

which implies

$$\sup_{1\le j\le\tilde{n}} \eta_j\mu_j(\pi j)^{-4} \le 1 + o(1), \quad n\to\infty,$$

with setting $\eta_j = \big(j/(j+2k-1/2)\big)^4$. □

References

[DJ95] D. L. Donoho and I. M. Johnstone, *Adapting to unknown smoothness via wavelet shrinkage*, J. Amer. Statist. Assoc. **90** (1995), no. 432, 1200–1224. MR1379464 (96k:62093)

[LG10] Z. Liu and W. Guo, *Data driven adaptive spline smoothing*, Statist. Sinica **20** (2010), no. 3, 1143–1163. MR2730177 (2011i:62077)

[Nus85] M. Nussbaum, *Spline smoothing in regression models and asymptotic efficiency in L_2*, Ann. Statist. **13** (1985), no. 3, 984–997, DOI 10.1214/aos/1176349651. MR803753 (87b:62050)

[Pin] M. S. Pinsker, *Optimal filtration of square-integrable signals in Gaussian noise*; Russian transl., Problems Inform. Transmission **16** (1980), no. 2, 120–133. MR624591 (82j:93048)

[PSH06] A. Pintore, P. Speckman, and C. C. Holmes, *Spatially adaptive smoothing splines*, Biometrika **93** (2006), no. 1, 113–125, DOI 10.1093/biomet/93.1.113. MR2277744

[Spe85] P. Speckman, *Spline smoothing and optimal rates of convergence in nonparametric regression models*, Ann. Statist. **13** (1985), no. 3, 970–983, DOI 10.1214/aos/1176349650. MR803752 (87a:62079)

[Sto82] C. J. Stone, *Optimal global rates of convergence for nonparametric regression*, Ann. Statist. **10** (1982), no. 4, 1040–1053. MR673642 (84b:62058)

[Utr88] F. I. Utreras, *Convergence rates for multivariate smoothing spline functions*, J. Approx. Theory **52** (1988), no. 1, 1–27, DOI 10.1016/0021-9045(88)90034-2. MR922591 (89a:41017)

[Wah90] G. Wahba, *Spline models for observational data*, CBMS-NSF Regional Conference Series in Applied Mathematics, vol. 59, Society for Industrial and Applied Mathematics (SIAM), Philadelphia, PA, 1990. MR1045442 (91g:62028)

[WDS13] X. Wang, P. Du, and J. Shen, *Smoothing splines with varying smoothing parameter*, Biometrika **100** (2013), no. 4, 955–970, DOI 10.1093/biomet/ast031.

Department of Mathematics, University of Science and Technology of China, Hefei, China

E-mail address: yangzw@ustc.edu.cn

Microsoft Research, Redmond, Washington 98052

E-mail address: kennyx@microsoft.com

School of Industrial and Systems Engineering, Georgia Institute of Technology, Atlanta, GA 30332, USA

E-mail address: xiaoming@isye.gatech.edu

Contemporary Mathematics
Volume **622**, 2014
http://dx.doi.org/10.1090/conm/622/12436

Variable selection for ultra-high-dimensional logistic models

Pang Du, Pan Wu, and Hua Liang

ABSTRACT. We propose a variable selection procedure through the optimization of a nonconcave penalized likelihood for logistic regression models with the dimension of covariates p diverging in an exponential rate of n. We first establish the oracle property of the procedure under such ultra-high-dimensional setting. Our optimization algorithm combines some recent developments, including the concave convex procedure and the coordinate descent algorithm, in solving regularization problems. Through extensive simulations, we show the promise of the proposed procedure in various high-dimensional logistic regression settings. An application to gene expression data from a breast cancer study illustrates the use of the method.

1. Introduction

This paper studies variable selection for logistic regression models with ultra-high-dimensional covariates. Let y be a binary response variable and $\mathbf{x} = (x_1, \ldots, x_p)^\top$ be the vector of covariate whose relationship with y can be described as

$$\text{logit}\{P(y = 1 \mid \mathbf{x})\} = \mathbf{x}^\top \boldsymbol{\beta}, \tag{1.1}$$

where $\boldsymbol{\beta} = (\beta_1, \ldots, \beta_p)^\top$ is a vector of unknown parameters. We are interested in the ultra-high-dimensional case when p can be much larger than the sample size n. Throughout the paper, we may use p_n when needed to emphasize the dependence of p on n.

Logistic models are standard and powerful tools to describe the relationship between a binary response variable and a set of covariates. Estimation and inference based on the maximum likelihood estimation in logistic regression have been well studied in theory and widely used in practice (**Lindsey, 1997**; **Nelder and Wedderburn, 1972**; **McCullagh and Nelder, 1983**; **Hosmer and Lemeshow, 1989**). Recently, logistic regression have been applied to analyze high-dimensional data where p may diverge with n (**Huang et al., 2008b**; **van de Geer, 2008**; **Fan and Song, 2010**; **Fan and Lv, 2011**; **Wang, 2011**). For instance, van de

---

2010 *Mathematics Subject Classification.* Primary 62G08.

Key words and phrases. Concave convex procedure, coordinate ascent, coordinate descent, LASSO, local linear approximation, local quadratic approximation, oracle property, penalized variable selection, SCAD.

The first authors research was partially supported by NSF grant DMS-1007126.

The third authors research was partially supported by NSF grants DMS-1007167 and DMS-1207444.

Geer (**2008**) studied estimation for high-dimensional logistic regression with emphasis on prediction error. Despite these developments, theoretical properties of the maximum likelihood estimator (MLE) in a general setting of high-dimensional $\boldsymbol{\beta}$ are not established yet. Fan and Lv (**2011**) offered some insights for a special case when all the components of the true $\boldsymbol{\beta}$ are nonzero and all the components of its MLE are not too close to zero.

Modern genetic and imaging processing studies often collect data on a large number of covariates while a relatively small set of them are believed to be important. Hence variable selection has become an important issue in both parametric and nonparametric regression models; see **Fan and Li (2006)** for a comprehensive overview. Traditional procedures, such as the Akaike information criterion (AIC) and the Bayesian information criterion (BIC), suffer from instability, costly computation, and lack of incorporating stochastic errors inherited in the stage of variable selection (**Breiman, 1996**). In the past decade, tremendous efforts have been devoted to developing various penalized variable selection procedures. Examples of relevant penalties include the bridge penalty (**Frank and Friedman, 1993**), the nonnegative garrote penalty (**Breiman, 1996**), the least absolute shrinkage and selection operator (LASSO) penalty (**Tibshirani, 1996**; **Zou and Hastie, 2005**), and the smoothly clipped absolute deviation (SCAD) penalty (**Fan and Li, 2001**). Among them, the SCAD procedure has been paid great attention since it has three desirable properties for all penalized variable selection procedures: unbiasedness, sparsity, and continuity. Some examples are **Fan and Li (2004)** for partially linear models with longitudinal observations, **Li and Liang (2008)** for generally semiparametric models, and **Liang and Li (2009)** for partially linear models with error-prone covariates. However, these papers can only deal with the case when the number of covariates is fixed. Generally this assumption is not satisfied in genetic and image processing studies where the number of subjects is small but the number of covariates is large.

Recent developments in variable selection focus on the case when p diverges with n. First attempts were made under the linear regression setting. For example, Fan and Peng (**2004**) and Zou and Zhang (**2009**) considered the case of $p = O(n^{\nu})$ for some $0 \leq \nu < 1$. For the ultra-high-dimensional case with $\log(p) = O(n^{\nu})$ for some $0 < \nu < 1$, Huang et al. (**2008a**) showed the oracle property for the adaptive LASSO (**Zou, 2006**) but their result required an initial consistent estimator which is often unavailable in high-dimensional problems. Fan and Lv (**2008**) proposed to select variables in two steps by first applying their sure independence screening (SIS) procedure to reduce the dimension to a much smaller scale such that a second-step procedure such as the LASSO or the low-dimensional SCAD can be applied. To achieve variable selection in one step, Kim et al. (**2008**) revisited the SCAD procedure and established its oracle property for the case of $\log(p) = O(n)$.

A few generalizations of high-dimensional variable selection procedures with $\log(p) = O(n^{\nu}), 0 < \nu < 1$ have been made recently to the logistic regression model (1.1). Extensions of LASSO were considered in **van de Geer (2008)** and **Huang et al. (2008b)**. The former focused only on the prediction aspect and did not address the variable selection problem. The latter established the oracle property for iterated LASSO but relied on the existence of a consistent initial estimate which may be unrealistic. Fan and Song (**2010**) and Fan et al. (**2009**) extend respectively the SIS and its variant, the iterated SIS (ISIS) procedures to generalized linear models. Moreover, Fan and Lv (**2011**) conducted a deep study on the estimation property of nonconcave penalized likelihood based estimators in generalized linear models with nonpolynomially growing dimensionality. For a wide class of penalty functions called the smooth integration of counting and absolute deviation (SICA) penalties, which include the Lasso, the SCAD, and the minimax concave penalties as special cases, they established non-asymptotic weak oracle properties and the oracle properties of the resulting estimators under general regularity conditions. Kwon and Kim (**2012**) studied large sample properties of the SCAD penalized maximum likelihood estimation for high-dimensional parameters. With the dimension of the parameters diverging at a polynomial order of n, they proved that the oracle MLE asymptotically becomes a local maximizer of the SCAD-penalized log-likelihood, and a global maximizer when the log-likelihood is strictly concave. In this paper, we focus on the SCAD procedure for logistic regression with exponentially growing dimensionality. Our method has the following features that distinguishes itself from the existing works: (i) Our proof relies on injective functions (**Chen et al., 1999**), which ensures our normality result under weaker assumptions than those of **Fan and Lv (2011)** and **Kwon and Kim (2012)**. Particularly, the number of nonzero covariates in our assumptions can diverge at the rate of $o(n^{c_1})$ for some $c_1 < 1$, while **Fan and Lv (2011)** requires $c_1 \leq 1/3$ and **Kwon and Kim (2012)** requires $c_1 \leq 1/5$. (ii) As discussed in the next paragraph, we implement the procedure using coordinate descent algorithm incorporating the concave convex procedure (CCCP) (**Le Thi Hoai An and Pham Dinh Tao, 1997**; **Yuille and Rangarajan, 2003**), while **Fan and Lv (2011)** employ the iterative coordinate ascent algorithm.

Although the SCAD penalty yields nice variable selection properties, its nonconcave feature often complicates the optimization of the relevant objective function. To deal with the non-concavity, Fan and Li (**2001**) proposed a local quadratic approximation (LQA) algorithm that iteratively approximates the penalty locally by a quadratic function. But the procedure requires a good initial estimator and manually sets small coefficients to zeros. Zou and Li **Zou and Li (2008)** proposed a local linear approximation (LLA) algorithm that iteratively works on adaptive LASSO type optimization after applying a local linear approximation. However, its requirement of a consistent initial estimator makes it applicable only in the fixed-dimension case. Kim et al. (**2008**) used the concave convex procedure (CCCP) (**Le Thi Hoai An and Pham Dinh Tao, 1997**; **Yuille and Rangarajan, 2003**) to transform the problem to iterative piecewise linear regularized optimization to which the algorithm in **Rosset and Zhu (2007)** applies. We use a similar algorithm here. First, a quadratic approximation to the log likelihood at the current estimate transforms the objective function to penalized weighted least squares. Then an application of the CCCP decomposes the penalty to the sum of two parts, whose tight convex upper bound consists of one part linear in $\boldsymbol{\beta}$ and the other an l_1-penalty on $\boldsymbol{\beta}$. The resulting objective function combining these two approximations is then solved by the coordinate descent algorithm in **Friedman et al. (2007)**. This algorithm was advocated to be less sensitive to the initial value, faster, more stable, and always guaranteed to converge to a minimum even with nonpolynomially diverging dimensionality. It is worth pointing out that Fan and Lv (**2011**) also used coordinate optimization in their implementation of the nonconcave penalized likelihood methods, an iterative coordinate ascent algorithm originated from

Lv and Fan (2009). The algorithm is fast, efficient, and has the ascent property. It also automatically produces the entire solution path without a need to pick an initial value (large enough regularization parameter leads to zero solution). We also tried the algorithm of **Rosset and Zhu (2007)** that was used by Kim et al. (**2008**) in their linear regression problem. But somehow it often exited early to yield severely under-fitted models. One possible reason is its requirement of inverting a hessian matrix computed from selected variables. This might have numeric difficulty when the number of selected variables grows. On the other hand, the coordinate descent algorithm does not have this problem and converges reasonably fast to suit our purpose.

The rest of the paper is organized as follows. Section 2 develops the SCAD procedure for high-dimensional logistic regression models, presents its oracle property, and describes the computation algorithm. The simulation results are presented in Section 3. In Section 4, we thoroughly analyze a gene expression data set from a breast cancer study to illustrate the use of the proposed method. We conclude the paper with some discussions in Section 5. All technical details are deferred to the Appendix.

2. High-dimensional SCAD

We now consider the variable selection problem for model (1.1) when the dimension $p = p_n$ is very high. Let $\mathcal{F}(v) = \{1 + \exp(-v)\}^{-1}$ be the logistic distribution function. Then the log-likelihood function is

$$\mathcal{L}(\boldsymbol{\beta}) = \sum_{i=1}^{n} \left[y_i \log\{\mathcal{F}(\mathbf{x}_i^\top \boldsymbol{\beta})\} + (1 - y_i) \log\{1 - \mathcal{F}(\mathbf{x}_i^\top \boldsymbol{\beta})\}\right]. \tag{2.1}$$

We estimate our coefficient vector $\boldsymbol{\beta}_0$ by minimizing the penalized likelihood

$$C(\boldsymbol{\beta}) = -\mathcal{L}(\boldsymbol{\beta}) + \sum_{j=1}^{p_n} J_\lambda(|\beta_j|), \tag{2.2}$$

where $J_\lambda(|\beta|)$ is the SCAD penalty proposed by Fan and Li (**2001**) and has the expression

$$J_\lambda(|\beta|) = \begin{cases} \lambda|\beta|, & \text{if } |\beta| < \lambda, \\ \dfrac{(a^2-1)\lambda^2 - (|\beta| - a\lambda)^2}{2(a-1)}, & \text{if } \lambda \le |\beta| < a\lambda, \\ \dfrac{(a+1)\lambda^2}{2}, & \text{if } |\beta| \ge a\lambda. \end{cases}$$

Here a is a constant often taking the value of 3.7 as suggested by the inventors. For lower-dimensional problems, several other penalties have been proposed in the literature. Examples include the bridge estimator $J_\lambda(|\beta|) = \lambda|\beta|^\gamma$, $\gamma > 0$ (**Frank and Friedman, 1993**) and the hard thresholding estimator $J_\lambda(|\beta|) = \lambda^2 - (|\beta| - \lambda)^2 I(|\beta| < \lambda)$ (**Fan and Li, 2001**). The SCAD penalty is considered better than them since it has all the three desirable properties of penalized variable selection estimators as we mentioned. The bridge estimator can only have one or two of such properties depending on the choice of γ. The hard thresholding estimator only has the unbiasedness and sparsity. Another penalized estimator that possesses all the three properties is the adaptive LASSO estimator (**Zou, 2006**). However, its requirement of a $\sqrt{n}$-consistent initial estimator makes it a less desired choice for ultra-high-dimensional problem where such an initial estimator is often hard to find. In contrast, Kim et al. (**2008**) have shown the oracle property for the SCAD estimator in high-dimensional linear regression problem.

2.1. The oracle property. Next we will show the oracle property for the ultra-high-dimensional SCAD in logistic regression model (1.1). Our claim is that the oracle estimator is asymptotically a local minimum of $C(\boldsymbol{\beta})$. Without loss of generality, we assume that the first q_n regression coefficients are nonzero and the remaining $p_n - q_n$ regression coefficients are 0s. Let $\mathbb{X} = (\mathbf{x}_1, \ldots, \mathbf{x}_n)^\top$ and $\mathbb{Y} = (y_1, \ldots, y_n)^\top$. Suppose $\mathbb{X} = (\mathbb{X}_{(1)}, \mathbb{X}_{(2)})$, where $\mathbb{X}_{(1)}$ is the first $n \times q_n$ submatrix and $\mathbb{X}_{(2)}$ is the last $n \times (p_n - q_n)$ submatrix of $\mathbb{X}$. Similarly, we write $\boldsymbol{\beta} = (\boldsymbol{\beta}_{(1)}, \boldsymbol{\beta}_{(2)})$. Then the oracle estimator is defined by $\widehat{\boldsymbol{\beta}}_{\mathrm{o}} = (\widehat{\boldsymbol{\beta}}_{(1)\mathrm{o}}, \mathbf{0}_{(2)})$, where $\widehat{\boldsymbol{\beta}}_{(1)\mathrm{o}}$ is the minimizer of $\mathcal{L}(\mathbb{X}_{(1)}, \boldsymbol{\beta}_{(1)})$ and $\mathbf{0}_{(2)}$ is $(p_n - q_n)$-dimensional 0 vector. The following theorem states that the oracle property holds even when p_n diverges at an exponential rate.

ASSUMPTION. (A1) There exists a positive constant M_1 such that

$$\frac{1}{n}(\mathbb{X}_{,j}^\top \mathbb{X}_{,j}) \le M_1 \quad \text{for all } j = 1, \ldots, p_n \text{ and all } n$$

(A2) $q_n = O(n^{c_1})$ for some $0 < c_1 < 1$.
(A3) There exist positive constants c_2 and M_3 such that $c_1 < c_2 \le 1$ and $n^{(1-c_2)/2} \min_{j=1,\ldots,q_n} |\beta_j| \ge M_3$.
(A4) $\max_{i,j}|x_{ij}| < \infty$ and there exist two positive constants $c_{\min}$ and $C_{\max}$ such that $c_{\min} \le \lambda_{\min}(\mathbb{X}_{(1)}^\top \mathbb{X}_{(1)}/n) \le \lambda_{\max}(\mathbb{X}_{(1)}^\top \mathbb{X}_{(1)}/n) \le C_{\max}$.

These conditions or their minor variants have been assumed in the model selection for high-dimensional models. For example, Kim et al. (**2008**) assumed Condition A1 for linear models and Kwon and Kim (**2012**) used a similar version of it for generalized linear models. Condition A2 is weaker than the corresponding conditions in both **Fan and Lv (2011)** and **Kwon and Kim (2012)**. Condition A3 specifies the minimal signal strength and was used in **Kim et al. (2008)** and **Kwon and Kim (2012)**. It may appear to be stronger than that in **Fan and Lv (2011)** but is probably necessary given the higher dimensionality we assume in A2. Condition A4, the sparse Riesz condition, was used in **Huang et al. (2008b)**. Similar versions of A4 were assumed in **Fan and Lv (2011)** and **Kwon and Kim (2012)**. Zhao and Yu **Zhao and Yu (2006)** imposed similar regularity conditions to prove the model selection consistency of their LASSO estimator.

THEOREM 2.1. *Denote the set of local minima of* (2.2) *with the SCAD penalty and a regularization parameter* λ_n *by*

$$\mathcal{A}_n(\lambda) = \{\boldsymbol{\beta}^* : \textit{there exists an } \epsilon > 0, \textit{ such that } C(\boldsymbol{\beta}) \ge C(\boldsymbol{\beta}^*) \quad \textit{whenever } |\boldsymbol{\beta} - \boldsymbol{\beta}^*| < \epsilon\}.$$

Under the Assumption, we have

$$P\{\widehat{\boldsymbol{\beta}}_{\mathrm{o}} \in \mathcal{A}_n(\lambda_n)\} \to 1$$

as $n \to \infty$ *provided that* $\lambda_n = o(n^{-(1-c_2+c_1)/2})$ *and* $\log(p_n)/n^{c_2-c_1} \to 0$, *where* $0 < c_1 < c_2 \le 1$ *are constants defined in the Assumption.*

Note that the exponential rate for p_n in Theorem 2.1 is much more general than the corresponding result in **Kim et al. (2008**, Theorem 1) for linear regression where p is only allowed to diverge in certain polynomial order. They could allow exponential rate for p_n only in the special case of Gaussian random errors (**Kim et al., 2008**, Theorem 2). Also of theoretical interest is the global optimality of the estimator, for which a theorem similar to Theorem 3 of **Kim et al. (2008)** can be established by following the steps in **Kim et al. (2008)** using (A.9) and (A.12).

2.2. Computation. We now introduce the algorithm for optimizing (2.2). A logistic regression model is often fitted through the method of iteratively reweighted least squares (IRLS). Given the current estimate $\widetilde{\boldsymbol{\beta}}$, the IRLS forms a quadratic approximation to $\mathcal{L}(\boldsymbol{\beta})$, which, after dropping some constant terms, is

$$\widetilde{\mathcal{L}}(\boldsymbol{\beta}) = \frac{1}{2n}\sum_{i=1}^{n} w_i(z_i - \mathbf{x}_i^\top \boldsymbol{\beta})^2. \tag{2.3}$$

Here

$$z_i = \mathbf{x}_i^\top \widetilde{\boldsymbol{\beta}} + \frac{y_i - \mathcal{F}(\mathbf{x}_i^\top \widetilde{\boldsymbol{\beta}})}{\mathcal{F}(\mathbf{x}_i^\top \widetilde{\boldsymbol{\beta}})\{1-\mathcal{F}(\mathbf{x}_i^\top \widetilde{\boldsymbol{\beta}})\}} \quad \text{and} \quad w_i = \mathcal{F}(\mathbf{x}_i^\top \widetilde{\boldsymbol{\beta}})\{1-\mathcal{F}(\mathbf{x}_i^\top \widetilde{\boldsymbol{\beta}})\}$$

are respectively the working responses and weights. Thus each IRLS iteration optimizes

$$\widetilde{C}(\boldsymbol{\beta}) = \widetilde{\mathcal{L}}(\boldsymbol{\beta}) + \sum_{j=1}^{p_n} J_\lambda(|\beta_j|). \tag{2.4}$$

Next, we describe the CCCP algorithm for solving (2.4), which updates the estimate with the minimizer of the tight convex upper bound of the non-convex objective function $\widetilde{C}(\boldsymbol{\beta})$ obtained at the current estimate. The CCCP algorithm always decreases the objective function at each iteration and thus guarantees to converge to a local minimum. Note that the SCAD penalty $J_\lambda(\beta)$ can be decomposed into the sum of a differentiable concave function $J_\lambda^*(\beta) = J_\lambda(\beta) - \lambda|\beta|$ and a convex function $\lambda|\beta|$. Hence (2.4) can be rewritten as

$$\widetilde{C}(\boldsymbol{\beta}) = \widetilde{\mathcal{L}}(\boldsymbol{\beta}) + \sum_{j=1}^{p_n} J_\lambda^*(|\beta_j|) + \lambda\sum_{j=1}^{p_n}|\beta_j|, \tag{2.5}$$

which is also the sum of concave and convex functions. Given the current estimate $\boldsymbol{\beta}^c$ at the CCCP iteration, the tight convex upper bound is given as

$$Q(\boldsymbol{\beta}) = \frac{1}{2n}\sum_{i=1}^{n} w_i(z_i - \mathbf{x}_i^\top \boldsymbol{\beta})^2 + \sum_{j=1}^{p_n} \nabla J_\lambda^*(|\beta_j^c|)\beta_j + \lambda\sum_{j=1}^{p_n}|\beta_j|. \tag{2.6}$$

The current estimate $\boldsymbol{\beta}^c$ is then updated by the minimizer of (2.6).

We minimize (2.6) by a coordinate descent algorithm modified from **Friedman et al. (2007)**. Define $S(\beta,\lambda) = \text{sign}(\beta)(|\beta|-\lambda)_+$, where $x_+ = \max(0,x)$. Given current solution $\breve{\boldsymbol{\beta}}$ to (2.6), let $\breve{y}_i^{(j)} = \sum_{k\neq j} x_{ik}\breve{\beta}_k$. Straightforward calculation similar to the derivation of (8) in **Friedman et al. (2007)** yields that the coordinate-wise updating equation for minimizing (2.6) is

$$\breve{\beta}_j^{\text{new}} \leftarrow \frac{n}{\sum_{i=1}^n x_{ij}^2} S\left\{\frac{1}{n}\sum_{i=1}^{n} x_{ij}(y_i - \breve{y}_i^{(j)}) - \nabla J_\lambda^*(|\beta_j^c|), \lambda\right\}. \tag{2.7}$$

Our algorithm needs to start with an initial estimator $\boldsymbol{\beta}_{\text{init}}$. While an arbitrary choice of $\boldsymbol{\beta}_{\text{init}}$ may give unstable results because of algorithm divergence, a number of reasonable choices of $\boldsymbol{\beta}_{\text{init}}$ are available, such as the zero vector and the LASSO estimator. Throughout the paper, we use the latter as our $\boldsymbol{\beta}_{\text{init}}$ since it performs better in our experience and easily available from the R package `glmnet`. The complete algorithm is

- Initialize $\widetilde{\boldsymbol{\beta}} = \boldsymbol{\beta}_{\text{init}}$.
- Do until convergence:
  - ∘ Initialize $\boldsymbol{\beta}^c = \boldsymbol{\beta}_{\text{init}}$.
  - ∘ Do until convergence:
    - ♠ Initialize $\breve{\boldsymbol{\beta}} = \boldsymbol{\beta}_{\text{init}}$.
    - ♠ Do until convergence:
      Update $\breve{\boldsymbol{\beta}}$ coordinate-wise by (2.7).

    Update $\boldsymbol{\beta}^c$ by converged $\breve{\boldsymbol{\beta}}$.

  Update $\widetilde{\boldsymbol{\beta}}$ by converged $\boldsymbol{\beta}^c$.

3. Simulation studies

In this section, we study the empirical performance of the SCAD estimator by comparing it with the SIS-SCAD and LASSO estimators in simulations in terms of prediction accuracy and variable selectivity. The SIS-SCAD estimator is obtained through a two-step procedure which first screens down the number of covariates by the SIS method and then applying the SCAD penalty. The LASSO estimator is obtained by a coordinate descent algorithm. We used the R packages `SIS` and `glmnet` to compute the SIS-SCAD and LASSO estimators respectively. Our SCAD estimator selected λ via the GCV score corresponding to the converged weighted least square problem in (2.4). For the SIS-SCAD estimator, the screening step selected $n/(4\log n)$ covariates, which is the value suggested by Fan et al.(**2009**) for logistic regression and the default value in the `SIS` package. The package does not have a default procedure for selecting λ, so we selected λ in the SIS-SCAD estimator by the AIC with number of nonzero $\hat{\beta}_j$'s being the degrees of freedom for the fitted model (**Zou et al., 2007**). λ for the LASSO estimator was selected by 10-fold cross-validation, the default method in the `glmnet` package.

We generated data from model (1.1). All the data had sample size $n = 100$ to match up with the sample size in our real example. The covariates $\mathbf{x} = (x_1, \ldots, x_p)^\top$ were multivariate normal with mean zero and $\text{cov}(x_i, x_j) = \rho^{|i-j|}$. The first q entries of $\boldsymbol{\beta}_0$ were nonzeros and the rest of entries were all zeros. We considered 12 settings with different combinations of $p = 500, 1000, q = 5, 15$ and $\rho = 0, 0.3, 0.6$. When $q = 5$, the first q entries of $\boldsymbol{\beta}_0$ were $(3.5, 3, 2.5, 2, 1.5)$. When $q = 15$, the first q entries of $\boldsymbol{\beta}_0$ decreased from 3.5 to 1.4 with step size -0.15. We simulated 100 data replicates for each setting.

To evaluate the prediction performance, we generated an independent test sample of size 10,000 for each setting. The estimates from each procedure were applied to this test sample such that each y_i was predicted to be $I(\mathcal{F}(\mathbf{x}_i^\top \hat{\boldsymbol{\beta}}) \geq 0.5)$. We then recorded the proportions of correct prediction and presented them in Table 1, where the numbers are means and standard deviations computed on 100 replicates. In terms of prediction, the SCAD estimator was better than the SIS-SCAD estimator and competitive to the LASSO in the independent covariate cases. In the cases of

TABLE 1. Comparison of proportions of correct predictions. Numbers are means and standard deviations across 100 data replicates.

p	q	ρ	Oracle	LASSO	SIS-SCAD	SCAD
500	5	0	0.9102 (.0005)	0.8072 (.0277)	0.7939 (.0501)	0.8149 (.0459)
		0.3	0.9293 (.0006)	0.8594 (.0202)	0.8837 (.0252)	0.8607 (.0401)
		0.6	0.9404 (.0005)	0.8873 (.0177)	0.9222 (.0120)	0.8674 (.0530)
500	15	0	0.9432 (.0006)	0.6623 (.0514)	0.6385 (.0371)	0.6552 (.0512)
		0.3	0.9576 (.0006)	0.7835 (.0324)	0.7395 (.0325)	0.7501 (.0514)
		0.6	0.9689 (.0005)	0.8589 (.0214)	0.8057 (.0241)	0.8031 (.0594)
1000	5	0	0.9071 (.0007)	0.7845 (.0321)	0.7748 (.0540)	0.8037 (.0531)
		0.3	0.9261 (.0006)	0.8500 (.0236)	0.8752 (.0312)	0.8643 (.0404)
		0.6	0.9406 (.0004)	0.8832 (.0201)	0.9198 (.0133)	0.8617 (.0487)
1000	15	0	0.9433 (.0007)	0.6226 (.0578)	0.6138 (.0470)	0.6177 (.0560)
		0.3	0.9579 (.0007)	0.7569 (.0425)	0.7302 (.0334)	0.7253 (.0519)
		0.6	0.9710 (.0006)	0.8416 (.0216)	0.8036 (.0238)	0.8013 (.0577)

medium correlation ($\rho = 0.3$), the three estimators were competitive to each other. In the high correlation cases ($\rho = 0.6$), the SCAD estimator was slightly worse than the other two estimators. Also, when the true model is not very sparse ($q = 15$), the SCAD estimator generally predicted better than the SIS-SCAD estimator when the covariates were not highly correlated. Although the prediction by the SCAD estimator is not as good as that of the LASSO estimator, the latter achieved this by including too many noise variables (see Tables 2 and 3).

Tables 2 and 3 respectively present the selection frequencies for the settings with $q = 5$ and $q = 15$, where the last column is the average number of noise variables selected. Clearly, although the LASSO estimator often had the highest selection frequencies for signal variables, it also tended to be conservative in selecting many noise variables. This suggests that the 10-fold cross-validation procedure used in the `glmnet` might be too conservative in favoring smaller λ. On the other hand, the two-step SIS-SCAD estimator were always the most aggressive one among the three procedures in selecting the least noise variables. However, it also tended to miss signal variables more often than the other two estimators. The SCAD estimator stroke a balance between these two approaches, having a high frequency of selecting the signal variables while including a reasonable number of noise variables.

When $q = 5$, the SCAD estimator was arguably the best among the three approaches when correlation between covariates were not too high ($\rho = 0$ and 0.3). In the high correlation case ($\rho = 0.6$), the SIS-SCAD outperformed the other two approaches. This is because that the SIS-SCAD uses marginal correlations between covariates and response to filter out specified number of covariates at the first step. When signal variables are highly correlated with each other, the probability of selecting all of them at the first step is very high for the SIS-SCAD and thus the resulting estimator can perform very well. However, when signal variables are not highly correlated, the first step of the SIS-SCAD may miss some signal variables and thus the resulting estimator suffers from the missing information. This is similar to what Kim et al. (**2008**) observed in the linear regression models.

When there were $q = 15$ signal variables, variable selection became a more difficult problem. All three approaches had lower selection frequencies than the case of $q = 5$. The SIS-SCAD estimator suffered the most, possibly because the small sample size $n = 100$ limited the number of covariates selected in the first screening stage. The LASSO estimator was still very conservative. The SCAD estimator appeared to provide reasonable selection performance in all the six settings.

In summary, our simulation results indicate that the SCAD is a promising method for high-dimensional logistic regression data when either the true model is not sparse or the signal variables are not strongly correlated. This echoes the finding in the simulations of **Kim et al. (2008)** for linear regression models. Following the suggestion of the Associate Editor, we also did simulations comparing our procedure with the SICA procedure in **Fan and Lv (2011)**, whose details are omitted here due to space concern. Our simulations revealed that while the SICA did slightly better in the independent cases, it did worse than our procedure in the correlated cases. A similar comparison was made in **Bradic et al. (2011)** under the Cox proportional hazards model setting. They also found that the SICA tends to lose some of TPs for the sake of smaller number of FPs.

4. Real data example

In this section, we apply the SCAD procedure together with the SIS-SCAD and LASSO procedures to a breast cancer study reported in **van 't Veer et al. (2002)**. The study had $n = 97$ lymph node-negative breast cancer patients who were 55 years old or younger. Among them, 46 developed distant metastases within

TABLE 2. Selection frequencies for simulations with $q = 5$ (Others are average number of noise variable selected).

Method	Signal variables					Others
	$p = 500$, $\rho = 0$					
SCAD	100	99	97	86	48	3.76
SIS-SCAD	98	90	73	44	12	1.83
LASSO	100	100	98	97	66	25.30
	$p = 500$, $\rho = 0.3$					
SCAD	99	98	92	82	45	1.95
SIS-SCAD	99	100	99	89	28	0.80
LASSO	100	100	100	96	79	25.11
	$p = 500$, $\rho = 0.6$					
SCAD	87	86	75	61	26	3.46
SIS-SCAD	100	99	100	99	87	0.12
LASSO	100	98	97	94	82	20.27
	$p = 1000$, $\rho = 0$					
SCAD	100	98	85	69	31	2.48
SIS-SCAD	97	91	64	32	13	2.03
LASSO	100	100	96	91	57	30.49
	$p = 1000$, $\rho = 0.3$					
SCAD	98	95	92	76	31	1.41
SIS-SCAD	99	99	98	73	29	0.98
LASSO	100	99	100	97	73	26.98
	$p = 1000$, $\rho = 0.6$					
SCAD	82	79	79	44	21	1.61
SIS-SCAD	99	100	100	99	69	0.30
LASSO	100	100	97	92	68	22.79

TABLE 3. Selection frequencies for simulations with $q = 15$ (Others are average number of noise variable selected).

Method	Signal variables															Others
	$p = 500,\ \rho = 0$															
SCAD	63	52	54	59	47	42	32	27	19	21	16	16	7	7	9	4.30
SIS-SCAD	51	40	37	44	23	30	21	16	14	9	12	6	4	4	1	1.88
LASSO	74	74	71	76	63	66	50	48	42	39	33	26	26	17	21	20.59
	$p = 500,\ \rho = 0.3$															
SCAD	61	62	64	63	56	49	39	38	39	27	24	23	10	18	9	3.87
SIS-SCAD	45	61	70	64	53	40	33	27	20	9	14	6	4	5	1	0.48
LASSO	86	95	89	92	85	86	78	74	77	59	59	56	41	52	34	24.85
	$p = 500,\ \rho = 0.6$															
SCAD	41	68	48	44	61	39	45	40	49	44	35	44	30	28	10	9.87
SIS-SCAD	18	47	66	62	70	60	54	31	34	20	7	10	2	0	0	0.03
LASSO	83	93	89	76	91	75	81	72	85	78	71	71	58	58	29	22.09
	$p = 1000,\ \rho = 0$															
SCAD	54	60	44	59	45	39	49	33	36	36	23	15	10	21	7	6.87
SIS-SCAD	37	36	35	42	17	22	18	9	12	6	5	5	5	5	1	2.45
LASSO	66	59	56	54	37	46	32	28	28	24	13	14	13	12	10	17.57
	$p = 1000,\ \rho = 0.3$															
SCAD	48	44	43	43	16	27	19	14	22	15	7	7	5	8	3	8.49
SIS-SCAD	32	66	61	64	48	44	30	24	20	18	11	7	3	3	0	0.69
LASSO	79	93	80	83	80	74	78	76	67	67	54	52	30	37	17	26.03
	$p = 1000,\ \rho = 0.6$															
SCAD	53	63	53	57	56	48	56	56	45	39	44	35	31	27	17	12.92
SIS-SCAD	7	50	63	73	72	56	56	45	23	16	10	6	4	1	0	0.03
LASSO	74	86	82	82	81	73	75	85	72	63	65	67	49	43	36	27.00

5 years (metastatic outcome coded as 1) and 51 remained metastases free for at least 5 years (metastatic outcome coded as 0). The data set is publicly available at `http://www.rii.com/publications/2002/vantveer.html`.

We first pre-process gene expression data as follows: (i) remove genes with more than 30% missing measurements; (ii) fill in missing gene expression measurements with median values across samples; (iii) normalize gene expressions to have zero means and unit variances; (iv) compute the simple correlation coefficients of gene expressions with the binary outcome; and (v) select p genes with the largest absolute values of correlation coefficients. We did the selection for $p = 500$ and 1000.

The detailed estimation results are displayed in Table 4. Besides estimation, we also did leave-one-out prediction to assess the prediction performance of the three approaches. When $p = 1000$, the LASSO estimator selected 55 genes and made 75 correct predictions, the SCAD estimator selected 12 genes and made 62 correct predictions, and the SIS-SCAD estimator selected 4 genes and made 59 correct predictions. When $p = 500$, the LASSO estimator selected 38 genes and made 85 correct predictions, the SCAD estimator selected 10 genes and made 79 correct predictions, and the SIS-SCAD estimator selected the same 4 genes and made 59 correct predictions. Interestingly, both the LASSO and the SCAD estimators could select fewer genes but achieve better prediction results.

TABLE 4. Estimation results for the breast cancer data set.

	$p = 500$			$p = 1000$		
Systematic Name	LASSO	SCAD	SIS-SCAD	LASSO	SCAD	SIS-SCAD
Contig28521_RC	0.000	0.000	0.000	0.003	0.000	0.000
X89657	0.000	0.000	0.000	0.158	0.000	0.000
Contig5816_RC	0.130	0.000	0.000	0.000	0.000	0.000
Contig35148_RC	−0.290	0.000	0.000	−0.337	−2.239	0.000
Contig45345_RC	0.000	0.000	0.000	−0.012	0.000	0.000
NM_003147	−0.371	−1.907	0.000	−0.196	0.000	0.000
Contig23356_RC	0.000	0.000	0.000	0.138	0.000	0.000
Contig35229_RC	−0.263	0.000	0.000	−0.059	0.000	0.000
Contig28433_RC	−0.004	0.000	0.000	0.000	0.000	0.000
Contig57801_RC	0.000	0.000	0.000	0.028	0.000	0.000
NM_003366	−0.311	0.000	0.000	−0.037	0.000	0.000
Contig20708_RC	0.000	0.000	0.000	0.026	0.000	0.000
NM_020120	0.000	0.000	0.533	0.000	0.000	0.533
NM_020123	0.333	2.596	0.000	0.269	1.244	0.000
Contig6238_RC	−0.276	0.000	0.000	−0.010	0.000	0.000
NM_004230	0.000	0.000	0.000	−0.212	−1.342	0.000
AF052087	0.000	0.000	0.000	−0.131	0.000	0.000
NM_005007	−0.078	0.000	0.000	0.000	0.000	0.000
AB018337	0.460	2.145	0.000	0.086	0.000	0.000
Contig47042	0.247	0.000	0.000	0.020	0.000	0.000
Contig38438_RC	−0.222	0.000	0.000	−0.193	0.000	0.000
X67055	−0.107	0.000	0.000	−0.058	0.000	0.000
NM_003862	0.000	0.000	−0.583	−0.017	0.000	−0.583
NM_003882	−0.243	0.000	0.000	−0.034	0.000	0.000
AF131819	0.465	2.138	0.000	0.110	0.000	0.000
NM_014003	0.146	0.000	0.000	0.213	0.000	0.000
NM_013376	0.000	0.000	0.000	−0.033	0.000	0.000
NM_005393	0.376	1.843	0.000	0.204	1.208	0.000
AJ011306	−0.185	0.000	0.000	−0.231	−1.806	0.000
Contig41613_RC	0.000	0.000	0.000	−0.005	0.000	0.000
NM_004953	0.000	0.000	0.000	0.093	0.000	0.000
NM_004994	0.129	0.000	0.000	0.015	0.000	0.000
AL080059	0.710	2.343	1.212	0.723	2.339	1.212
U17327	0.000	0.000	0.000	0.233	0.000	0.000
Contig42563_RC	0.169	0.000	0.000	0.036	0.000	0.000
AL080109	0.000	0.000	0.000	0.020	0.000	0.000
NM_006544	0.170	0.000	0.000	0.161	0.000	0.000
Contig64861_RC	0.202	0.000	0.000	0.135	0.000	0.000
AF055033	0.686	1.642	0.000	0.466	1.438	0.000
Contig14086_RC	0.000	0.000	0.000	0.195	0.000	0.000
NM_016017	0.000	0.000	0.689	0.000	0.000	0.689
Contig47544_RC	0.693	2.545	0.000	0.546	1.855	0.000
Contig48697_RC	0.059	0.000	0.000	0.000	0.000	0.000
Contig2947_RC	0.000	0.000	0.000	0.318	2.112	0.000

Continued on next page

TABLE 4 (continued)

	$p = 500$			$p = 1000$		
Systematic Name	LASSO	SCAD	SIS-SCAD	LASSO	SCAD	SIS-SCAD
NM_016361	−0.084	0.000	0.000	−0.087	0.000	0.000
AL137584	0.000	0.000	0.000	−0.239	0.000	0.000
Contig412_RC	−0.605	−1.930	0.000	−0.416	−1.906	0.000
NM_016564	0.581	1.397	0.000	0.154	0.000	0.000
NM_015849	0.000	0.000	0.000	−0.029	0.000	0.000
Contig48208_RC	−0.016	0.000	0.000	0.000	0.000	0.000
NM_018089	0.091	0.000	0.000	0.213	0.000	0.000
U79298	−0.215	0.000	0.000	−0.133	0.000	0.000
NM_000127	0.280	0.000	0.000	0.361	1.111	0.000
NM_019018	−0.101	0.000	0.000	−0.005	0.000	0.000
NM_000232	0.000	0.000	0.000	0.059	0.000	0.000
NM_001061	0.000	0.000	0.000	−0.066	0.000	0.000
BE739817_RC	0.000	0.000	0.000	0.148	0.000	0.000
Contig41716_RC	0.106	0.000	0.000	0.079	0.000	0.000
NM_000765	0.000	0.000	0.000	−0.078	0.000	0.000
NM_002200	0.175	0.000	0.000	0.241	0.390	0.000
NM_000801	0.173	0.000	0.000	0.008	0.000	0.000
Contig3920_RC	−0.010	0.000	0.000	0.000	0.000	0.000
NM_018964	0.000	0.000	0.000	0.002	0.000	0.000

5. Discussion

For ultra-high-dimensional logistic regression models with the dimension of covarates diverging at an exponential rate, we have proposed a regularization based method for the selection of signal covariates and showed the oracle property for the method. It is also worthwhile to mention that our ultra-high-dimensional setting covers more general cases than the polynomial rate assumption imposed on linear regression models by **Kim et al. (2008)**. We believe that the procedure can be extended to other generalized linear models and similar theoretical results can be established for them. Other potential extensions are to the Cox model, robust regression, and procedures based on quasi-likelihood functions. Further effort is needed to build up similar procedure and theoretical results under these settings. The major difficulty for such extensions would be establishing an exponential inequality like (A.14) in our proof for the logistic regression model.

Appendix A

We first prove a technical lemma for proving Theorem 2.1.

LEMMA A.1. *Let $W_1, \ldots, W_n$ be independent random variables with variances $\sigma_1^2, \ldots, \sigma_n^2$. Define $Z_k = (W_k - EW_k)/\sigma_k$, and let $M_k(u) = E\{\exp(uZ_k)\}$ be the generating function of Z_k for $k = 1, \ldots, n$. Assume that there is a positive constant t_0 such that $E\{\exp(t|W_k|)\} < \infty$ for $0 \le t \le t_0/\sigma_k$. Moreover, let A, A_k, a_{nk}, $1 \le k \le n$, be constants satisfying $A_n \ge \sum_{k=1}^n a_{nk}^2 \sigma_k^2$ and $A \ge \max_k |a_{nk}\sigma_k|/A_n$.*

If

$$M \equiv \sup_{1\le k\le n}\sup_{0\le u\le t_0}\left|\frac{\mathrm{d}^3\log M_k(u)}{\mathrm{d}t^3}\right| < \infty, \tag{A.1}$$

then for $0<\zeta<t_0/A$, *we have*

$$P\left\{\left|\sum_{k=1}^n a_{nk}(W_k-EW_k)\right|>\zeta\right\}\le 2\exp\left\{-\frac{\zeta^2}{2A_n}\left(1-\frac{1}{3}AM\zeta\right)\right\}. \tag{A.2}$$

PROOF. Define $t_\zeta=\zeta/A_n$ for $\zeta\le t_0/A$. Then $|a_{nk}\sigma_k t_\zeta| = |a_{nk}\sigma_k\zeta/A_n|\le A\zeta\le t_0$. Applying a Taylor expansion to $\log\{M_k(u)\}$ at $u=0$, we have, for $0\le u\le t_0$,

$$\log M_k(u) = \log M_k(0) + u\frac{\mathrm{d}\log M_k(u)}{\mathrm{d}u}\bigg|_{u=0} + \frac{u^2}{2}\frac{\mathrm{d}^2\log M_k(u)}{\mathrm{d}u^2}\bigg|_{u=0} + \frac{u^3}{6}\frac{\mathrm{d}^3\log M_k(u)}{\mathrm{d}u^3}\bigg|_{u=u_*},$$

where $u_*\in(0,u)$. Note that $\log M_k(0)=0$,

$$\frac{\mathrm{d}\log M_k(u)}{\mathrm{d}u}\bigg|_{u=0} = E(Z_k)=0, \qquad \frac{\mathrm{d}^2\log M_k(u)}{\mathrm{d}u^2}\bigg|_{u=0}=1,$$

and

$$\left|\frac{\mathrm{d}^3\log M_k(u)}{\mathrm{d}u^3}\bigg|_{u=u_*}\right|\le M.$$

It follows that

$$\begin{aligned}\log\{M_k(a_{nk}\sigma_k t_\zeta)\} &\le 2^{-1}\left(\frac{a_{nk}\sigma_k\zeta}{A_n}\right)^2+\frac{1}{6}\left|\frac{a_{nk}\sigma_k\zeta}{A_n}\right|^3\cdot M\\ &\le \frac{a_{nk}^2\sigma_k^2\zeta^2}{2A_n^2}\left(1+\frac{1}{3}AM\zeta\right).\end{aligned}$$

Some algebra yields

$$\begin{aligned}\log P\left\{\sum_{k=1}^n a_{nk}(W_k-EW_k)>\zeta\right\} &= \log P\left\{\sum_{k=1}^n a_{nk}\sigma_k Z_k>\zeta\right\}\\ &\le \log E\left[\exp\left\{t_\zeta\left(\sum_{k=1}^n a_{nk}\sigma_k Z_k-\zeta\right)\right\}\right]\\ &= -\zeta t_\zeta+\sum_{k=1}^n\log M_k(a_{nk}\sigma_k t_\zeta)\\ &\le -\frac{\zeta^2}{A_n}+\sum_{k=1}^n\frac{a_{nk}^2\sigma_k^2\zeta^2}{2A_n}\left(1+\frac{1}{3}AM\zeta\right)\\ &\le -\frac{\zeta^2}{A_n}+\frac{\zeta^2}{2A_n}\left(1+\frac{1}{3}AM\zeta\right)\\ &= -\frac{\zeta^2}{2A_n}\left(1-\frac{1}{3}AM\zeta\right).\end{aligned}$$

Hence

$$\text{(A.3)} \qquad P\left\{\sum_{k=1}^{n} a_{nk}(W_k - EW_k) > \zeta\right\} \le \exp\left\{-\frac{\zeta^2}{2A_n}\left(1 - \frac{1}{3}AM\zeta\right)\right\}.$$

Similarly we can show that

$$\text{(A.4)} \qquad P\left\{\sum_{k=1}^{n} a_{nk}(W_k - EW_k) < -\zeta\right\} \le \exp\left\{-\frac{\zeta^2}{2A_n}\left(1 - \frac{1}{3}AM\zeta\right)\right\}.$$

Combining (A.3) and (A.4) completes the proof. □

PROOF OF THEOREM 2.1. Let $L_\lambda(\boldsymbol{\beta}) = \mathcal{L}(\boldsymbol{\beta}) + \sum_{j=1}^{p} \tilde{J}_\lambda(|\beta_j|)$ and $S_j(\boldsymbol{\beta}) = -(1/n)\sum_{i=1}^{n} x_{ij}\{y_i - \mathcal{F}(\mathbf{x}_i^\top \boldsymbol{\beta})\}$. Note that

$$\frac{\partial L_\lambda(\boldsymbol{\beta})}{\partial \beta_j} = \begin{cases} S_j(\boldsymbol{\beta}) - \lambda \operatorname{sign}(\beta_j) & \text{if } |\beta_j| \ge a\lambda \\ S_j(\boldsymbol{\beta}) + \dfrac{a\lambda - |\beta_j|}{(a-1)} \operatorname{sign}(\beta_j) - \lambda \operatorname{sign}(\beta_j) & \text{if } \lambda \le |\beta_j| < a\lambda \\ S_j(\boldsymbol{\beta}) & \text{if } |\beta_j| \le \lambda. \end{cases}$$

By the second-order sufficiency of the Karush–Kuhn–Tucker condition (see **Bertsekas (1999**, p. 320)), any $\boldsymbol{\beta}$ satisfying

$$\text{(A.5)} \qquad S_j(\boldsymbol{\beta}) = 0 \quad \text{and} \quad |\beta_j| \ge a\lambda \qquad \text{for } j = 1, \dots, q_n$$

$$\text{(A.6)} \qquad |S_j(\boldsymbol{\beta})| \le \lambda \quad \text{and} \quad |\beta_j| \le a\lambda \qquad \text{for } j = q_n + 1, \dots, p_n$$

is an element of $A(\lambda)$. Thus it suffices to show that $\widehat{\boldsymbol{\beta}}_{\mathrm{o}}$ satisfies (A.5) and (A.6) with $\lambda = \lambda_n$.

Step 1. Note that $S_j(\widehat{\boldsymbol{\beta}}_{\mathrm{o}}) = 0$ for $j \le q_n$ holds trivially by the definition of the oracle estimator. So to prove (A.5), it suffices to prove that as $n \to \infty$,

$$\text{(A.7)} \qquad P(|\widehat{\boldsymbol{\beta}}_{j\mathrm{o}}| \ge a\lambda_n \text{ for } j = 1, \dots, q_n) \to 1$$

Let $\varepsilon_i = y_i - \mathcal{F}(\mathbf{x}_i^\top \boldsymbol{\beta})$. Note that

$$\frac{1}{n}\sum_{i=1}^{n} \mathbb{X}_{(1)i}\{y_i - \mathcal{F}(\mathbb{X}_{(1)i}^\top \widehat{\boldsymbol{\beta}}_{(1)\mathrm{o}})\} = \mathbf{0}.$$

The left-hand side can be expressed as

$$\begin{aligned} &\frac{1}{n}\sum_{j=1}^{n} \mathbb{X}_{(1)i}\varepsilon_i + \frac{1}{n}\sum_{i=1}^{n} \mathbb{X}_{(1)i}\{\mathcal{F}(\mathbb{X}_{(1)i}^\top \boldsymbol{\beta}_{(1)\mathrm{o}}) - \mathcal{F}(\mathbb{X}_{(1)i}^\top \widehat{\boldsymbol{\beta}}_{(1)\mathrm{o}})\} \\ &\qquad = \frac{1}{n}\sum_{j=1}^{n} \mathbb{X}_{(1)i}\varepsilon_i + \frac{1}{n}\sum_{i=1}^{n} \mathbb{X}_{(1)i}\mathbb{X}_{(1)i}^\top (\widehat{\boldsymbol{\beta}}_{(1)\mathrm{o}} - \boldsymbol{\beta}_{(1)\mathrm{o}})\mathcal{F}^{(1)}(\xi_i^*), \end{aligned}$$

where ξ_i^* lies between $\mathbb{X}_{(1)i}^\top \boldsymbol{\beta}_{(1)\mathrm{o}}$ and $\mathbb{X}_{(1)i}^\top \widehat{\boldsymbol{\beta}}_{(1)\mathrm{o}}$.

Denote $\operatorname{diag}_1\{\boldsymbol{\mathcal{F}}^{(1)}\} = \operatorname{diag}\{\mathcal{F}^{(1)}(\xi_1^*), \dots, \mathcal{F}^{(1)}(\xi_n^*)\}$. It follows that

$$\text{(A.8)} \qquad \begin{aligned} \mathbf{u}^\top(\widehat{\boldsymbol{\beta}}_{(1)\mathrm{o}} - \boldsymbol{\beta}_{(1)}) &= \mathbf{u}^\top \left[\frac{1}{n}\mathbb{X}_{(1)}^\top \operatorname{diag}_1\{\boldsymbol{\mathcal{F}}^{(1)}\}\mathbb{X}_{(1)}\right]^{-1} \mathbb{X}_{(1)}^\top \boldsymbol{\varepsilon} \\ &\overset{\text{def}}{=} \mathbf{u}^\top \Sigma_n \mathbb{X}_{(1)}^\top \boldsymbol{\varepsilon}. \end{aligned}$$

Note that $|\widehat{\beta}_{jo}| \geq |\beta_j| - |\widehat{\beta}_{jo} - \beta_j|$. Because $\min_{j \leq q_n} |\beta_j| = O(n^{-(1-c_2)/2})$ and $\lambda_n = o(n^{-(1+c_2-c_1)/2})$, it suffices to show that

$$\max_{j \leq q_n} |\widehat{\beta}_{jo} - \beta_j| = o_p(n^{-(1-c_2)/2}). \tag{A.9}$$

Let $z_j = \sqrt{n}(\widehat{\beta}_{jo} - \beta_j)$. For (A.9), we show that

$$\max_{j \leq q_n} |z_j| = o_p(n^{c_2/2}). \tag{A.10}$$

Let $\boldsymbol{\xi}_j$ be a q_n-vector, whose jth element is 1 and 0 otherwise. A direct calculation yields that, for any $C > 0$,

$$\sum_{j=1}^{q_n} P\{|\sqrt{n}\boldsymbol{\xi}_j^\top (\widehat{\boldsymbol{\beta}}_{(1)o} - \boldsymbol{\beta}_{(1)})| > Cn^{c_2/2}\} \leq C \sum_{j=1}^{q_n} \frac{nE(\boldsymbol{\xi}_j^\top \Sigma_n^{-1} \mathbb{X}_{(1)}^\top \boldsymbol{\varepsilon}\boldsymbol{\varepsilon}^\top \mathbb{X}_{(1)} \Sigma_n^{-1} \boldsymbol{\xi}_j)}{n^{c_2}}.$$

A tedious but straightforward derivation shows that the numerator of the jth term is proportional to $n\boldsymbol{\xi}_j^\top (\mathbb{X}_{(1)}^\top \mathbb{X}_{(1)})^{-1} \boldsymbol{\xi}_j$, which is of order $O(1)$. These evidences indicate that

$$\max_{1 \leq j \leq q_n} \left|\sqrt{n}(\widehat{\beta}_{jo} - \beta_j)\right| = o_p(n^{c_2/2}). \tag{A.11}$$

Step 2. To show (A.6), we have $|\widehat{\beta}_{jo}| \leq \lambda_n$ for $j > q_n$, because $\widehat{\beta}_{jo} = 0$ by the definition. Thus it suffices to show that

$$P(|S_j(\widehat{\boldsymbol{\beta}}_o)| > \lambda_n \text{ for some } j = 1 + q_n, \ldots, p_n). \tag{A.12}$$

Note that for $j = q_n + 1, \ldots, n$,

(A.13)

$$\begin{aligned} S_j(\widehat{\boldsymbol{\beta}}_o) &= -\frac{1}{n}\mathbb{X}_{(2)}^\top \{\mathbf{Y} - \mathcal{F}(\mathbb{X}_{(1)}^\top \widehat{\boldsymbol{\beta}}_{(1)o})\} \\ &= -\frac{1}{n}\mathbb{X}_{(2)}^\top \{\mathbf{Y} - \mathcal{F}(\mathbb{X}_{(1)}^\top \boldsymbol{\beta}_{(1)o})\} - \frac{1}{n}\mathbb{X}_{(2)}^\top \{\mathcal{F}(\mathbb{X}_{(1)}^\top \boldsymbol{\beta}_{(1)o}) - \mathcal{F}(\mathbb{X}_{(1)}^\top \widehat{\boldsymbol{\beta}}_{(1)o})\} \\ &= -\frac{1}{n}\mathbb{X}_{(2)}^\top \boldsymbol{\varepsilon} + \frac{1}{n}\mathbb{X}_{(2)}^\top \operatorname{diag}_1\{\boldsymbol{\mathcal{F}}^{(1)}\}\mathbb{X}_{(1)}^\top \{\widehat{\boldsymbol{\beta}}_{(1)o} - \boldsymbol{\beta}_{(1)o}\}. \end{aligned}$$

Substituting the expression of $(\widehat{\boldsymbol{\beta}}_{(1)o} - \boldsymbol{\beta}_{(1)o})$ given in (A.8) indicates that (A.13) can be expressed as

$$-\frac{1}{n}\mathbb{X}_{(2)}^\top \left[\mathbb{I} - \frac{1}{n}\operatorname{diag}_1\{\boldsymbol{\mathcal{F}}^{(1)}\}\mathbb{X}_{(1)}\Sigma_n \mathbb{X}_{(1)}^\top\right]\boldsymbol{\varepsilon}.$$

Let a_{nk} be the kth element of the vector $-(1/n)\mathbf{u}_j^\top \mathbb{X}_{(2)}^\top [\mathbb{I} - \operatorname{diag}_1\{\boldsymbol{\mathcal{F}}^{(1)}\}\mathbb{X}_{(1)}\Sigma_n \mathbb{X}_{(1)}^\top]$, where $\mathbf{u}_j^\top$ is a $(p_n - q_n)$-vector with the jth element being 1 and zeros otherwise. Note that

$$\begin{aligned} \sum_{k=1}^{n} a_{nk}^2 = \frac{1}{n^2} \sum_{k=1}^{n} \mathbf{u}_j^\top \mathbb{X}_{(2)}^\top &\left[\mathbb{I} - \frac{1}{n}\operatorname{diag}_1\{\boldsymbol{\mathcal{F}}^{(1)}\}\mathbb{X}_{(1)}\Sigma_n \mathbb{X}_{(1)}^\top\right] \\ &\times \left[\mathbb{I} - \frac{1}{n}\mathbb{X}_{(1)}\Sigma_n \mathbb{X}_{(1)}^\top \operatorname{diag}_1\{\boldsymbol{\mathcal{F}}^{(1)}\}\right]\mathbb{X}_{(2)}\mathbf{u}_j. \end{aligned}$$

A direct simplification results in

$$\sum_{k=1}^{n} a_{nk}^2 \leq \frac{1}{n^2}\mathbf{u}_j^\top \mathbb{X}_{(2)}^\top \mathbb{X}_{(2)} \mathbf{u}_j = O(n^{-1}) \quad \text{and} \quad \max_{1 \leq k \leq n} |a_{nk}| = O(n^{-1}).$$

Take $A_n = 1/n$ and $A = 1$. We can verify that the conditions in Lemma A.1 are ensured. Recall $\log(p_n)/n^{c_2-c_1} \to 0$ and $\lambda_n = o(n^{-(1-c_2+c_1)/2})$. It follows that

$$\text{(A.14)} \quad \sum_{j=q_n+1}^{p_n} P\left(\left|\sum_{k=1}^{n} a_{nk}\varepsilon_k\right| > \lambda_n\right) \leq 2(p_n - q_n)\exp\left\{-\frac{\lambda_n^2(1 - A\lambda_n/3)}{2A_n}\right\} \leq 2(p_n - q_n)\exp(-2n\lambda_n^2) \to 0.$$

This completes the proof. □

References

Le Thi Hoai An and Pham Dinh Tao. 1997. *Solving a class of linearly constrained indefinite quadratic problems by d.c. algorithms*, J. Global Optim. **11**, no. 3, 253–285, DOI 10.1023/A:1008288411710. MR1469128 (98k:90068)

Bertsekas, D. P. 1999. *Nonlinear programming*, 2nd ed., Athena Scientific, Belmount, MA.

Bradic, J., J. Fan, and J. Jiang. 2011. *Regularization for Cox's proportional hazards model with NP-dimensionality*, Ann. Statist. **39**, no. 6, 3092–3120, DOI 10.1214/11-AOS911. MR3012402

Breiman, L. 1996. *Heuristics of instability and stabilization in model selection*, Ann. Statist. **24**, no. 6, 2350–2383, DOI 10.1214/aos/1032181158. MR1425957 (97j:62093)

Chen, K., I. Hu, and Z. Ying. 1999. *Strong consistency of maximum quasi-likelihood estimators in generalized linear models with fixed and adaptive designs*, Ann. Statist. **27**, no. 4, 1155–1163, DOI 10.1214/aos/1017938919. MR1740117 (2001d:62025)

Fan, J. and R. Li. 2001. *Variable selection via nonconcave penalized likelihood and its oracle properties*, J. Amer. Statist. Assoc. **96**, no. 456, 1348–1360, DOI 10.1198/016214501753382273. MR1946581 (2003k:62160)

Fan, J. and R. Li. 2004. *New estimation and model selection procedures for semiparametric modeling in longitudinal data analysis*, J. Amer. Statist. Assoc. **99**, no. 467, 710–723, DOI 10.1198/016214504000001060. MR2090905 (2005d:62053)

Fan, J. and R. Li. 2006. *Statistical challenges with high dimensionality: feature selection in knowledge discovery*, International Congress of Mathematicians. Vol. III, Eur. Math. Soc., Zürich, pp. 595–622. MR2275698 (2008m:62099)

Fan, J. and J. Lv. 2008. *Sure independence screening for ultrahigh dimensional feature space*, J. R. Stat. Soc. Ser. B Stat. Methodol. **70**, no. 5, 849–911, DOI 10.1111/j.1467-9868.2008.00674.x. MR2530322

Fan, J. and J. Lv. 2011. *Nonconcave penalized likelihood with NP-dimensionality*, IEEE Trans. Inform. Theory **57**, no. 8, 5467–5484, DOI 10.1109/TIT.2011.2158486. MR2849368 (2012k:62211)

Fan, J. and H. Peng. 2004. *Nonconcave penalized likelihood with a diverging number of parameters*, Ann. Statist. **32**, no. 3, 928–961, DOI 10.1214/009053604000000256. MR2065194 (2005g:62047)

Fan, J., R. Samworth, and Y. Wu. 2009. *Ultrahigh dimensional feature selection: beyond the linear model*, J. Mach. Learn. Res. **10**, 2013–2038. MR2550099 (2010j:62174)

Fan, J. and R. Song. 2010. *Sure independence screening in generalized linear models with NP-dimensionality*, Ann. Statist. **38**, no. 6, 3567–3604, DOI 10.1214/10-AOS798. MR2766861 (2012a:62201)

Frank, I. E. and J. H. Friedman. 1993. *A statistical view of some chemometrics regression tools (with discussion)*, Technometrics **35**, no. 2, 109–148, DOI 10.1080/00401706.1993.10485033.

Friedman, J., T. Hastie, H. Höfling, and R. Tibshirani. 2007. *Pathwise coordinate optimization*, Ann. Appl. Stat. **1**, no. 2, 302–332, DOI 10.1214/07-AOAS131. MR2415737

Hosmer, D. W. and S. Lemeshow. 1989. *Applied logistic regression*, Wiley, New York.

Huang, J., S. Ma, and C.-H. Zhang. 2008a. *Adaptive Lasso for sparse high-dimensional regression models*, Statist. Sinica **18**, no. 4, 1603–1618. MR2469326 (2010a:62214)

Huang, J., S. Ma, and C.-H. Zhang. 2008b. *The iterated LASSO for high-dimensional logistic regression*, Technical Report 392, Department of Statistics and Actuarial Science, The University of Iowa.

Kim, Y., H. Choi, and H.-S. Oh. 2008. *Smoothly clipped absolute deviation on high dimensions*, J. Amer. Statist. Assoc. **103**, no. 484, 1665–1673, DOI 10.1198/016214508000001066. MR2510294 (2010i:62183)

Kwon, S. and Y. Kim. 2012. *Large sample properties of the SCAD-penalized maximum likelihood estimation on high dimensions*, Statist. Sinica **22**, no. 2, 629–653, DOI 10.5705/ss.2010.027. MR2954355

Li, R. and H. Liang. 2008. *Variable selection in semiparametric regression modeling*, Ann. Statist. **36**, no. 1, 261–286, DOI 10.1214/009053607000000604. MR2387971 (2009g:62051)

Liang, H. and R. Li. 2009. *Variable selection for partially linear models with measurement errors*, J. Amer. Statist. Assoc. **104**, no. 485, 234–248, DOI 10.1198/jasa.2009.0127. MR2504375 (2010d:62175)

Lindsey, J. K. 1997. *Applying generalized linear models*, Springer Texts Statist., vol. 13, Springer, New York.

Lv, J. and Y. Fan. 2009. *A unified approach to model selection and sparse recovery using regularized least squares*, Ann. Statist. **37**, no. 6A, 3498–3528, DOI 10.1214/09-AOS683. MR2549567 (2010m:62219)

McCullagh, P. and J. A. Nelder. 1983. *Generalized linear models*, Monographs on Statistics and Applied Probability, Chapman & Hall, London. MR727836 (85k:62161)

Nelder, J. A. and R. W. M. Wedderburn. 1972. *Generalized linear models*, J. Roy. Statist. Soc. Ser. A **135**, no. 3, 370–384.

Rosset, S. and J. Zhu. 2007. *Piecewise linear regularized solution paths*, Ann. Statist. **35**, no. 3, 1012–1030, DOI 10.1214/009053606000001370. MR2341696 (2009b:62140)

Tibshirani, R. 1996. *Regression shrinkage and selection via the lasso*, J. Roy. Statist. Soc. Ser. B **58**, no. 1, 267–288. MR1379242 (96j:62134)

van de Geer, S. A. 2008. *High-dimensional generalized linear models and the lasso*, Ann. Statist. **36**, no. 2, 614–645, DOI 10.1214/009053607000000929. MR2396809 (2009h:62048)

van 't Veer, L. J., H. Dai, M. J. van de Vijver, Y. D. He, A. A. M. Hart, M. Mao, H. L. Peterse, K. van der Kooy, M. J. Marton, A. T. Witteveen, G. J. Schreiber, R. M. Kerkhoven, C. Roberts, P. S. Linsley, R. Bernards, and S. H. Friend. 2002. *Gene expression profiling predicts clinical outcome of breast cancer*, Nature **415**, no. 6871, 530–536, DOI 10.1038/415530a.

Wang, L. 2011. *GEE analysis of clustered binary data with diverging number of covariates*, Ann. Statist. **39**, no. 1, 389–417, DOI 10.1214/10-AOS846. MR2797851 (2012b:62085)

Yuille, A. L. and A. Rangarajan. 2003. *The concave-convex procedure*, Neural Computation **15**, no. 4, 915–936, DOI 10.1162/08997660360581958.

Zhao, P. and B. Yu. 2006. *On model selection consistency of Lasso*, J. Mach. Learn. Res. **7**, 2541–2563. MR2274449

Zou, H. 2006. *The adaptive lasso and its oracle properties*, J. Amer. Statist. Assoc. **101**, no. 476, 1418–1429, DOI 10.1198/016214506000000735. MR2279469 (2008d:62024)

Zou, H. and T. Hastie. 2005. *Regularization and variable selection via the elastic net*, J. R. Stat. Soc. Ser. B Stat. Methodol. **67**, no. 2, 301–320, DOI 10.1111/j.1467-9868.2005.00503.x. MR2137327

Zou, H., T. Hastie, and R. Tibshirani. 2007. *On the "degrees of freedom" of the lasso*, Ann. Statist. **35**, no. 5, 2173–2192, DOI 10.1214/009053607000000127. MR2363967 (2009d:62096)

Zou, H. and R. Li. 2008. *One-step sparse estimates in nonconcave penalized likelihood models*, Ann. Statist. **36**, no. 4, 1509–1533, DOI 10.1214/009053607000000802. MR2435443 (2010a:62222)

Zou, H. and H. H. Zhang. 2009. *On the adaptive elastic-net with a diverging number of parameters*, Ann. Statist. **37**, no. 4, 1733–1751, DOI 10.1214/08-AOS625. MR2533470 (2010j:62210)

Department of Statistics, Virginia Polytechnic Institute and State University, Blacksburg, Virginia 24061
E-mail address: pangdu@vt.edu

Department of Biostatistics and Computational Biology, University of Rochester, Rochester, New York 14642
E-mail address: pan_wu@urmc.rochester.edu

Department of Statistics, The George Washington University, Washington, DC 20052
E-mail address: hliang@gwu.edu

Contemporary Mathematics
Volume **622**, 2014
http://dx.doi.org/10.1090/conm/622/12432

Shrinkage estimation and selection for a logistic regression model

Shakhawat Hossain and S. Ejaz Ahmed

Abstract. This paper considers the problem of variable selection and the estimation for a logistic regression model via shrinkage and three penalty methods. We develop a large sample theory for the shrinkage estimators including asymptotic distributional bias and risk. We show that if the shrinkage dimension exceeds two, the asymptotic risk of the shrinkage estimator is strictly less than the classical estimators for a wide class of models. This reduction holds globally in the parameter space. Furthermore, we consider three different penalty estimators: the LASSO, adaptive LASSO, and SCAD and compare their relative performance with the shrinkage estimators numerically. A Monte Carlo simulation study is conducted for different combinations of inactive predictors and the performance of each method is evaluated in terms of a simulated mean squared error. This study indicates that shrinkage method is comparable to the LASSO, adaptive LASSO, and SCAD when the number of inactive predictors in the model is relatively large. A real data example is presented to illustrate the proposed methodologies.

1. Introduction

Logistic regression is a widely used statistical tool to model independent binary response data in medical and epidemiologic studies. This model assumes that the logit of the response variable can be modeled by a linear combination of unknown parameters. Many researchers prefer to work directly with this model. For this reason, it is treated as an independent model in its own right. For detailed information on logistic regression we refer to [**H**] and [**HL**], and among others.

Model selection plays an important role as the presence of a number of candidate predictors occur in a wide variety of scientific field. By selecting a subset of active predictors, one wants to achieve accurate predictions and interpretable models. In this paper, a shrinkage approach is proposed, which focus on improved estimation technique for estimating the regression parameters based on the information of inactive predictors in the model for binary data. This paper also compare

---

2010 *Mathematics Subject Classification.* Primary 62J07.

Key words and phrases. Penalty estimators, shrinkage estimators, asymptotic distributional bias and risk, Monte Carlo simulation, logistic regression, likelihood ratio test.

The research of both authors was supported by the Natural Sciences and Engineering Research Council of Canada.

The authors would like to thank the referees and the editor for their valuable suggestions in the revision of this paper.

the performance of shrinkage estimators to three popular penalty methods recently developed for simultaneously variable selection and parameter estimation.

We consider the problem of estimating the coefficients of a logistic regression model when there are many predictors and some of them may be less relevant for the response of interest. In this situation, we consider the information from inactive predictors and use either the full model or the sub-model. The shrinkage method which combines the full and sub model estimators, is inspired by Stein's result that in a dimension greater than two, efficient estimates can be obtained by shrinking a full model estimator in the direction of a sub model estimator. Existing literature (see [**AHD**] and [**ADHY**]) shows that the shrinkage estimators improve upon the penalty [**T**] and other classical estimators.

Several authors developed the shrinkage estimation strategy for parametric, semiparametric, and nonparametric linear models for censored and uncensored data, see [**AHD**, **ADHY**, **AHS**, **HDEA**], and others. In this paper we extend the shrinkage estimation method for modeling binary data by amalgamating the ideas from the recent literature on sparsity patterns and compares the resulting estimator to the full and sub model estimators as well as to a version of the penalty estimators. We also derive the asymptotic distributional biases and risks for the shrinkage estimators.

Motivating example. Hosmer and Lemeshow [**HL**] considered the low birth weight data. This data were collected at Baystate Medical Center in Springfield, Massachusetts in 1996, USA and the low birth weight is a response variable that has been a concern for physicians. The goal of this study was to identify risk factors associated with giving birth to a low birth weight baby (weighing less than 2500 grams). Data were collected on 189 women, 59 of whom had the low birth weight babies and 130 of whom had the normal birth weight babies. The observed predictor variables are age, weight of the mother at the last menstrual period, race, smoking status, history of premature labor, history of hypertension, presence of uterine irritability, and the number of physician visits during the first trimester of pregnancy. The shrinkage method uses a two-step approach to estimate the coefficients of active predictors. In the first step, AIC or BIC criterion is used to form a subset of the total set of predictors. This criterion shows that the history of premature labor, history of hypertension, weight of the mother at the last menstrual period, smoking status, and the race of the mother are the active predictors, and the effects of the other inactive predictors may be ignored. In this situation we can partition the full parameter vector $\boldsymbol{\beta}$ into the active and inactive parameter sub-vectors as $\boldsymbol{\beta} = (\boldsymbol{\beta}_1', \boldsymbol{\beta}_2')'$, where $\boldsymbol{\beta}_1$ and $\boldsymbol{\beta}_2$ are assumed to have dimensions $p_1 \times 1$ and $p_2 \times 1$, respectively, such that $p = p_1 + p_2$. Our interest lies in the estimation of the parameter sub-vector $\boldsymbol{\beta}_1$ when the information on $\boldsymbol{\beta}_2$ is available. The information about the inactive parameters may be used to estimate $\boldsymbol{\beta}_1$ when their values are near some specified value which, without loss of generality, may be set to a null vector,

$$\boldsymbol{\beta}_2 = \boldsymbol{\beta}_2^0 = \mathbf{0}. \tag{1.1}$$

In the second step, we combine the sub and full model estimators in an optimal way in order to achieve an improved estimator for the remaining active predictors. This approach can be implemented for moderate values of p.

On the other hand, when p is large and n is small, a common goal is to find the genetic explanations that are responsible for observable traits in biomedical studies. So understanding the genetic associations of diseases helps medical researchers to further investigate the diseases and to develop the corresponding treatment methods. Suppose a medical researcher was measured about 600 microRNA (miRNA) expressions in serum samples from two groups of participants. One group consisted of 30 oral cancer patients and the other group consisted of 26 individuals without cancer. The question is whether these miRNA readings can be used to distinguish the cancer patients from the others. If the treatment method is successful, the genetic information might be further used to predict whether an oral-cancer patient will progress from a minor tumor to a serious one. Using all 600 miRNAs for the classification leads to a poor predictive value because of the high level of noise. Consequently, it is important to select those that make the greatest contribution to identifying oral-cancer patients. A logistic regression of the tumor type on the miRNA readings can be used to identify the relevant miRNAs by selecting the most important predictors. However, the number of predictors p is 600, and the number of participants n is just 56. This large-p-small-n situation places this problem outside the domain of classical model selection methods. We need penalization methods, such as LASSO, ALASSO, and SCAD to deal with the high dimensionality.

Penalty likelihood methods are studied for simultaneous variable selection and parameter estimation. These methods deal with ill-defined regression problems in the classical frameworks. One of the penalty methods is LASSO [**T**], which shrinks regression coefficients toward 0 and produces some coefficients that are exactly zero and, hence, implement variable selection. It has been shown that this procedure produces biases even in the simple regression setting [**FL**] due to its linear increase of penalty on regression coefficients. To avoid this bias issue, the adaptive LASSO and SCAD penalties were proposed recently in the literature. Fan and Li [**FL**] is considered the SCAD penalty to overcome the inefficiency of traditional variable selection procedures including the stepwise regression and the best subset selection. This SCAD estimator has many desirable properties, including continuity, sparsity, and unbiasedness. It also has the (asymptotically) oracle property when the dimension of predictors is fixed or diverges more slowly than the sample size. Zou [**Z**] is modified the LASSO penalty so that different amounts of shrinkage are allowed for different regression coefficients. It has been shown theoretically that the adaptive LASSO estimator is able to identify the true model consistently, and the resulting estimator is as efficient as oracle. Meier, van de Geer, and Bühlmann [**MvdGB**] studied the group LASSO for logistic regression. They showed that the group LASSO is consistent under the certain conditions, and they proposed a block coordinate descent algorithm that can handle high-dimensional data. Zou [**Z**] studied a one-step approach in nonconcave penalized likelihood methods in models with fixed p. This approach is closely related to the adaptive LASSO. Park and Hastie [**PH**] proposed an algorithm for computing the entire solution path of the L_1 regularized maximum likelihood estimates, which facilitates the choice of tuning parameter. This algorithm does both the shrinkage and variable selection due to the nature of the constraint region, which leads to exact zeros for some of the coefficients. It does not satisfy oracle properties i.e., it does not yield unbiased estimates [**FL**]. Zhu and Hastie [**ZH**] used L_2-penalized method for logistic regression to pursue classification in the context of microarray cancer studies with categorical outcomes. The above three methods have been extensively studied in literature; for example, [**RJ**, **WSW**, **HMZ**, **WL**, **YL**, **EHJT**], and others.

The rest of the paper is organized as follows. In Section 2, we present the multiple logistic regression model and derive the resulting maximum likelihood and other estimators. Section 3 is devoted to develop the asymptotic properties of the non-penalty estimators and their asymptotic distributional biases and risks. In Section 4, an extensive simulation study is conducted to compare the shrinkage estimators with the penalty estimators. A real data set from a low birth weight study is used to illustrate the proposed methodology in Section 5. Finally, the concluding remarks are given in Section 6.

2. Estimation strategies

Suppose that $y_1, y_2, \dots, y_n$ are independent binary response variables that take a value of 0 or 1, and that $\boldsymbol{x}_i = (x_{i1}, x_{i2}, \dots, x_{ip})'$ is a $p \times 1$ vector of predictors for the ith subject. Define $\pi(z) = \mathrm{e}^z/1+\mathrm{e}^z$. The logistic regression model assumes that

$$\mathrm{P}(y = 1 \mid \boldsymbol{x}_i) = \pi(\boldsymbol{x}_i'\boldsymbol{\beta}) = \frac{\exp(\boldsymbol{x}_i'\boldsymbol{\beta})}{1+\exp(\boldsymbol{x}_i'\boldsymbol{\beta})}, \quad 1 \le i \le n,$$

where $\boldsymbol{\beta}$ is a $p \times 1$ vector of regression parameters. The log-likelihood is given by

$$l(\boldsymbol{\beta}) = \sum_{i=1}^{n} \big[y_i \ln \pi(\boldsymbol{x}_i'\boldsymbol{\beta}) + (1-y_i)\ln\big(1-\pi(\boldsymbol{x}_i'\boldsymbol{\beta})\big)\big]. \tag{2.1}$$

The log-likelihood function depends on $\boldsymbol{\beta}$. The derivative of the log-likelihood with respect to $\boldsymbol{\beta}$ is obtained by using the chain rule:

$$\frac{\partial l}{\partial \boldsymbol{\beta}} = \sum_{i=1}^{n} [y_i - \pi(\boldsymbol{x}_i'\boldsymbol{\beta})]\boldsymbol{x}_i = 0. \tag{2.2}$$

2.1. The unrestricted and restricted maximum likelihood estimators. The unrestricted maximum likelihood estimator (UE) $\widehat{\boldsymbol{\beta}}$ of $\boldsymbol{\beta}$ is obtained by solving the score equation (2.2). This equation is non-linear in parameter $\boldsymbol{\beta}$. To solve this equation, we use an iterative procedure such as Newton–Raphson to determine the value of $\widehat{\boldsymbol{\beta}}$ of $\boldsymbol{\beta}$ that maximize the log-likelihood function $l(\boldsymbol{\beta})$.

Under the usual regularity conditions, it can be shown that $\widehat{\boldsymbol{\beta}}$ is consistent and asymptotically normal with variance-covariance matrix $\big(\boldsymbol{I}(\boldsymbol{\beta})\big)^{-1}$, where

$$\boldsymbol{I}(\boldsymbol{\beta}) = \sum_{i=1}^{n} \pi(\boldsymbol{x}_i'\boldsymbol{\beta})(1-\pi(\boldsymbol{x}_i'\boldsymbol{\beta}))\boldsymbol{x}_i\boldsymbol{x}_i'.$$

The matrix $\boldsymbol{I}(\widehat{\boldsymbol{\beta}})$ can be partitioned as

$$\begin{pmatrix} \boldsymbol{I}_{11} & \boldsymbol{I}_{12} \\ \boldsymbol{I}_{21} & \boldsymbol{I}_{22} \end{pmatrix},$$

when $\boldsymbol{\beta} = (\boldsymbol{\beta}_1', \boldsymbol{\beta}_2')'$.

The restricted maximum likelihood estimator (RE), $\widetilde{\boldsymbol{\beta}}$ of $\boldsymbol{\beta}$ can be obtained by maximizing the log-likelihood function (2.1) under the linear restriction $\boldsymbol{\beta}_2 = \boldsymbol{0}$.

2.2. The shrinkage and positive shrinkage estimators. The shrinkage estimator (SE) that shrinks the unrestricted estimator $\widehat{\boldsymbol{\beta}}$ towards $\widetilde{\boldsymbol{\beta}}$ is

$$\widehat{\boldsymbol{\beta}}^S = \widetilde{\boldsymbol{\beta}} + \left(1 - (p_2 - 2)D^{-1}\right)(\widehat{\boldsymbol{\beta}} - \widetilde{\boldsymbol{\beta}}), \quad p_2 \geq 3,$$

where D is the likelihood ratio statistic for testing $H_0 : \boldsymbol{\beta}_2 = \mathbf{0}$. If $l(\widehat{\boldsymbol{\beta}})$ and $l(\widetilde{\boldsymbol{\beta}})$ are the values of the log-likelihood at the unrestricted and restricted estimates respectively, then

$$\begin{aligned} D &= 2[l(\widehat{\boldsymbol{\beta}}; y_1, \ldots, y_n) - l(\widetilde{\boldsymbol{\beta}}; y_1, \ldots, y_n)], \\ &= n\widehat{\boldsymbol{\beta}}_2'(\boldsymbol{I}_{22} - \boldsymbol{I}_{21}\boldsymbol{I}_{11}^{-1}\boldsymbol{I}_{12})\widehat{\boldsymbol{\beta}}_2 + o_p(1). \end{aligned}$$

Under H_0, the distribution of D converges to a χ^2 distribution with p_2 degrees of freedom as $n \to \infty$.

The shrinkage estimator is not a convex combination of unrestricted and restricted estimators and also changes the sign of the unrestricted estimator. This could happen if $(p_2 - 2)D^{-1}$ is larger than unity. From the practical point of view, the change of sign would affect its interpretability. However, this behavior does not adversely affect the risk performance of SE. To overcome the sign problem, it is better to define a truncated estimator called a positive-part shrinkage estimator (PSE). This estimator can be defined as

$$\widehat{\boldsymbol{\beta}}^{S+} = \widetilde{\boldsymbol{\beta}} + (1 - (p_2 - 2)D^{-1})^+(\widehat{\boldsymbol{\beta}} - \widetilde{\boldsymbol{\beta}}),$$

where $z^+ = \max(0, z)$.

2.3. Penalty estimators. Tibshirani [**T**] proposed the LASSO method that uses an L_1 penalty under the log-likelihood setting, the objective function of LASSO is

$$\begin{aligned} \widehat{\boldsymbol{\beta}}_\lambda^{\text{LASSO}} &= \arg\min_{\boldsymbol{\beta}}\{-l(\boldsymbol{\beta}) + \lambda\|\boldsymbol{\beta}\|_1\} \\ &= \arg\min_{\boldsymbol{\beta}}\left[-\sum_{i=1}^{n}\Big(y_i \ln \pi(\boldsymbol{x}_i'\boldsymbol{\beta}) + (1 - y_i)\ln\big(1 - \pi(\boldsymbol{x}_i'\boldsymbol{\beta})\big)\Big) + \lambda\|\boldsymbol{\beta}\|_1\right], \end{aligned}$$

where the tuning parameter $\lambda > 0$ controls the strength of penalty, which shrinks the coefficients $\boldsymbol{\beta}$ towards zero vector. Hence, the LASSO produces sparse solutions. It has been shown that the LASSO penalty produces biases in the simple regression setting [**FL**] due to the linear increase of the penalty on the regression coefficients. To remedy this bias issue, the following two penalties were proposed recently.

Zou [**Z**] proposed an adaptive version of the LASSO method. It is essentially an L_1 penalization method where the degree of penalty is adaptively chosen. This method penalizes large weights to inactive variables, and thus heavily shrinks their associated coefficients. This is done by introducing the individual weights w_i in the following penalized likelihood function. The adaptive LASSO is the solution of

$$\widehat{\boldsymbol{\beta}}_\lambda^{\text{ALASSO}} = \arg\min_{\boldsymbol{\beta}}\left[-\sum_{i=1}^{n}\Big(y_i \ln \pi(\boldsymbol{x}_i'\boldsymbol{\beta}) + (1 - y_i)\ln\big(1 - \pi(\boldsymbol{x}_i'\boldsymbol{\beta})\big)\Big) + \lambda\sum_{i=1}^{k}|\beta_i|w_i\right],$$

where $w_i = |\widehat{\boldsymbol{\beta}}_i|^{-\gamma}$ for some positive γ. The adaptive LASSO is computationally easy to handle because the optimization problem is convex. It satisfies the oracle property, that is, as the sample size $n \to \infty$, it selects with probability one the correct regression model and the nonzero estimates are asymptotically normal with the same covariance matrix as if the nonzero coefficients were known a priori.

Fan and Li [**FL**] introduced the SCAD penalty method, which complies to the mathematical conditions of sparsity, continuity, and unbiasedness. This method selects variables and estimate parameters $\boldsymbol{\beta}$ simultaneously by minimizing the penalized likelihood function

$$\widehat{\boldsymbol{\beta}}_{\lambda}^{\mathrm{SCAD}} = \arg\min_{\boldsymbol{\beta}}\left[-\sum_{i=1}^{n}\Big(y_i \ln \pi(\boldsymbol{x}_i'\boldsymbol{\beta}) + (1-y_i)\ln\big(1-\pi(\boldsymbol{x}_i'\boldsymbol{\beta})\big)\Big) + \lambda\sum_{i=1}^{k} p_{\lambda}(|\beta_i|)\right],$$

where $p_\lambda(\cdot)$ is the smoothly clipped absolute deviation penalty with a tuning parameter λ. The penalty function $p_\lambda(\cdot)$ corresponds to a quadratic spline function with knots at λ and $a\lambda$, which leaves large values of the parameter $\boldsymbol{\beta}$ not excessively penalized and makes the solution continuous. It satisfies $p_\lambda(0) = 0$, and its first-order derivative

$$p_\lambda'(\theta) = \lambda\left[I(\theta \le \lambda) + \frac{(a\lambda - \theta)_+}{(a-1)\lambda} I(\theta > \lambda)\right],$$

where a is some constant usually taken to be 3.7 and $(t)_+ = tI\{t > 0\}$, see [**FL**].

In common among these penalization methods is the necessity of determining the value of the penalty parameter, λ, that minimizes the penalized likelihood function. We use cross-validation to determine the value of this penalty parameter λ. Specifically, k-fold cross-validation splits the available observations into a fixed number of set, k, each with a roughly equal number of observations. In the next step, it determines the prediction error in each set by using the observations in the $k-1$ of other sets to fit the regression model. We select the value of λ that gives the smallest prediction error.

The output of the above three penalty methods looks like a shrinkage method by both shrinking and deleting coefficients. However, it is different from the shrinkage estimation procedure in that it weighs all the predictor coefficients equally. It does not use a specified linear subspace with $\boldsymbol{\beta}_2 = \mathbf{0}$.

3. Asymptotic results and comparison

We consider the properties of the shrinkage estimators for a logistic regression model where the subspace $\boldsymbol{\beta}_2 = \mathbf{0}$ may be wrong, that is, $\boldsymbol{\beta}_2 = \mathbf{0} + \boldsymbol{d}$. Because the statistic D converges to ∞ for fixed $\boldsymbol{d} \neq \mathbf{0}$, the SE and PSE will be asymptotically equivalent in probability to $\widehat{\boldsymbol{\beta}}$ for such $\boldsymbol{d}$. This leads us to consider the usual local alternatives:

$$K_{(n)}\colon \boldsymbol{\beta}_2 = \frac{\boldsymbol{\delta}}{\sqrt{n}}, \tag{3.1}$$

where $\boldsymbol{\delta} = (\delta_1, \dots, \delta_{p_2}) \in \Re^{p_2}$ is a fixed vector. The vector $\boldsymbol{\delta}/n^{1/2}$ is a measure of how much the local alternatives $K_{(n)}$ differs from the subspace (1.1). For such local alternatives, the SE and PSE may not be asymptotically unbiased estimators of $\boldsymbol{\beta}$. With that in mind, we introduce the following loss function:

$$L(\boldsymbol{\beta}^*, \boldsymbol{\beta}; \boldsymbol{W}) = \big[\sqrt{n}(\boldsymbol{\beta}^* - \boldsymbol{\beta})\big]'\boldsymbol{W}\big[\sqrt{n}(\boldsymbol{\beta}^* - \boldsymbol{\beta})\big], \tag{3.2}$$

where $\boldsymbol{W}$ is a positive semidefinite weight matrix and $\boldsymbol{\beta}^*$ is any one of $\widehat{\boldsymbol{\beta}}$, $\widetilde{\boldsymbol{\beta}}$, $\widehat{\boldsymbol{\beta}}^S$, or $\widehat{\boldsymbol{\beta}}^{S+}$. A common choice of $\boldsymbol{W}$ is the identity matrix $\boldsymbol{I}$, which is used in the simulation study. Now we introduce the asymptotic distribution function of $\boldsymbol{\beta}^*$ under $K_{(n)}$ by

$$G(\boldsymbol{y}) = \lim_{n\to\infty} \mathrm{P}\big[\sqrt{n}(\boldsymbol{\beta}^* - \boldsymbol{\beta}) \le \boldsymbol{y} \mid K_{(n)}\big],$$

where $G(\boldsymbol{y})$ is a nondegenerate distribution function. Using the distribution of $G(\boldsymbol{y})$ and taking the expected value in both sides of (3.2), we get the expected loss that would be called the quadratic risk.

Now we define the *asymptotic distributional risk* (ADR) by

$$R(\boldsymbol{\beta}^*; \boldsymbol{W}) = \int \cdots \int \boldsymbol{y}'\boldsymbol{W}\boldsymbol{y}\,\mathrm{d}G(\boldsymbol{y}) = \operatorname{trace}(\boldsymbol{W}\boldsymbol{\Sigma}^*), \tag{3.3}$$

where $\boldsymbol{\Sigma}^* = \int \cdots \int \boldsymbol{y}\boldsymbol{y}'\,\mathrm{d}G(\boldsymbol{y})$ is the dispersion matrix for the distribution $G(\boldsymbol{y})$.

We define *asymptotic distributional bias* (ADB) of an estimator $\boldsymbol{\beta}^*$ as

$$\mathrm{ADB}(\boldsymbol{\beta}^*) = \lim_{n\to\infty} E\{n^{1/2}(\boldsymbol{\beta}^* - \boldsymbol{\beta})\} = \int \cdots \int \boldsymbol{y}\,\mathrm{d}G(\boldsymbol{y}),$$

where second equality can be established under our model assumptions.

Two key results to the study of ADR and ADB of the shrinkage estimators are given in the following theorem.

THEOREM 3.1. *Under the local alternatives $K_{(n)}$ in (3.1) and the usual regularity conditions, as $n \to \infty$,*

(1) $\sqrt{n}\widehat{\boldsymbol{\beta}}_2 \xrightarrow{L} N(\boldsymbol{\delta}, \boldsymbol{I}_{22.1})$

(2) *The test statistic D converges to a non-central chi-squared distribution $\chi^2_{p_2}(\Delta)$ with p_2 degrees of freedom and non-centrality parameter $\Delta = \boldsymbol{\delta}'\boldsymbol{I}_{22.1}\boldsymbol{\delta}$, where $\boldsymbol{I}_{22.1} = \boldsymbol{I}_{22} - \boldsymbol{I}_{21}\boldsymbol{I}_{11}^{-1}\boldsymbol{I}_{12}$ is a positive definite matrix.*

Using this theorem, we can obtain the main results of this section.

THEOREM 3.2. *Under the local alternatives $K_{(n)}$ and the condition of Theorem 3.1, the ADBs of the estimators are*

$$\begin{aligned}
\mathrm{ADB}(\widehat{\boldsymbol{\beta}}) &= \boldsymbol{0},\\
\mathrm{ADB}(\widetilde{\boldsymbol{\beta}}) &= \boldsymbol{M}\boldsymbol{\delta}, \quad \boldsymbol{M} = \boldsymbol{I}_{11}^{-1}\boldsymbol{I}_{12},\\
\mathrm{ADB}(\widehat{\boldsymbol{\beta}}^S) &= (p_2 - 2)\boldsymbol{M}\boldsymbol{\delta}E\big(\chi^{-2}_{p_2+2}(\Delta)\big),\\
\mathrm{ADB}(\widehat{\boldsymbol{\beta}}^{S+}) &= \mathrm{ADB}(\widehat{\boldsymbol{\beta}}^S) + \boldsymbol{M}\boldsymbol{\delta}\Psi_{p_2+2}(p_2 - 2, \Delta)\\
&\quad - (p_2 - 2)\boldsymbol{M}\boldsymbol{\delta}E\Big(\chi^{-2}_{p_2+2}(\Delta)I\big(\chi^2_{p_2+2}(\Delta) < (p_2 - 2)\big)\Big),
\end{aligned}$$

where the notation $\Psi_\nu(x, \Delta)$ is a non-central chi-square distribution function with ν degrees of freedom and non-centrality parameter Δ, and

$$E\big(x_\nu^{-2j}(\Delta)\big) = \int_0^\infty x^{-2j}\,\mathrm{d}\Psi_g(x, \Delta)$$

PROOF. See the Appendix. □

We now turn to the ADRs of the estimators.

THEOREM 3.3. *Under the local alternatives $K_{(n)}$ and the assumptions of Theorem 3.1, the ADRs of the estimators are*

$$R(\widehat{\boldsymbol{\beta}}; \boldsymbol{W}) = \operatorname{trace}[\boldsymbol{W}\boldsymbol{I}_{11.2}^{-1}],$$

$$
\begin{aligned}
R(\widetilde{\boldsymbol{\beta}};\boldsymbol{W}) &= R(\widehat{\boldsymbol{\beta}};\boldsymbol{W}) - \operatorname{trace}[\boldsymbol{I}_{22.1}^{-1}\boldsymbol{M}'\boldsymbol{W}\boldsymbol{M}] + \boldsymbol{\delta}'(\boldsymbol{M}'\boldsymbol{W}\boldsymbol{M})\boldsymbol{\delta},\\
R(\widehat{\boldsymbol{\beta}}^{S};\boldsymbol{W}) &= R(\widehat{\boldsymbol{\beta}};\boldsymbol{W}) + (p_2-2)\operatorname{trace}[\boldsymbol{I}_{22.1}^{-1}\boldsymbol{M}'\boldsymbol{W}\boldsymbol{M}]\big((p_2-2)E(Z_1^2) - 2E(Z_1)\big)\\
&\quad - (p_2-2)\boldsymbol{\delta}'(\boldsymbol{M}'\boldsymbol{W}\boldsymbol{M})\boldsymbol{\delta}[2E(Z_1) - (p_2-2)E(Z_2^2) + 2E(Z_2)],\\
R(\widehat{\boldsymbol{\beta}}^{S+};\boldsymbol{W}) &= R(\widehat{\boldsymbol{\beta}}^{S};\boldsymbol{W}) + 2\boldsymbol{\delta}'(\boldsymbol{M}'\boldsymbol{W}\boldsymbol{M})\boldsymbol{\delta}E\left[(1-(p_2-2)Z_1)I((p_2-2)Z_1>1)\right]\\
&\quad - \operatorname{trace}[\boldsymbol{I}_{22.1}^{-1}\boldsymbol{M}'\boldsymbol{W}\boldsymbol{M}]E[(1-(p_2-2)Z_1)^2 I((p_2-2)Z_1>1)]\\
&\quad - \boldsymbol{\delta}'(\boldsymbol{M}'\boldsymbol{W}\boldsymbol{M})\boldsymbol{\delta}E[(1-(p_2-2)Z_2)^2 I((p_2-2)Z_2>1)],
\end{aligned}
$$

where $Z_1 = \chi_{p_2+2}^{-2}(\Delta)$, $Z_2 = \chi_{p_2+4}^{-2}(\Delta)$, *and* $\boldsymbol{I}_{11.2} = \boldsymbol{I}_{11} - \boldsymbol{I}_{12}\boldsymbol{I}_{22}^{-1}\boldsymbol{I}_{21}$.

PROOF. See the Appendix. □

In order to elucidate the performance of the estimators reviewed in this section, we provide an extensive simulation study in the next section.

4. Simulation results

In this section, we have carried out a Monte Carlo simulation study to examine the risk (namely MSE) performance of the proposed estimators under consideration. Our simulation is based on a logistic regression model with (a) sample size $n = 250$ for the low dimensional setting ($n \geq p$); and (b) sample sizes $n = 10,\ 20$ for the high dimensional setting ($n \leq p$). A binary response is generated from the following model:

$$
\ln\left(\frac{p_i}{1-p_i}\right) = \eta_i = \boldsymbol{x}_i'\boldsymbol{\beta}, \quad i = 1, \ldots, n,
$$

where $p_i = \mathrm{P}(Y = 1 \mid x_i)$ and the predictor values $\boldsymbol{x}_i' = (x_{i1}, x_{i2}, \ldots, x_{in})$ have been drawn from a multivariate standard normal distribution.

Since we are interested in testing the hypothesis $H_0 : \boldsymbol{\beta}_2 = \mathbf{0}$ and so our objective is to estimate $\boldsymbol{\beta}_1$ here. We partition the regression coefficient vector as $\boldsymbol{\beta} = (\boldsymbol{\beta}_1', \boldsymbol{\beta}_2')' = (\boldsymbol{\beta}_1', \mathbf{0})$. We consider 24 predictors, of which only 4 are active, and consider different number p_2 of 20 inactive predictors. We set the true values of $\boldsymbol{\beta}$ at $\boldsymbol{\beta} = (\boldsymbol{\beta}_1', \mathbf{0})'$ with $\boldsymbol{\beta}_1 = (1.5, -1.5, 1.8, -1.9)$ and the weight matrix $\boldsymbol{W} = \boldsymbol{I}$. The number of replications was initially varied and each trial was repeated 2000 times to obtain stable results. We define the parameter

$$
\Delta = \left\|\boldsymbol{\beta} - \boldsymbol{\beta}^{(0)}\right\|^2,
$$

where $\boldsymbol{\beta}^{(0)} = (\boldsymbol{\beta}_1', \mathbf{0}')'$, $\boldsymbol{\beta}$ is the true parameter in the model and $\|\cdot\|$ is the Euclidian norm. Samples were generated using Δ between 0 and 1.41.

We evaluated the performance of the estimators by using a mean square error (MSE) relative to the MSE of $\widehat{\boldsymbol{\beta}}$. The simulated relative efficiency (SRE) of an estimator $\boldsymbol{\beta}^*$ to $\widehat{\boldsymbol{\beta}}$ is defined by

$$
\mathrm{SRE}(\widehat{\boldsymbol{\beta}} : \boldsymbol{\beta}^*) = \frac{\text{Simulated MSE}(\widehat{\boldsymbol{\beta}})}{\text{Simulated MSE}(\boldsymbol{\beta}^*)}.
$$

Here the amount SRE is larger than 1 which indicates how much more efficient $\boldsymbol{\beta}^*$ than $\widehat{\boldsymbol{\beta}}$.

TABLE 1. Simulated relative efficiencies of RE, SE, PSE, LASSO, adaptive LASSO, and SCAD with respect to $\widehat{\boldsymbol{\beta}}$ when the restricted parameter space $\boldsymbol{\beta}_2 = \mathbf{0}$. is correct ($\Delta = 0$).

	$n = 250$					
Method	$p_2 = 3$	$p_2 = 5$	$p_2 = 9$	$p_2 = 13$	$p_2 = 16$	$p_2 = 20$
RE	1.59	2.10	3.14	4.67	6.18	8.68
SE	1.15	1.42	2.04	2.69	3.18	3.77
PSE	1.17	1.47	2.11	2.77	3.28	3.88
LASSO	1.31	1.50	1.92	2.23	2.69	3.26
Adaptive LASSO	1.33	1.51	1.95	2.28	2.89	3.29
SCAD	1.44	1.77	2.34	2.92	3.14	3.56

4.1. Relative efficiency when $\beta_2 = 0$ is correct ($\Delta = 0$) and $n \geq p$. In this sub-section, where $\boldsymbol{\beta}_1 \neq \mathbf{0}$, $\boldsymbol{\beta}_2 = \mathbf{0}$, and $n \geq p$, the penalty estimators (LASSO, adaptive LASSO, and SCAD) can be expected to estimate $\boldsymbol{\beta}_1$ well by selecting λ so that many of the components of $\boldsymbol{\beta}_2$ will be set to zero. On the other hand, the shrinkage estimators can be expected to do well by placing almost all weights on the restricted estimator $\widetilde{\boldsymbol{\beta}}$. Here, we evaluate how these two methods work for low dimensional data.

In Table 1, we present the simulated relative efficiencies of restricted, shrinkage, and three penalty estimators with respect to the unrestricted estimator for $\Delta = 0$ and $n = 250$. We used the 10-fold cross validation method for choosing the best value of λ. Based on this λ, we computed penalty estimators.

First, we note that the relative efficiency of all the estimators increases as the number of inactive variables increases. Moreover, at $\Delta = 0$, as we would expect, the RE is the best and all the estimators perform better than the UE. Table 1 reveals that the penalty methods performs better than the shrinkage strategy when the number of inactive predictors p_2 in the model is small. On the other hand, the shrinkage estimators outshine the penalty estimators for larger values of p_2.

Generally speaking, in the presence of a relatively large number of inactive predictors in the model, the shrinkage strategy does well relative to the penalty estimators. However, the adaptive shrinkage estimators are preferable when the number of inactive predictors is relatively large.

4.2. Relative efficiencies when $\beta_2 = 0$ is correct and incorrect ($\Delta \geq 0$) and $n \geq p$. The penalty estimators are not included in the $\Delta \geq 0$ case, because these estimators do not take advantage of the fact that the regression parameter lies in a subspace $\boldsymbol{\beta}_2 = \mathbf{0}$. The shrinkage estimators are expected do well by adapting to the $\boldsymbol{\beta}_2 \neq \mathbf{0}$ case. Here we investigate how the shrinkage estimators improve on the unrestricted estimates for low dimensional data. The simulation results for the case where the active component of parameters $\boldsymbol{\beta}_1 = (1.5, -1.5, 1.8, -1.9)$ and the inactive component of parameters $\boldsymbol{\beta}_2 = (\beta_3, \boldsymbol{a})$, where β_3=$0, 0.55, 0.77, 0.89, 1.00, 1.10, 1.22$, and 1.41 and $\boldsymbol{a}$ is a zero vector with different dimensions, are presented in Figure 1 and Table 2. Note that there are $m = p_2 - 1$ inactive variables and that $\Delta = \beta_3^2$ in this framework. The findings are summarized as follows:

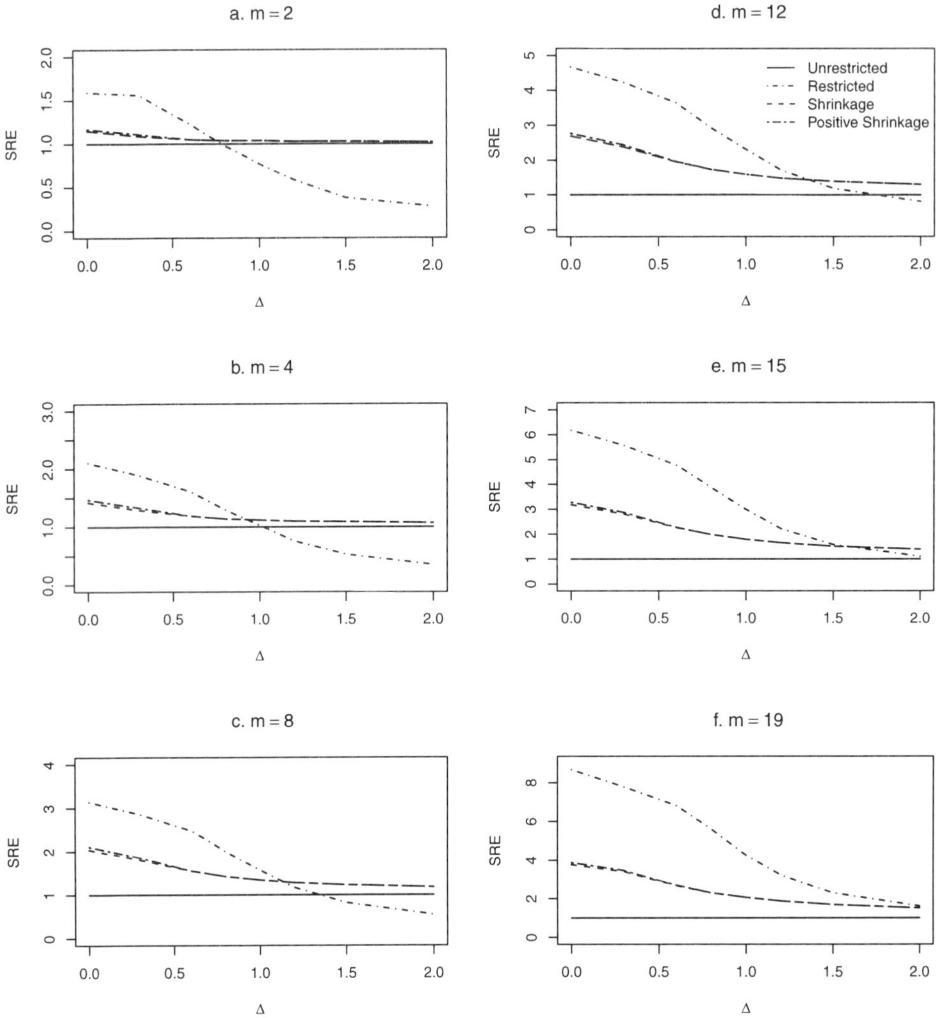

FIGURE 1. Simulated relative efficiency with respect to $\widehat{\boldsymbol{\beta}}$ of the estimators when the subspace misspecifies β_3 as zero as a function of $\Delta = \beta_3^2$. Here $m = p_2 - 1$ the number of inactive predictors and $n = 250$.

(i) The restricted estimator $\widetilde{\boldsymbol{\beta}}$ is better than all the estimators at and near $\Delta = 0$. On the contrary, as Δ gets larger than zero, the relative efficiency of RE decreases sharply, crosses the reference line at a certain level of Δ, and converges to zero as $\Delta \to \infty$ whereas the relative efficiency of all the other estimators remain bounded and approaches 1. This is in agreement with the asymptotic results of Section 3.

(ii) Keeping n and Δ fixed, the relative efficiency of the shrinkage estimator is an increasing function of the number of inactive variables m. For $m = 15$, the SRE of the PSE can be as large as 3.28 (Figure 1) for small Δ, indicating a remarkable performance of the PSE. The figures also reveal that for small Δ the

TABLE 2. Simulated MSE of RE, SE, and PSE with respect to $\widehat{\boldsymbol{\beta}}$ when the restricted parameter space is correct and incorrect ($\Delta \geq 0$) for $n = 250$ and $p_2 = 13$.

Δ	$\widetilde{\boldsymbol{\beta}}$	$\boldsymbol{\beta}^S$	$\boldsymbol{\beta}^{S+}$
0.00	4.67	2.69	2.77
0.30	4.23	2.38	2.44
0.60	3.64	1.95	1.96
0.80	2.92	1.73	1.73
1.00	2.31	1.59	1.59
1.20	1.71	1.48	1.48
1.50	1.19	1.39	1.39
2.00	0.81	1.30	1.30

TABLE 3. Simulated RMSE of adaptive LASSO and SCAD with respect to LASSO when $n \leq p$.

	$n = 10$		
Method	$p = 30$	$p = 40$	$p = 50$
Adaptive LASSO	1.29	1.34	1.43
SCAD	1.42	1.47	1.57
	$n = 20$		
Adaptive LASSO	1.10	1.18	1.27
SCAD	1.13	1.24	1.32

shrinkage estimator offers a large improvement in SRE as the number m of inactive predictors is increased.

4.3. Penalty estimators when $n \leq p$. In this sub-section, we consider the penalty estimators when the number of predictors is larger than the sample size ($n \leq p$). The standard maximum likelihood method for logistic regression is not applicable here for the high-dimensional data. For this reason, the shrinkage estimators are not included here. Here we investigate the relative MSE's of adaptive LASSO and SCAD with respect to LASSO when the sample sizes are $n = 10, 20$ and the predictors are $p = 30, 40, 50$.

Table 3 shows the relative MSE's of the adaptive LASSO and SCAD estimators when 4 predictor variables are active and all other predictors are inactive. The relative MSE's of adaptive LASSO and SCAD are larger than the LASSO estimator, which implies that the adaptive LASSO and SCAD outperform the LASSO estimator.

5. Example: low birth weight data

To further illustrate the proposed methodology, we consider the motivating example in the Introduction Section. The response variable is low birth weight ($0 =$ birth weight ≥ 2500 grams, $1 =$ birth weight < 2500 grams). The predictors variables are age of mother in years, weight of mother at last menstrual period, race ($1 =$ white, $2 =$ black, $3 =$ other), smoking status during pregnancy ($0 =$ no, $1 =$ yes), history of premature labor ($0 =$ none, $1 =$ one, etc.), history of hypertension

TABLE 4. Estimates (first row) and standard errors (second row) of the coefficients for the effect of history of premature labor (β_1), history of hypertension (β_2), weight of the mother at her last menstrual period (β_3), race (black) (β_4), smoking status (β_5) on the low birth weight. The SRE column gives the relative mean square error of the estimators with respect to the unrestricted maximum likelihood estimator.

Estimators	β_1	β_2	β_3	β_4	β_5	SRE
UE	−1.362	2.084	−0.017	−1.230	0.843	1.000
	0.546	1.535	0.009	0.638	0.486	
RE	−1.284	1.989	−0.0173	−1.034	0.934	2.178
	0.423	0.934	0.009	0.432	0.356	
SE	−1.341	2.063	−0.017	−1.180	0.869	1.110
	0.526	1.522	0.009	0.571	0.472	
PSE	−1.341	2.063	−0.017	−1.183	0.869	1.118
	0.527	1.522	0.009	0.567	0.471	
LASSO	−1.170	1.536	−0.012	−0.811	0.646	1.941
	0.493	0.849	0.008	0.546	0.429	
Adaptive LASSO	−1.177	1.555	−0.013	−0.821	0.656	1.958
	0.489	0.845	0.008	0.552	0.426	
SCAD	−1.131	1.483	−0.008	−0.863	0.614	2.116
	0.462	0.821	0.008	0.506	0.383	

(0 = no, 1 = yes), presence of uterine irritability (0 = no, 1 = yes), and number of physician visits (0 = none, 1 = 1).

We started analysis this data based on AIC and BIC criteria. Based on AIC, a sub model has been selected with five predictors: history of premature labor, history of hypertension, weight of the mother at her last menstrual period, race (black), and smoking status. That is, these predictors are the most important risk factors for the low birth weight, and the other 4 are inactive predictors for the low birth weight. Similarly we applied BIC criterion in this data set and obtained similar result as AIC. Based on this result, a natural linear subspace would omit 4 variables. In this example, our null hypothesis has $\boldsymbol{\beta}_2 = (\beta_6, \beta_7, \beta_8, \beta_9) = (0, 0, 0, 0)$, $p = 9$, $p_1 = 5$, $p_2 = 4$, and $n = 189$.

The point estimates, the standard errors, and relative efficiency based on a parametric bootstrap size of 1000 are displayed in Table 4. The findings in this table are consistent with the simulation results. Thus, the penalty estimators perform well compared to the shrinkage when the number of inactive variables is relatively small. On the other hand, the shrinkage estimators perform well when there are moderate or relatively large numbers of inactive predictors in the model.

However, the restricted estimator outperforms the penalty methods because the inactive predictors that are deleted in the sub-model are indeed irrelevant or nearly irrelevant for the response.

6. Discussion and conclusion

This paper have compared the shrinkage estimation method with the penalty method for a logistic regression model in low dimensional setting when it is suspected that some of the parameters are under linear constraint. We studied the risk properties of the shrinkage estimators via asymptotic distributional risk and a Monte Carlo simulation study. Our analytical and simulation results demonstrate that shrinkage estimators provide excellent prediction accuracy and outperform the maximum likelihood estimator uniformly. The properties of the penalty estimators were evaluated in terms of Monte Carlo simulation study and it was found that it is competitive when there are many inactive predictors in the model for $n \geq p$. On the other hand, the adaptive shrinkage estimator performs better when the number of inactive predictors is relatively large. In fact, if the shrinkage dimension is 3 or larger, the asymptotic risk of the shrinkage estimator is strictly less than that of the maximum likelihood estimator.

In the high-dimensional simulation setting, we compare the relative performance of adaptive LASSO and SCAD with the LASSO estimator. It shows improvement of the adaptive LASSO and SCAD over the LASSO.

Finally, we applied the shrinkage, LASSO, Adaptive LASSO, and SCAD to a real data set to evaluate the relative performance of the suggested estimators. These results corroborate with our analytical and simulated findings. The theoretical and numerical results can be extended to the entire class of generalized linear models. We are currently working on this project.

Appendix A. Proof of Theorems 3.2 and 3.3

To prove the theorems, we use $\boldsymbol{\beta} = (\boldsymbol{\beta}_1', \mathbf{0})'$. Under the local alternative $K_{(n)}$ and the assumed regularity conditions, the joint distributions are:

$$\text{(i)} \qquad \begin{pmatrix} \boldsymbol{\eta}_1 \\ \boldsymbol{\eta}_2 \end{pmatrix} \sim N_{p_1+p_2}\left[\begin{pmatrix} 0 \\ -\boldsymbol{M}\boldsymbol{\delta} \end{pmatrix}, \begin{pmatrix} \boldsymbol{I}_{11.2}^{-1} & \boldsymbol{M}\boldsymbol{I}_{22.1}^{-1}\boldsymbol{M}' \\ \boldsymbol{M}'\boldsymbol{I}_{22.1}^{-1}\boldsymbol{M} & \boldsymbol{M}\boldsymbol{I}_{22.1}^{-1}\boldsymbol{M}' \end{pmatrix}\right]$$

$$\text{(ii)} \qquad \begin{pmatrix} \boldsymbol{\eta}_2 \\ \boldsymbol{\eta}_3 \end{pmatrix} \sim N_{p_1+p_2}\left[\begin{pmatrix} -\boldsymbol{M}\boldsymbol{\delta} \\ \boldsymbol{M}\boldsymbol{\delta} \end{pmatrix}, \begin{pmatrix} \boldsymbol{M}\boldsymbol{I}_{22.1}^{-1}\boldsymbol{M}' & \mathbf{0} \\ \mathbf{0} & \boldsymbol{\Phi} \end{pmatrix}\right],$$

where $\boldsymbol{\eta}_1 = \sqrt{n}(\widehat{\boldsymbol{\beta}} - \boldsymbol{\beta})$, $\boldsymbol{\eta}_2 = \sqrt{n}(\widehat{\boldsymbol{\beta}} - \widetilde{\boldsymbol{\beta}})$, $\boldsymbol{\eta}_3 = \sqrt{n}(\widetilde{\boldsymbol{\beta}} - \boldsymbol{\beta})$, $\boldsymbol{M} = \boldsymbol{I}_{11}^{-1}\boldsymbol{I}_{12}$, and $\boldsymbol{\Phi} = \boldsymbol{I}_{11.2}^{-1} - \boldsymbol{I}_{11}^{-1}\boldsymbol{I}_{12}\boldsymbol{I}_{22.1}^{-1}\boldsymbol{I}_{21}\boldsymbol{I}_{11}^{-1}$.

Justification of (i) and (ii). Since $\boldsymbol{\eta}_2$ and $\boldsymbol{\eta}_3$ are linear functions of $\widehat{\boldsymbol{\beta}}$, they are also asymptotically normally distributed

$$\begin{aligned}
E(\boldsymbol{\eta}_1) &= E[\sqrt{n}(\widehat{\boldsymbol{\beta}} - \boldsymbol{\beta})] = \mathbf{0} \\
E(\boldsymbol{\eta}_3) &= E[\sqrt{n}(\widetilde{\boldsymbol{\beta}} - \boldsymbol{\beta})] \\
&= E[\sqrt{n}(\widehat{\boldsymbol{\beta}}_1 - \boldsymbol{\beta}_1 + \boldsymbol{I}_{11}^{-1}\boldsymbol{I}_{12}\widehat{\boldsymbol{\beta}}_2)], \quad \text{see } [\mathbf{LS}] \\
&= \mathbf{0} + \boldsymbol{I}_{11}^{-1}\boldsymbol{I}_{12} \cdot \sqrt{n}\frac{\boldsymbol{\delta}}{\sqrt{n}} = \boldsymbol{M}\boldsymbol{\delta} \\
E(\boldsymbol{\eta}_2) &= E[\sqrt{n}(\widehat{\boldsymbol{\beta}} - \widetilde{\boldsymbol{\beta}})] \\
&= E\big[\sqrt{n}\big((\widehat{\boldsymbol{\beta}} - \boldsymbol{\beta}) - (\widetilde{\boldsymbol{\beta}} - \boldsymbol{\beta})\big)\big] \\
&= \mathbf{0} - (\boldsymbol{M}\boldsymbol{\delta}) = -\boldsymbol{M}\boldsymbol{\delta}
\end{aligned}$$

$$
\begin{aligned}
\operatorname{Var}(\boldsymbol{\eta}_1) &= \boldsymbol{I}_{11.2}^{-1} \\
\operatorname{Var}(\boldsymbol{\eta}_3) &= \operatorname{Var}\big[\sqrt{n}(\widetilde{\boldsymbol{\beta}} - \boldsymbol{\beta})\big] \\
&= \operatorname{Var}\big[\sqrt{n}(\widehat{\boldsymbol{\beta}}_1 - \boldsymbol{\beta}_1) + \boldsymbol{I}_{11}^{-1}\boldsymbol{I}_{12}\widehat{\boldsymbol{\beta}}_2\big] \\
&= \operatorname{Var}\big[\sqrt{n}(\widehat{\boldsymbol{\beta}}_1 - \boldsymbol{\beta}_1)\big] + \boldsymbol{I}_{11}^{-1}\boldsymbol{I}_{12}\operatorname{Var}\big[\sqrt{n}\widehat{\boldsymbol{\beta}}_2\big]\boldsymbol{I}_{21}\boldsymbol{I}_{11}^{-1} \\
&\qquad\qquad + 2\operatorname{Cov}\big[\sqrt{n}(\widehat{\boldsymbol{\beta}}_1 - \boldsymbol{\beta}_1), \boldsymbol{I}_{11}^{-1}\boldsymbol{I}_{12}\sqrt{n}\widehat{\boldsymbol{\beta}}_2\big] \\
&= \boldsymbol{I}_{11.2}^{-1} + \boldsymbol{I}_{11}^{-1}\boldsymbol{I}_{12}\boldsymbol{I}_{22.1}^{-1}\boldsymbol{I}_{21}\boldsymbol{I}_{11}^{-1} + 2\operatorname{Cov}\big[\sqrt{n}(\widehat{\boldsymbol{\beta}}_1 - \boldsymbol{\beta}_1), \sqrt{n}\widehat{\boldsymbol{\beta}}_2\big](\boldsymbol{I}_{11}^{-1}\boldsymbol{I}_{12})' \\
&= \boldsymbol{I}_{11.2}^{-1} + \boldsymbol{I}_{11}^{-1}\boldsymbol{I}_{12}\boldsymbol{I}_{22.1}^{-1}\boldsymbol{I}_{21}\boldsymbol{I}_{11}^{-1} - 2\boldsymbol{I}_{11}^{-1}\boldsymbol{I}_{12}\boldsymbol{I}_{22.1}^{-1}\boldsymbol{I}_{21}\boldsymbol{I}_{11}^{-1} \\
&= \boldsymbol{I}_{11.2}^{-1} - \boldsymbol{I}_{11}^{-1}\boldsymbol{I}_{12}\boldsymbol{I}_{22.1}^{-1}\boldsymbol{I}_{21}\boldsymbol{I}_{11}^{-1} \\
&= \boldsymbol{I}_{11.2}^{-1} - \boldsymbol{M}\boldsymbol{I}_{22.1}^{-1}\boldsymbol{M}' = \boldsymbol{\Phi} \\
\operatorname{Var}(\boldsymbol{\eta}_2) &= \operatorname{Var}\big[\sqrt{n}(\widehat{\boldsymbol{\beta}} - \widetilde{\boldsymbol{\beta}})\big] \\
&= \operatorname{Var}\big[\sqrt{n}\big((\widehat{\boldsymbol{\beta}} - \boldsymbol{\beta}) - (\widetilde{\boldsymbol{\beta}} - \boldsymbol{\beta})\big)\big] \\
&= \operatorname{Var}\big[\boldsymbol{\eta}_1 - \boldsymbol{\eta}_3\big] \\
&= \operatorname{Var}\big[\boldsymbol{\eta}_1\big] + \operatorname{Var}\big[\boldsymbol{\eta}_3\big] - 2\operatorname{Cov}\big[\boldsymbol{\eta}_1, \boldsymbol{\eta}_3\big] \\
&= \boldsymbol{I}_{11.2}^{-1} + \boldsymbol{I}_{11.2}^{-1} - \boldsymbol{I}_{11}^{-1}\boldsymbol{I}_{12}\boldsymbol{I}_{22.1}^{-1}\boldsymbol{I}_{21}\boldsymbol{I}_{11}^{-1} - 2\boldsymbol{I}_{11.2}^{-1} + 2\boldsymbol{I}_{11}^{-1}\boldsymbol{I}_{12}\boldsymbol{I}_{22.1}^{-1}\boldsymbol{I}_{21}\boldsymbol{I}_{11}^{-1} \\
&= \boldsymbol{I}_{11}^{-1}\boldsymbol{I}_{12}\boldsymbol{I}_{22.1}^{-1}\boldsymbol{I}_{21}\boldsymbol{I}_{11}^{-1} \\
&= \boldsymbol{M}\boldsymbol{I}_{22.1}^{-1}\boldsymbol{M}' \\
\operatorname{Cov}(\boldsymbol{\eta}_1, \boldsymbol{\eta}_2) &= \operatorname{Cov}\big[\sqrt{n}(\widehat{\boldsymbol{\beta}} - \boldsymbol{\beta}), \sqrt{n}(\widehat{\boldsymbol{\beta}} - \widetilde{\boldsymbol{\beta}})\big] \\
&= \operatorname{Cov}\big[\sqrt{n}(\widehat{\boldsymbol{\beta}} - \boldsymbol{\beta}), \sqrt{n}\big((\widehat{\boldsymbol{\beta}} - \boldsymbol{\beta}) - (\widetilde{\boldsymbol{\beta}} - \boldsymbol{\beta})\big)\big] \\
&= \operatorname{Var}\big[\sqrt{n}(\widehat{\boldsymbol{\beta}} - \boldsymbol{\beta})\big] - \operatorname{Cov}\big[\sqrt{n}(\widehat{\boldsymbol{\beta}} - \boldsymbol{\beta}), \sqrt{n}(\widetilde{\boldsymbol{\beta}} - \boldsymbol{\beta})\big] \\
&= \boldsymbol{I}_{11.2}^{-1} + \operatorname{Cov}\big[\sqrt{n}(\widehat{\boldsymbol{\beta}}_1 - \boldsymbol{\beta}_1), \boldsymbol{I}_{11}^{-1}\boldsymbol{I}_{12}\sqrt{n}\widehat{\boldsymbol{\beta}}_2\big] \\
&= \boldsymbol{I}_{11.2}^{-1} - \boldsymbol{I}_{11.2}^{-1} + \operatorname{Cov}\big[\sqrt{n}(\widehat{\boldsymbol{\beta}}_1 - \boldsymbol{\beta}_1), \sqrt{n}\widehat{\boldsymbol{\beta}}_2\big](\boldsymbol{I}_{11}^{-1}\boldsymbol{I}_{12})' \\
&= \boldsymbol{I}_{11}^{-1}\boldsymbol{I}_{12}\boldsymbol{I}_{22.1}^{-1}\boldsymbol{I}_{21}\boldsymbol{I}_{11}^{-1} \\
&= \boldsymbol{M}\boldsymbol{I}_{22.1}^{-1}\boldsymbol{M}' \\
\operatorname{Cov}(\boldsymbol{\eta}_1, \boldsymbol{\eta}_3) &= \operatorname{Cov}\big[\sqrt{n}(\widehat{\boldsymbol{\beta}} - \boldsymbol{\beta}), \sqrt{n}(\widetilde{\boldsymbol{\beta}} - \boldsymbol{\beta})\big] \\
&= \boldsymbol{I}_{11.2}^{-1} - \boldsymbol{M}\boldsymbol{I}_{22.1}^{-1}\boldsymbol{M}' \\
\operatorname{Cov}(\boldsymbol{\eta}_2, \boldsymbol{\eta}_3) &= \operatorname{Cov}\big[\sqrt{n}(\widehat{\boldsymbol{\beta}} - \widetilde{\boldsymbol{\beta}}), \sqrt{n}(\widetilde{\boldsymbol{\beta}} - \boldsymbol{\beta})\big] \\
&= \operatorname{Cov}\big[\sqrt{n}((\widehat{\boldsymbol{\beta}} - \boldsymbol{\beta}) - (\widetilde{\boldsymbol{\beta}} - \boldsymbol{\beta})), \sqrt{n}(\widetilde{\boldsymbol{\beta}} - \boldsymbol{\beta})\big] \\
&= \operatorname{Cov}\big[\sqrt{n}((\widehat{\boldsymbol{\beta}} - \boldsymbol{\beta}), \sqrt{n}(\widetilde{\boldsymbol{\beta}} - \boldsymbol{\beta})\big] - \operatorname{Var}[\sqrt{n}(\widetilde{\boldsymbol{\beta}} - \boldsymbol{\beta})\big] \\
&= \boldsymbol{I}_{11.2}^{-1} - \boldsymbol{I}_{11}^{-1}\boldsymbol{I}_{12}\boldsymbol{I}_{22.1}^{-1}\boldsymbol{I}_{21}\boldsymbol{I}_{11}^{-1} - \boldsymbol{I}_{11.2}^{-1} + \boldsymbol{I}_{11}^{-1}\boldsymbol{I}_{12}\boldsymbol{I}_{22.1}^{-1}\boldsymbol{I}_{21}\boldsymbol{I}_{11}^{-1} \\
&= \boldsymbol{0}.
\end{aligned}
$$

We present the following lemma below, which will help us to derive the results of Theorems 3.2 and 3.3.

LEMMA. *Let $\boldsymbol{x}$ be an p_2 vector that follows normal distribution with mean $\boldsymbol{\mu}_x$ vector and covariance matrix $\boldsymbol{\Sigma}_{p_2}$ as $\boldsymbol{x} \sim N_{p_2}(\boldsymbol{\mu}_x, \boldsymbol{\Sigma}_{p_2})$. Then, for a measurable* *function of* ϕ, *we have*

$$E[\boldsymbol{x}\phi(\boldsymbol{x}'\boldsymbol{x})] = \boldsymbol{\mu}_x E\big[\phi\big(\chi^2_{p_2+2}(\Delta)\big)\big]$$
$$E[\boldsymbol{x}\boldsymbol{x}'\phi(\boldsymbol{x}'\boldsymbol{x})] = \boldsymbol{\Sigma}_{p_2} E\big[\phi\big(\chi^2_{p_2+2}(\Delta)\big)\big] + \boldsymbol{\mu}_x\boldsymbol{\mu}_x' E\big[\phi\big(\chi^2_{p_2+4}(\Delta)\big)\big],$$

where $\Delta = \boldsymbol{\mu}_x'\boldsymbol{\Sigma}_{p_2}^{-1}\boldsymbol{\mu}_x$.

The proof can be found in [**JB**].

PROOF OF THEOREM 3.2. Here, we provide the proof of the bias expressions. It is obvious that $\text{ADB}(\widehat{\boldsymbol{\beta}}) = 0$. The ADB of the reduced, shrinkage, and positive shrinkage estimators are as follows:

$$\begin{aligned}
\text{ADB}(\widehat{\boldsymbol{\beta}}) &= E\Big[\lim_{n\to\infty}\sqrt{n}(\widehat{\boldsymbol{\beta}}-\boldsymbol{\beta})\Big] = \mathbf{0}\\
\text{ADB}(\widetilde{\boldsymbol{\beta}}) &= E\left[\lim_{n\to\infty}\sqrt{n}(\widetilde{\boldsymbol{\beta}}-\boldsymbol{\beta})\right]\\
&= E\Big[\lim_{n\to\infty}\sqrt{n}(\widehat{\boldsymbol{\beta}}_1-\boldsymbol{\beta}_1)\Big] + \boldsymbol{I}_{11}^{-1}\boldsymbol{I}_{12}\sqrt{n}\boldsymbol{\delta}/\sqrt{n})\\
&= \boldsymbol{I}_{11}^{-1}\boldsymbol{I}_{12}\boldsymbol{\delta} = \boldsymbol{M}\boldsymbol{\delta}\\
\text{ADB}(\widehat{\boldsymbol{\beta}}^S) &= E\Big[\lim_{n\to\infty}\sqrt{n}(\widehat{\boldsymbol{\beta}}^S-\boldsymbol{\beta})\Big]\\
&= E\Big[\lim_{n\to\infty}\big(\sqrt{n}(\widehat{\boldsymbol{\beta}}-\boldsymbol{\beta}) - (\widehat{\boldsymbol{\beta}}-\widetilde{\boldsymbol{\beta}})\big)(p_2-2)D_n^{-1}\Big]\\
&= \mathbf{0} - E\Big[\lim_{n\to\infty}\big(\sqrt{n}(\widehat{\boldsymbol{\beta}}-\boldsymbol{\beta}) - \sqrt{n}(\widetilde{\boldsymbol{\beta}}-\boldsymbol{\beta})\big)(p_2-2)D_n^{-1}\Big]\\
&= (p_2-2)E[\boldsymbol{\eta}_3 D_n^{-1}]\\
&= (p_2-2)\boldsymbol{M}\boldsymbol{\delta}E(Z_1), \quad \text{where } Z_1 = \chi^{-2}_{p_2+2}(\Delta).\\
\text{ADB}(\widehat{\boldsymbol{\beta}}^{S+}) &= E\Big[\lim_{n\to\infty}\sqrt{n}(\widehat{\boldsymbol{\beta}}^{S+}-\boldsymbol{\beta})\Big]\\
&= E\Big[\lim_{n\to\infty}\sqrt{n}(\widehat{\boldsymbol{\beta}}^S-\boldsymbol{\beta})\\
&\qquad - \lim_{n\to\infty}\sqrt{n}(\widehat{\boldsymbol{\beta}}-\widetilde{\boldsymbol{\beta}})(1-(p_2-2)D_n^{-1})I(D_n < (p_2-2)\Big]\\
&= \text{ABD}(\widehat{\boldsymbol{\beta}}^S) - E[\boldsymbol{\eta}_2(1-(p_2-2)D_n^{-1})I(D_n < (p_2-2)]\\
&= \text{ABD}(\widehat{\boldsymbol{\beta}}^S) + \boldsymbol{M}\boldsymbol{\delta}E[(1-(p_2-2)Z_1)I((p_2-2)Z_1 > 1)]\\
&= \text{ABD}(\widehat{\boldsymbol{\beta}}^S)\\
&\qquad + \boldsymbol{M}\boldsymbol{\delta}\big[\Psi_{(p_2+2)}\big((p_2-2),\Delta\big) - (p_2-2)E\big(Z_1 I(Z_1>1)\big)\big]. \quad \square
\end{aligned}$$

PROOF OF THEOREM 3.3. In order to prove this theorem, we first derive the asymptotic covariance matrices for the four estimators. The covariance matrix $\boldsymbol{\Sigma}^*(\widehat{\boldsymbol{\beta}}^*)$ of any estimator $\boldsymbol{\beta}^*$ is defined as:

$$\boldsymbol{\Sigma}^*(\boldsymbol{\beta}^*) = E\left[\lim_{n\to\infty} n(\boldsymbol{\beta}^*-\boldsymbol{\beta}))(\boldsymbol{\beta}^*-\boldsymbol{\beta})'\right].$$

First, we derive the covariance matrices of UE and RE:

$$\begin{aligned}
\boldsymbol{\Sigma}^*(\widehat{\boldsymbol{\beta}}) &= E\Big[\lim_{n\to\infty}\sqrt{n}(\widehat{\boldsymbol{\beta}}-\boldsymbol{\beta})\sqrt{n}(\widehat{\boldsymbol{\beta}}-\boldsymbol{\beta})'\Big]\\
&= E(\boldsymbol{\eta}_1\boldsymbol{\eta}_1') = \text{Var}(\boldsymbol{\eta}_1) + E(\boldsymbol{\eta}_1)E(\boldsymbol{\eta}_1')\\
&= \text{Var}(\boldsymbol{\eta}_1) = \boldsymbol{I}_{11.2}^{-1}
\end{aligned}$$

$$\begin{aligned}\mathbf{\Sigma}^*(\widetilde{\boldsymbol{\beta}}) &= E\Big[\lim_{n\to\infty}\sqrt{n}(\widetilde{\boldsymbol{\beta}}-\boldsymbol{\beta})\sqrt{n}(\widetilde{\boldsymbol{\beta}}-\boldsymbol{\beta})'\Big] \\ &= E(\boldsymbol{\eta}_3\boldsymbol{\eta}_3') = \mathrm{Var}(\boldsymbol{\eta}_3) + E(\boldsymbol{\eta}_3)E(\boldsymbol{\eta}_3') \\ &= \mathbf{\Phi} + \boldsymbol{M}\boldsymbol{\delta}\delta'\boldsymbol{M}'.\end{aligned}$$

Second, we derive the covariance matrices of the shrinkage estimators:

$$\begin{aligned}\mathbf{\Sigma}^*(\widehat{\boldsymbol{\beta}}^S) &= E\Big[\lim_{n\to\infty}\sqrt{n}(\widehat{\boldsymbol{\beta}}^S-\boldsymbol{\beta})\sqrt{n}(\widehat{\boldsymbol{\beta}}^S-\boldsymbol{\beta})'\Big] \\ &= E\Big[\lim_{n\to\infty}\sqrt{n}\big(\widehat{\boldsymbol{\beta}}-\boldsymbol{\beta}+(p_2-2)D_n^{-1}(\widehat{\boldsymbol{\beta}}-\widetilde{\boldsymbol{\beta}})\big) \\ &\qquad\qquad \times\sqrt{n}\big(\widehat{\boldsymbol{\beta}}-\boldsymbol{\beta}+(p_2-2)D_n^{-1}(\widehat{\boldsymbol{\beta}}-\widetilde{\boldsymbol{\beta}})\big)'\Big] \\ &= E\Big[\lim_{n\to\infty}(\widehat{\boldsymbol{\beta}}-\boldsymbol{\beta})(\widehat{\boldsymbol{\beta}}-\boldsymbol{\beta})' - 2(p_2-2)D_n^{-1}(\widehat{\boldsymbol{\beta}}-\widetilde{\boldsymbol{\beta}})(\widehat{\boldsymbol{\beta}}-\boldsymbol{\beta})' \\ &\qquad\qquad + (p_2-2)^2D_n^{-2}(\widehat{\boldsymbol{\beta}}-\widetilde{\boldsymbol{\beta}})(\widehat{\boldsymbol{\beta}}-\widetilde{\boldsymbol{\beta}})'\Big] \\ &= E[\boldsymbol{\eta}_1\boldsymbol{\eta}_1' - 2(p_2-2)D_n^{-1}\boldsymbol{\eta}_2\boldsymbol{\eta}_1' + (p_2-2)^2D_n^{-2}\boldsymbol{\eta}_2\boldsymbol{\eta}_2'].\end{aligned}$$

Using the conditional mean of bivariate normal, the second term of $\mathbf{\Sigma}^*(\widehat{\boldsymbol{\beta}}^S)$ without $-2(p_2-2)$ is equal to

$$\begin{aligned}E[\boldsymbol{\eta}_2\boldsymbol{\eta}_1'D_n^{-1}] &= E[E(\boldsymbol{\eta}_2\boldsymbol{\eta}_1'D_n^{-1}\mid\boldsymbol{\eta}_2)] \\ &= E[\boldsymbol{\eta}_2E(\boldsymbol{\eta}_1'D_n^{-1}\mid\boldsymbol{\eta}_2)] \\ &= E\big[\boldsymbol{\eta}_2\big(E(\boldsymbol{\eta}_1)+(\boldsymbol{\eta}_2-\boldsymbol{M}\boldsymbol{\delta})\big)'D_n^{-1}\big] \\ &= E[\boldsymbol{\eta}_2(\boldsymbol{\eta}_2-\boldsymbol{M}\boldsymbol{\delta})'D_n^{-1}] \\ &= E[\boldsymbol{\eta}_2\boldsymbol{\eta}_2'D_n^{-1}] - E[\boldsymbol{\eta}_2D_n^{-1}]\boldsymbol{\delta}'\boldsymbol{M}' \\ &= \mathrm{Var}(\boldsymbol{\eta}_2)E(Z_1) + E(\boldsymbol{\eta}_2)E(\boldsymbol{\eta}_2)'E(Z_2) - E(\boldsymbol{\eta}_2)\boldsymbol{\delta}'\boldsymbol{M}'E(Z_1) \\ &= \boldsymbol{M}\boldsymbol{I}_{22.1}^{-1}\boldsymbol{M}'E(Z_1) + \boldsymbol{M}\boldsymbol{\delta}\delta'\boldsymbol{M}'E(Z_2) + \boldsymbol{M}\boldsymbol{\delta}\delta'\boldsymbol{M}'E(Z_1),\end{aligned}$$

where $Z_2 = \chi_{p_2+4}^{-2}(\Delta)$. Therefore,

$$\begin{aligned}\mathbf{\Sigma}^*(\widehat{\boldsymbol{\beta}}^S) &= \boldsymbol{I}_{11.2}^{-1} - 2(p_2-2)[\boldsymbol{M}\boldsymbol{I}_{22.1}^{-1}\boldsymbol{M}'E(Z_1) \\ &\qquad + \boldsymbol{M}\boldsymbol{\delta}\delta'\boldsymbol{M}'E(Z_2) - \boldsymbol{M}\boldsymbol{\delta}\delta'\boldsymbol{M}'E(Z_1)] \\ &\qquad + (p_2-2)^2[\boldsymbol{M}\boldsymbol{I}_{22.1}^{-1}\boldsymbol{M}'E(Z_1^2) + \boldsymbol{M}\boldsymbol{\delta}\delta'\boldsymbol{M}'E(Z_2^2)] \\ &= \boldsymbol{I}_{11.2}^{-1} + (p_2-2)\boldsymbol{M}\boldsymbol{I}_{22.1}^{-1}\boldsymbol{M}'[(p_2-2)E(Z_1^2) - 2E(Z_1)] \\ &\qquad - (p_2-2)\boldsymbol{M}\boldsymbol{\delta}\delta'\boldsymbol{M}'[2E(Z_1) - (p_2-2)E(Z_2^2) + 2E(Z_2)].\end{aligned}$$

Again,

$$\begin{aligned}&\mathbf{\Sigma}^*(\widehat{\boldsymbol{\beta}}^{S+}) \\ &\quad = E\Big[\lim_{n\to\infty}\sqrt{n}(\widehat{\boldsymbol{\beta}}^{S+}-\boldsymbol{\beta})\sqrt{n}(\widehat{\boldsymbol{\beta}}^{S+}-\boldsymbol{\beta})'\Big] \\ &\quad = \mathbf{\Sigma}^*(\widehat{\boldsymbol{\beta}}^S) - 2E\Big[\lim_{n\to\infty}n\Big((\widehat{\boldsymbol{\beta}}-\widetilde{\boldsymbol{\beta}})(\widetilde{\boldsymbol{\beta}}-\boldsymbol{\beta})'(1-(p_2-2)D_n^{-1})I(D_n<(p_2-2))\Big)\Big] \\ &\qquad\quad - E\Big[\lim_{n\to\infty}n(\widehat{\boldsymbol{\beta}}-\widetilde{\boldsymbol{\beta}})(\widehat{\boldsymbol{\beta}}-\widetilde{\boldsymbol{\beta}})'(1-(p_2-2)D_n^{-1})^2I(D_n<(p_2-2))\Big] \\ &\quad = \mathbf{\Sigma}^*(\widehat{\boldsymbol{\beta}}^S) - 2E\big[\boldsymbol{\eta}_2\boldsymbol{\eta}_3'(1-(p_2-2)D_n^{-1})I\big(D_n<(p_2-2)\big)\big] \\ &\qquad\quad - E\big[\boldsymbol{\eta}_2\boldsymbol{\eta}_2'(1-(p_2-2)D_n^{-1})^2I\big(D_n<(p_2-2)\big)\big].\end{aligned}$$

Consider the second term without -2 and use the rule of conditional expectation

$$\begin{aligned}
&E\big[\boldsymbol{\eta}_2\boldsymbol{\eta}_3'(1-(p_2-2)D_n^{-1})I\big(D_n<(p_2-2)\big)\big]\\
&\quad= E\big[\boldsymbol{\eta}_2 E\big(\boldsymbol{\eta}_3'(1-(p_2-2)D_n^{-1})I\big(D_n<(p_2-2)\big)\mid \boldsymbol{\eta}_2\big)\big]\\
&\quad= E\big[\boldsymbol{\eta}_2\big((\boldsymbol{M\delta})'+\underbrace{\mathrm{Cov}(\boldsymbol{\eta}_2,\boldsymbol{\eta}_3)}_{\mathbf{0}}\cdot\boldsymbol{\Phi}\cdot(\boldsymbol{\eta}_2-\boldsymbol{M\delta})\big)'\\
&\qquad\qquad\times(1-(p_2-2)D_n^{-1})I\big(D_n<(p_2-2)\big)\big]\\
&\quad= E\big[\boldsymbol{\eta}_2(\boldsymbol{M\delta})'(1-(p_2-2)D_n^{-1})I\big(D_n<(p_2-2)\big)\big]\\
&\quad= -\boldsymbol{M\delta}\delta'\boldsymbol{M}'E\big[(1-(p_2-2)Z_1)I\big((p_2-2)Z_1>1\big)\big].
\end{aligned}$$

Therefore,

$$\begin{aligned}
\boldsymbol{\Sigma}^*(\widehat{\boldsymbol{\beta}}^{S+}) = \boldsymbol{\Sigma}^*(\widehat{\boldsymbol{\beta}}^{S}) &+ 2\boldsymbol{M\delta}\delta'\boldsymbol{M}'E\big[(1-(p_2-2)Z_1)I\big((p_2-2)Z_1>1\big)\big]\\
&- \boldsymbol{M}\boldsymbol{I}_{22.1}^{-1}\boldsymbol{M}'E\big[(1-(p_2-2)Z_1)^2 I\big((p_2-2)Z_1>1\big)\big]\\
&- \boldsymbol{M\delta}\delta'\boldsymbol{M}'E\big[(1-(p_2-2)Z_2)^2 I\big((p_2-2)Z_2>1\big)\big].
\end{aligned}$$

The proof of Theorem 3.3 now follows using (3.3) and the above covariance matrices. □

References

[ADHY] S. E. Ahmed, K. A. Doksum, S. Hossain, and J. You, *Shrinkage, pretest and absolute penalty estimators in partially linear models*, Aust. N. Z. J. Stat. **49** (2007), no. 4, 435–454, DOI 10.1111/j.1467-842X.2007.00493.x. MR2394591 (2009b:62139)

[AHD] S. E. Ahmed, S. Hossain, and K. A. Doksum, *LASSO and shrinkage estimation in Weibull censored regression models*, J. Statist. Plann. Inference **142** (2012), no. 6, 1273–1284, DOI 10.1016/j.jspi.2011.12.027. MR2891480

[AHS] S. E. Ahmed, A. A. Hussein, and P. K. Sen, *Risk comparison of some shrinkage M-estimators in linear models*, J. Nonparametr. Stat. **18** (2006), no. 4-6, 401–415, DOI 10.1080/10485250601046752. MR2284191

[EHJT] B. Efron, T. Hastie, I. Johnstone, and R. Tibshirani, *Least angle regression*, Ann. Statist. **32** (2004), no. 2, 407–499, DOI 10.1214/009053604000000067. With discussion, and a rejoinder by the authors. MR2060166 (2005d:62116)

[FL] J. Fan and R. Li, *Variable selection via nonconcave penalized likelihood and its oracle properties*, J. Amer. Statist. Assoc. **96** (2001), no. 456, 1348–1360, DOI 10.1198/016214501753382273. MR1946581 (2003k:62160)

[H] J. M. Hilbe, *Logistic regression models*, Chapman & Hall/CRC Texts in Statistical Science Series, CRC Press, Boca Raton, FL, 2009. MR2554479 (2010i:62001)

[HDEA] S. Hossain, K. A. Doksum, and S. Ejaz Ahmed, *Positive shrinkage, improved pretest and absolute penalty estimators in partially linear models*, Linear Algebra Appl. **430** (2009), no. 10, 2749–2761, DOI 10.1016/j.laa.2008.12.015. MR2509855 (2010f:62190)

[HL] D. W. Hosmer and S. Lemeshow, *Applied logistic regression*, 2nd ed., Wiley, New York, 2000.

[HMZ] J. Huang, S. Ma, and C.-H. Zhang, *Adaptive Lasso for sparse high-dimensional regression models*, Statist. Sinica **18** (2008), no. 4, 1603–1618. MR2469326 (2010a:62214)

[JB] G. G. Judge and M. E. Bock, *The statistical implications of pre-test and Stein-rule estimators in econometrics*, North-Holland Publishing Co., Amsterdam, 1978. Studies in Mathematical and Managerial Economics, Vol. 25. MR0483199 (58 #3217)

[LS] J. F. Lawless and K. Singhal, *Efficient screening of nonnormal regression models*, Biometrics **34** (1978), no. 2, 318–327.

[MvdGB] L. Meier, S. van de Geer, and P. Bühlmann, *The group Lasso for logistic regression*, J. R. Stat. Soc. Ser. B Stat. Methodol. **70** (2008), no. 1, 53–71, DOI 10.1111/j.1467-9868.2007.00627.x. MR2412631

[PH] M. Y. Park and T. Hastie, *L_1-regularization path algorithm for generalized linear models*, J. R. Stat. Soc. Ser. B Stat. Methodol. **69** (2007), no. 4, 659–677, DOI 10.1111/j.1467-9868.2007.00607.x. MR2370074

[RJ] P. Radchenko and G. M. James, *Improved variable selection with forward-Lasso adaptive shrinkage*, Ann. Appl. Stat. **5** (2011), no. 1, 427–448, DOI 10.1214/10-AOAS375. MR2810404 (2012d:62225)

[T] R. Tibshirani, *Regression shrinkage and selection via the lasso*, J. Roy. Statist. Soc. Ser. B **58** (1996), no. 1, 267–288. MR1379242 (96j:62134)

[WL] H. Wang and C. Leng, *Unified LASSO estimation by least squares approximation*, J. Amer. Statist. Assoc. **102** (2007), no. 479, 1039–1048, DOI 10.1198/016214507000000509. MR2411663

[WSW] M. Wang, L. Song, and X. Wang, *Bridge estimation for generalized linear models with a diverging number of parameters*, Statist. Probab. Lett. **80** (2010), no. 21-22, 1584–1596, DOI 10.1016/j.spl.2010.06.012. MR2684004 (2011e:62212)

[YL] M. Yuan and Y. Lin, *Model selection and estimation in regression with grouped variables*, J. R. Stat. Soc. Ser. B Stat. Methodol. **68** (2006), no. 1, 49–67, DOI 10.1111/j.1467-9868.2005.00532.x. MR2212574

[Z] H. Zou, *The adaptive lasso and its oracle properties*, J. Amer. Statist. Assoc. **101** (2006), no. 476, 1418–1429, DOI 10.1198/016214506000000735. MR2279469 (2008d:62024)

[ZH] J. Zhu and T. Hastie, *Classification of gene microarrays by penalized logistic regression*, Biostatistics **5** (2004), no. 3, 427–444, DOI 10.1093/biostatistics/kxg046.

Department of Mathematics and Statistics, University of Winnipeg, Winnipeg, MB R3B 2E9, Canada

E-mail address: `sh.hossain@uwinnipeg.ca`

Department of Mathematics, Brock University, St. Catharines, ON L2S 3A1, Canada

E-mail address: `sahmed5@brocku.ca`

Contemporary Mathematics
Volume **622**, 2014
http://dx.doi.org/10.1090/conm/622/12429

Manifold unfolding by Isometric Patch Alignment with an application in protein structure determination

Pooyan Khajehpour Tadavani, Babak Alipanahi, and Ali Ghodsi

ABSTRACT. We propose Isometric Patch Alignment (IPA), a novel dimensionality reduction method which has a low computational cost. This method is inspired by the observation that, reasonably large chunks of a high-dimensional dataset can be approximated by low-dimensional patches over its underlying manifold. Thus, in our proposed approach, first the input data is clustered, such that the neighboring clusters have some overlap. Then, each cluster is modeled by a low-dimensional patch on the underlying manifold of the data. Conceptually, the input is now reduced to a set of patches. IPA rearranges these patches such that the shared points between the neighboring ones match accordingly. We show that this rearrangement can be computed by solving a relatively small semi-definite program. In addition, we prove that attaching two neighboring patches on their overlapping parts aligns them together. Based on this, the manifold resulted from the convex optimization of IPA will be unfolded and thus, can be represented in a low-dimensional subspace. The proposed method is stable, fast, and scalable; our experimental results demonstrate its capability for dimensionality reduction and data visualization, particularly on input datasets with complex structures, for example where the underlying manifold is not convex. Moreover, as a distinct application, IPA is employed for protein structure determination.

1. Introduction

The past decade has witnessed enormous emphasis on the need to analyze large volumes of high-dimensional data. Despite the best efforts of industrial and academic research, the 'curse of dimensionality' continues to severely challenge machine learning and data mining algorithms. Thus there is sustained interest in dimensionality reduction methods insightful enough to discover meaningful nonlinear relationships in the data, and scalable enough to handle very large datasets.

Many such dimensionality reduction algorithms have been proposed, balancing various motivations, tradeoffs, and limitations. This paper proposes *Isometric Patch Alignment (IPA)*, a novel method that is motivated by two key observations: (i) high-dimensional data can be modeled by a number of low-dimensional patches on its underlying manifold, and (ii) attaching the neighboring patches and aligning them, results in unfolding the underlying manifold of the data.

This paper is organized as follows. The rest of Section 1 reviews some of the most popular dimensionality reduction methods, and explains subspace clustering.

2010 *Mathematics Subject Classification.* Primary 68T99.

Section 2 introduces a novel idea for dimensionality reduction, and based on that, proposes an efficient algorithm. Finally, Section 3 illustrates some experimental results to demonstrate the quality of embedding by IPA on synthetic and real-world datasets, with a notable application in structural biochemistry at the end.

1.1. Dimensionality reduction methods. Dimensionality reduction algorithms attempt to reduce the dimensionality of the data, while they preserve some properties. One approach is a geometric approach in which it is assumed that the data is on or close to a low-dimensional subspace or manifold. In this approach, dimensionality reduction is the process of projecting the high-dimensional data on its underlying subspace, as Principal Component Analysis (PCA) [**7**] does. In the more challenging case that the data is close to an underlying manifold rather than an affine subspace, the process of dimensionality reduction proceeds through two steps: projecting the high-dimensional data on the low-dimensional underlying manifold, and then unfolding the manifold.

Among prominent algorithms, some attempt to preserve the local geometry of the manifold throughout the unfolding, for example Locally Linear Embedding (LLE) [**11**], and Local Tangent Space Alignment (LTSA) [**21**]. There are other algorithms that preserve the local structure of the manifold by preserving the local distances explicitly, such as Isomap [**15**], and Maximum Variance Unfolding (MVU) [**19**]. Kernel PCA [**12**] is a non-linear version of PCA, which is simply the original PCA applied to the projected data in a high-dimensional feature space. However, common closed-form kernels, such as Gaussians, usually perform poorly when kernel PCA is being used for unfolding. Most dimensionality reduction techniques have been discovered to be instances of kernel PCA [**5**]; In fact, the properties and behavior of each algorithm are encoded in its kernel. Considering that kernels are positive semi-definite matrices, one approach could be to use Semi-definite Programming (SDP) to learn a kernel with a desired functionality. For example, a kernel could be constructed to preserve local pairwise distances as in [**13, 19**], or just to preserve the neighborhood as in [**14**].

The quality of embedding of these methods is usually satisfactory, but they suffer from the computational complexity of SDP. Unfortunately, the existing methods can only be applied to problems with small number of data points. Although some large-scale variations of these methods have been proposed (e.g., Landmark MVU [**18**] and Fast MVU [**20**]), they use approximation schemes to reduce the complexity, which results in suboptimal solutions that require post-processing.

After learning a kernel matrix with SDP, kernel PCA is applied for obtaining a low-dimensional representation for the points. Crucially, if the learned kernel is not low-rank, the desired properties which have been considered in learning the kernel matrix, will be disturbed. This is another problem with using SDP for dimensionality reduction: there is no direct way to minimize the rank of the kernel. The rank function is not convex, and the closest convex objective that approximates the rank of a matrix is its trace [**3**]. However, in practice using the trace of the kernel matrix is not helpful enough and therefore, to circumvent the rank optimization problem in dimensionality reduction, the rank function is generally relaxed by some other convex objective functions, such as the total variance of the data points (as employed by the MVU-inspired methods). Unfortunately, rank relaxation usually does not result in a desired low-rank solution, and thus, some information will be lost in the final step, where kernel PCA is applied.

In this paper, we propose IPA, a scalable method for non-linear dimensionality reduction based on semi-definite programming, which does not have the aforementioned issues. IPA in its first step clusters the data points and then, the problem is solved based on these clusters, thereby reducing the complexity of the SDP. The proposed method creates an embedding for the input data points, in which while the local distances are preserved, the underlying manifold of the data is unfolded. In this way, applying kernel PCA for obtaining the final low-dimensional representation, does not disturb the local patterns of the data.

For partitioning the data any clustering technique can be used. One good choice, however, is subspace clustering. Subspace clustering algorithms model the data as a configuration of affine subspaces. This can be performed by partitioning the data into clusters, such that the points of each cluster lie on a low-dimensional subspace. In this approach the output is a set of low-rank clusters[1]; this is why this approach is also called *local* dimensionality reduction (for more detail see [**10**]).

The two most popular methods for local dimensionality reduction are Mixture of Probabilistic PCA (MPPCA) [**16**] and Mixture of Factor Analyzers (MFA) [**6**], which soft-partition the data points such that the rank of each partition is minimized. There are other methods, such as Vector Quantization PCA (VQPCA) [**8**] and K-Subspace Clustering [**17**], which find clusters with the minimum reconstruction error by firm-partitioning. In general, the reconstruction error of these firm-partitioning methods is less than the mixture methods (MPPCA and MFA); however, they tend to generate clusters that are not as localized. For IPA, we would suggest using a recent subspace clustering method, called *Low-Dimensional Localized Clustering (LDLC)* [**9**]. This technique forms low-rank clusters of the data points which are localized on the underlying manifold of the data.

2. The proposed method

Our proposed method, IPA, reduces the dimensionality of the data by unfolding its underlying manifold. During unfolding, in most cases, the structure of reasonably large clusters of the data can be preserved as whole, rather than by dividing them into small pairwise neighborhoods over the data points. Therefore, in the first step, we partition the input data into a number of clusters. For each cluster, we consider a flat *patch* on the underlying manifold of data. Intuitively speaking, IPA rearranges these patches to construct the unfolded underlying manifold.

First, we explain how the patches are built, and also, we define the meaning of rearranging for the patches. Then, we will discuss how IPA reconstructs a manifold by rearranging its patches. Finally, we prove that attaching the neighboring patches will align them, which results in unfolding the manifold and thereby reducing the dimensionality of the data over that.

2.1. Patch alignment method. Assume the input data is partitioned into c preferably low-rank clusters. If the manifold is smooth or there are enough clusters, a simple clustering algorithm such as k-means can be applied for this partitioning. In case of complex manifolds, subspace clustering algorithms should be used to obtain low-rank clusters.

---

[1]The rank of a cluster is the affine rank of the set of its points, which is defined as the maximum number of affinely independent points in that cluster.

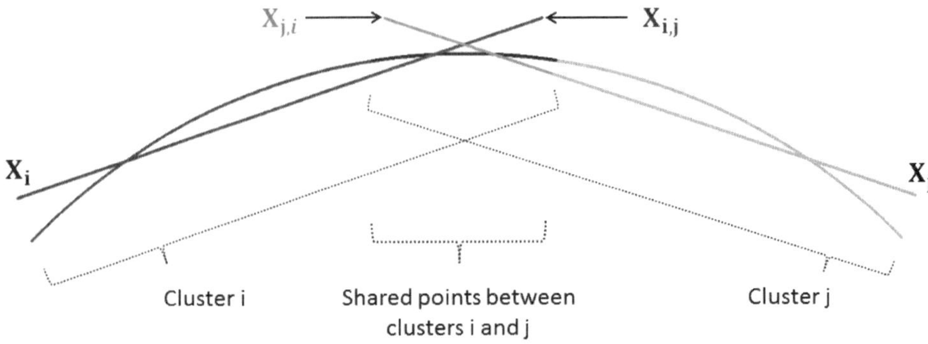

FIGURE 1. A two-dimensional manifold which consists of two clusters i and j. It is clear that the coordinates of the shared points (i.e., $\mathbf{X}_{i,j}$ and $\mathbf{X}_{j,i}$) are different in their one-dimensional patches.

First of all, we expand each cluster to have some points in common with its neighboring clusters. A simple approach for expanding a cluster is to add the k-nearest neighbors of each point to the cluster. After expanding, if two clusters have any point in common, we call them neighbors. Based on this notion, we create a connectivity graph $\mathcal{G}$ over c vertices with the edge set of $\mathcal{E}$. The number of neighboring clusters to the cluster i is denoted by d_i. In addition, for each $(i, j) \in \mathcal{E}$, the number of points shared between the clusters i and j is indicated by $n_{i,j} > 0$.

Since the clusters are low-rank, each of them can approximately be represented in an r-dimensional space, where r is the target[2] dimensionality. We refer to the r-dimensional representation of the points of a cluster, as the patch of that cluster. So in the second step, we embed the clusters into their r-dimensional patches[3] $\mathbf{X}_1, \mathbf{X}_2, \ldots, \mathbf{X}_c$ by a simple distance preserving dimensionality reduction method such as PCA.

Since the clusters are separately embedded into the patches, each shared point may have different coordinates in different patches (For example in Figure 1). Here $\mathbf{X}_{i,j}$ represents the shared points between the clusters i and j, which are embedded into the patch $\mathbf{X}_i$. Note that $\mathbf{X}_{j,i}$ can be different, but it is $r \times n_{i,j}$ too.

In order to construct an unfolded manifold, IPA rearranges the patches of the underlying manifold of the data, such that the neighboring patches can be attached together. First, we mathematically define the meaning of rearranging for the patches:

DEFINITION 2.1 (Rearranging the patches). Moving the patches into a common Euclidean space of dimensionality d, using isometric transformations. Each r-dimensional point $\mathbf{x} \in \mathbf{X}_i$ is transferred to the d-dimensional space, using a distance-preserving mapping which can be modeled by a $d \times r$ orthonormal[4] matrix $\mathbf{R}_i$ and a d-dimensional translation vector $\mathbf{t}_i$, such that $\mathbf{y} = \mathbf{R}_i\mathbf{x} + \mathbf{t}_i$ represents its new coordinate.

[2]IPA attempts to reduce the dimensionality of the input data to r. The target dimensionality can be given by a user, or set based on the rank of the clusters.

[3]Throughout this paper, we represent a set of points by a matrix in which each column represents one point.

[4]A matrix $\mathbf{\Psi}$ is orthonormal, if and only if $\mathbf{\Psi}^\top \mathbf{\Psi} = \mathbf{I}$

After rearranging, in order to attach the neighboring patches, their shared points should have been matched accordingly. Therefore, we define the following error term, by which we can minimize the distances between the shared points of two neighboring patches:

DEFINITION 2.2 (Matching error of two patches). For two neighboring patches $\mathbf{X}_i$ and $\mathbf{X}_j$, matching error is defined as follows:

$$
\begin{aligned}
(2.1)\quad e_{i,j}^2 &= \left\|\mathbf{R}_i\mathbf{X}_{i,j} + \mathbf{t}_i\mathbf{1}_{n_{i,j}}^\top - \mathbf{R}_j\mathbf{X}_{j,i} - \mathbf{t}_j\mathbf{1}_{n_{i,j}}^\top\right\|_F^2 \\
&= \mathbf{Tr}\big\{\big(\mathbf{R}_i\mathbf{X}_{i,j} - \mathbf{R}_j\mathbf{X}_{j,i} + (\mathbf{t}_i - \mathbf{t}_j)\mathbf{1}_{n_{i,j}}^\top\big)^\top \\
&\qquad\qquad \times \big(\mathbf{R}_i\mathbf{X}_{i,j} - \mathbf{R}_j\mathbf{X}_{j,i} + (\mathbf{t}_i - \mathbf{t}_j)\mathbf{1}_{n_{i,j}}^\top\big)\big\} \\
&= \mathbf{Tr}\big\{\mathbf{X}_{i,j}^\top\mathbf{X}_{i,j} + \mathbf{X}_{j,i}^\top\mathbf{X}_{j,i} - 2\,\mathbf{X}_{i,j}^\top\mathbf{R}_i^\top\mathbf{R}_j\mathbf{X}_{j,i}\big\} \\
&\qquad + n_{i,j}\,(\mathbf{t}_i - \mathbf{t}_j)^\top(\mathbf{t}_i - \mathbf{t}_j) + 2\,n_{i,j}\,(\mathbf{t}_i - \mathbf{t}_j)^\top(\mathbf{R}_i\overline{\mathbf{x}}_{i,j} - \mathbf{R}_j\overline{\mathbf{x}}_{j,i})
\end{aligned}
$$

where $\|.\|_F$ is the Frobenius norm, and $\overline{\mathbf{x}}_{i,j} = (1/n_{i,j})\mathbf{X}_{i,j}\mathbf{1}_{n_{i,j}}$ is the mean of $\mathbf{X}_{i,j}$. The total normalized matching error is also defined as: $e^2 = \sum_{(i,j)\in\mathcal{E}}(1/n_{i,j})e_{i,j}^2$.

In order to construct the unfolded manifold, the total normalized matching error needs to be minimized in rearranging. To minimize this error, first we take the derivative with respect to $\mathbf{t}_i$ and set it to zero:

$$
\begin{aligned}
(2.2)\quad \frac{\partial e^2}{\partial \mathbf{t}_i} &= 4\sum_{j:(i,j)\in\mathcal{E}}\big(\mathbf{t}_i - \mathbf{t}_j + \mathbf{R}_i\overline{\mathbf{x}}_{i,j} - \mathbf{R}_j\overline{\mathbf{x}}_{j,i}\big) = \mathbf{0} \\
&\Longrightarrow d_i\mathbf{t}_i - \sum_{j:\,(i,j)\in\mathcal{E}}\mathbf{t}_j = \sum_{j:\,(i,j)\in\mathcal{E}}\mathbf{R}_j\overline{\mathbf{x}}_{j,i} - \mathbf{R}_i\sum_{j:(i,j)\in\mathcal{E}}\overline{\mathbf{x}}_{i,j}
\end{aligned}
$$

Suppose $\mathbf{T} = [\mathbf{t}_1, \mathbf{t}_2, \ldots, \mathbf{t}_c]$ as an unknown $d\times c$ translation matrix, and an unknown $d\times rc$ matrix $\mathbf{R} = [\mathbf{R}_1, \mathbf{R}_2, \ldots, \mathbf{R}_c]$, whose blocks are the orthonormal $\mathbf{R}_i$ matrices. Using these notations, we can simplify (2.2) for each i, and by putting them together, we will have the following equation:

$$
(2.3)\qquad \mathbf{T}\mathbf{l}_i = \mathbf{R}\mathbf{z}_i \Longrightarrow \mathbf{T}\mathbf{L} = \mathbf{R}\mathbf{Z} \Longrightarrow \mathbf{T} = -\mathbf{R}\mathbf{Z}\mathbf{L}^\dagger
$$

where $\mathbf{l}_i$ is a c-dimensional vector, whose ith element is d_i, and jth element is -1 if $(i,j)\in\mathcal{E}$, and zero otherwise. It is easy to show that $\mathbf{L} = [\mathbf{l}_1, \mathbf{l}_2, \ldots, \mathbf{l}_c]$ is the $c\times c$ Laplacian matrix of the connectivity graph $\mathcal{G}$. With a similar structure, $\mathbf{z}_i$ is an (rc)-dimensional vector, consisting of c blocks of size r. The ith block is equal to $-\sum_{j:(i,j)\in\mathcal{E}}\overline{\mathbf{x}}_{i,j}$, and the jth block is equal to $\overline{\mathbf{x}}_{j,i}$ if $(i,j)\in\mathcal{E}$ and zero otherwise. Based on this, we have the $rc\times c$ matrix $\mathbf{Z} = [\mathbf{z}_1, \mathbf{z}_2, \ldots, \mathbf{z}_ct]$. Note that, this matrix can be calculated from $\overline{\mathbf{x}}_{i,j}$ of the neighboring patches, and so it is known.

Clearly, the last step in (2.3) explains that $\mathbf{T}$ is linearly related to $\mathbf{R}$. By substituting $\mathbf{t}_i$ and $\mathbf{t}_j$, (2.1) can be rewritten based on the unknown matrix $\mathbf{R}$, and so we will have:

$$
(2.4)\qquad \frac{e_{i,j}^2}{n_{i,j}} = \|\mathbf{R}\mathbf{B}_{i,j}\|_F^2 = \mathbf{Tr}\{\mathbf{B}_{i,j}^\top\mathbf{R}^\top\mathbf{R}\mathbf{B}_{i,j}\} = \mathbf{Tr}\{\mathbf{R}^\top\mathbf{R}\mathbf{K}_{i,j}\}
$$

where $\mathbf{B}_{i,j}$ is a known $rc\times n_{i,j}$ matrix, which is calculated based on $\mathbf{X}_{i,j}, \mathbf{X}_{j,i}, \mathbf{Z}$, and $\mathbf{L}^\dagger$. Based on this, the $rc\times rc$ positive semi-definite matrix $\mathbf{K}_{i,j} = \mathbf{B}_{i,j}\mathbf{B}_{i,j}^\top$ can be calculated. Finally, the total normalized matching error will have the following quadratic form:

$$e^2 = \sum_{(i,j)\in\mathcal{E}} \mathbf{Tr}\{\mathbf{R}^\top \mathbf{R}\mathbf{K}_{i,j}\} = \mathbf{Tr}\{\mathbf{R}^\top \mathbf{R}\mathbf{K}\} = \mathbf{Tr}\{\mathbf{A}\mathbf{K}\} \tag{2.5}$$

where $\mathbf{K} = \sum \mathbf{K}_{i,j}$ is a known $rc \times rc$ positive semi-definite matrix. Now instead of $\mathbf{R}$, we have an unknown $rc \times rc$ positive semi-definite matrix $\mathbf{A} = \mathbf{R}^\top \mathbf{R}$. Suppose $\mathbf{A}$ is partitioned into c by c blocks of $r \times r$ matrices, such that the block (i, j) is $\mathbf{R}_i^\top \mathbf{R}_j$. To impose the orthonomality constraint on $\mathbf{R}_i$, it is sufficient to have $\mathbf{I}_r$ in the block (i, i). That is, $\mathbf{A}$ has c blocks of $r \times r$ identity matrices on its main diagonal to guarantee that each $\mathbf{R}_i$ is an orthonormal matrix. Thus, to calculate the set of isometric mappings which minimize the total normalized matching error, one can solve the following standard semi-definite programming problem:

$$\min_{\mathbf{A}\succeq 0} \mathbf{Tr}\{\mathbf{A}\mathbf{K}\} \quad s.t. \mathbf{A} = \begin{pmatrix} \mathbf{I}_r & & ? \\ & \mathbf{I}_r & \\ & & \ddots & \\ ? & & & \mathbf{I}_r \end{pmatrix} \tag{2.6}$$

After solving this SDP optimization, $\mathbf{R}$ can be calculated by applying eigen-decomposition to $\mathbf{A}$. Since $\mathbf{A}$ is $rc \times rc$, the computed $\mathbf{R}$ will have rc rows and columns. This means that in order to achieve the minimum error in rearranging the patches, they are mapped to an rc-dimensional space, and therefore setting $d = rc$ provides enough degrees of freedom to reconstruct the structure of the manifold[5]. The coordinate of point $\mathbf{x}$ after rearranging can be obtained from the following linear relation to $\mathbf{R}$:

$$\mathbf{y} = \frac{1}{m_{\mathbf{x}}} \sum_{\forall i:\mathbf{x}\in\mathbf{X}_i} \mathbf{R}_i \mathbf{x}^i + \mathbf{t}_i = \mathbf{R}\mathbf{f}_i \implies \mathbf{Y} = \mathbf{R}\mathbf{F} \tag{2.7}$$

where $\mathbf{f}_i$ is a known dc-dimensional vector, and so $\mathbf{F} = [\mathbf{f}_1, \mathbf{f}_2, \ldots, \mathbf{f}_n]$ can be calculated for all n data points. Here, $m_{\mathbf{x}}$ is the number of patches that include $\mathbf{x}$, and $\mathbf{x}^i$ is the coordinate of $\mathbf{x}$ in patch $\mathbf{X}_i$ after embedding in r dimensions. Since $\mathbf{Y}$, the unfolded version of $\mathbf{X}$, is rc-dimensional, PCA should be applied at the end to remove its non-informant dimensions, and embed it into a low-dimensional space.

2.2. Alignment of the attached patches. In the previous section, we proposed a method for rearranging the patches of data such that the total matching error is minimized. We showed that the dimensionality of the data after rearranging is rc. Now, we explain why the rank of the rearranged data is much less than its dimensionality, very close to the target dimensionality r.

DEFINITION 2.3 (Attaching two patches). Rearranging two neighboring patches such that their matching error is minimized.

[5]Since each patch uses r orthonormal vectors as its basis, the maximum possible rank of the configuration of all c patches will be rc, considering that $\mathbf{T}$ is linearly related to $\mathbf{R}$ and therefore, the translations do not increase the final rank. It is important to note that the patches can not be rearranges in a space of dimensionality r. This is because in many cases, an r-dimensional space does not provide enough degree of freedom for rearranging the patches in order to preserve the topology of the underlying manifold. For example, if a circle is partitioned into one-dimensional patches, rearranging them in a one-dimensional space destroys the topological structure of the circle, and therefore, at least two dimensions are needed for rearranging.

One might imagine that the union of two attached patches has a higher rank than that of each of them separately. For example, if two one-dimensional patches (i.e., lines) are attached based on their single shared point, they can form a V-shape configuration which is obviously of rank two. In the same way, two attached planar patches can form a configuration of rank four, three or two. By the following theorem, we will prove that under certain conditions, matching aligns the patches, and does not increase the overall rank.

THEOREM 2.4. *Suppose two r-dimensional patches are attached in a space of dimensionality $d \geq r$, and form a configuration there. If the cross-covariance matrix of their shared points is full-rank, the rank of the union of their point in this configuration will not exceed r.*

PROOF. To minimize the matching error of two patches $\mathbf{X}_i$ and $\mathbf{X}_j$, first we can calculate $(\mathbf{t}_i - \mathbf{t}_j)$ from (2.1), by taking the derivative and setting it to zero:

$$\frac{\partial e_{i,j}^2}{\partial(\mathbf{t}_i - \mathbf{t}_j)} = 0 \implies \mathbf{t}_i - \mathbf{t}_j = \mathbf{R}_j\overline{\mathbf{x}}_{j,i} - \mathbf{R}_i\overline{\mathbf{x}}_{i,j} \tag{2.8}$$

By substituting (2.8) in (2.1) we will have:

$$\begin{aligned} e_{i,j}^2 &= \left\|\mathbf{R}_i\mathbf{X}_{i,j} - \mathbf{R}_j\mathbf{X}_{j,i} + (\mathbf{t}_i - \mathbf{t}_j)\mathbf{1}_{n_{i,j}}^\top\right\|_F^2 = \left\|\mathbf{R}_i\overline{\mathbf{X}}_{i,j} - \mathbf{R}_j\overline{\mathbf{X}}_{j,i}\right\|_F^2 \\ &= \mathbf{Tr}\{\overline{\mathbf{X}}_{i,j}^\top\overline{\mathbf{X}}_{i,j} + \overline{\mathbf{X}}_{j,i}^\top\overline{\mathbf{X}}_{j,i}\} - 2\,\mathbf{Tr}\{\overline{\mathbf{X}}_{i,j}^\top\mathbf{R}_i^\top\mathbf{R}_j\overline{\mathbf{X}}_{j,i}\} \end{aligned} \tag{2.9}$$

where $\overline{\mathbf{X}}_{i,j}$ is the centered version of the shared points of the two patches in $\mathbf{X}_i$. Since the first term is constant, to minimize the error it is only necessary to maximize the trace in the second term. This trace term can be rewritten as:

$$\begin{aligned} \mathbf{Tr}\{\overline{\mathbf{X}}_{i,j}^\top\mathbf{R}_i^\top\mathbf{R}_j\overline{\mathbf{X}}_{j,i}\} &= \mathbf{Tr}\{\mathbf{R}_i^\top\mathbf{R}_j\mathbf{M}\} = \mathbf{Tr}\{\mathbf{R}_i^\top\mathbf{R}_j\mathbf{U}\mathbf{S}\mathbf{V}^\top\} \\ &= \mathbf{Tr}\{(\mathbf{R}_i\mathbf{V})^\top(\mathbf{R}_j\mathbf{U})\mathbf{S}\} = \mathbf{Tr}\{\mathbf{Q}_i^\top\mathbf{Q}_j\mathbf{S}\} \\ &= \sum \mathbf{q}_{i_\alpha}^\top\mathbf{q}_{j_\alpha}\, s_{\alpha,\alpha} \end{aligned} \tag{2.10}$$

where $\mathbf{M} = \overline{\mathbf{X}}_{j,i}\overline{\mathbf{X}}_{i,j}^\top$ is the cross-covariance matrix of the shared points. Singular Value Decomposition (SVD) has been applied to $\mathbf{M}$ for obtaining the $r \times r$ orthonormal matrices $\mathbf{U}$ and $\mathbf{V}$, and also the diagonal matrix $\mathbf{S}$. It is clear that $\mathbf{Q}_i = \mathbf{R}_i\mathbf{V}$ and $\mathbf{Q}_j = \mathbf{R}_j\mathbf{U}$ are orthonormal $d \times r$ matrices, and we represent their columns by $\mathbf{q}_{i_\alpha}$ and $\mathbf{q}_{j_\alpha}$ respectively.

Given that $\mathbf{M}$ is full-rank[6], all of the diagonal elements of $\mathbf{S}$ are greater than zero ($s_{\alpha,\alpha} > 0$). Since $\mathbf{q}_{i_\alpha}$ and $\mathbf{q}_{j_\alpha}$ are normal vectors, we have $\mathbf{q}_{i_\alpha}^\top\mathbf{q}_{j_\alpha}\, s_{\alpha,\alpha} \leq s_{\alpha,\alpha}$. The equality case only happens if $\mathbf{q}_{i_\alpha} = \mathbf{q}_{j_\alpha}$, which means to maximize the summation in (2.10), $\mathbf{Q}_i$ and $\mathbf{Q}_j$ should be equal and so, we can show both of them by a single matrix $\mathbf{Q}$ consisting of r orthonormal columns.

Suppose $\mathbf{X}_i$ and $\mathbf{X}_j$ are rearranged to be attached together. We put all of their points in a single d-dimensional matrix: $[\mathbf{R}_i\mathbf{X}_i + \mathbf{t}_i\mathbf{1}^\top \quad \mathbf{R}_j\mathbf{X}_j + \mathbf{t}_j\mathbf{1}^\top]$. Although the dimensionality of this set is greater than r, we can show that its points are placed on an r-dimensional subspace, and so their affine rank is at most r. First, we translate all of the points by subtracting $\mathbf{t}_i$: $[\mathbf{R}_i\mathbf{X}_i \quad \mathbf{R}_j\mathbf{X}_j - (\mathbf{t}_i - \mathbf{t}_j)\mathbf{1}^\top]$. Since

[6]The clusters should be expanded such that each pair of neighboring patches has more than r points in common to form a shared point matrix of rank r. In practice, it is always possible to choose more than r linearly independent points for each pair of neighboring clusters such that the cross-covariance matrix $\mathbf{M}$ becomes full-rank.

$\mathbf{U}$ and $\mathbf{V}$ are orthonormal square matrices, we have $\mathbf{U}\mathbf{U}^\top = \mathbf{V}\mathbf{V}^\top = \mathbf{I}_r$, so we can rewrite the term $\mathbf{t}_1 - \mathbf{t}_2$ from (2.8):

$$\mathbf{t}_i - \mathbf{t}_j = \mathbf{R}_j\mathbf{U}\mathbf{U}^\top\overline{\mathbf{x}}_{j,i} - \mathbf{R}_i\mathbf{V}\mathbf{V}^\top\overline{\mathbf{x}}_{i,j} = \mathbf{Q}(\mathbf{U}^\top\overline{\mathbf{x}}_{j,i} - \mathbf{V}^\top\overline{\mathbf{x}}_{i,j}) = \mathbf{Q}\boldsymbol{\delta} \tag{2.11}$$

Using the same approach, now we can rewrite the set of all points as:

$$[\mathbf{Q}\mathbf{V}^\top\mathbf{X}_i \;\; \mathbf{Q}\mathbf{U}^\top\mathbf{X}_j - \mathbf{Q}\boldsymbol{\delta}\mathbf{1}^\top] = \mathbf{Q}\boldsymbol{\Delta} \tag{2.12}$$

in which it is clear that the rearranged points are on an r-dimensional subspace, which passes through $\mathbf{t}_i$ (and also $\mathbf{t}_j$) and is spanned by the r columns of the orthonormal matrix $\mathbf{Q}$. In other words, the attached patches are now aligned in an r-dimensional subspace. □

Intuitively speaking, when the patches of manifold are rearranged such that the neighboring ones are attached together, it is expected to see them aligned in a linear subspace of a rank close to that of the patches. Therefore, the final configuration can be viewed as the underlying manifold of the input data, unfolded in a low-dimensional subspace. This is the main reason that the proposed method reduces the dimensionality of the input dataset as low as the intrinsic dimensionality[7] of its underlying manifold.

2.3. The algorithm. A simple pseudo-code for the proposed method is presented in Algorithm 1. As previously discussed, first the input data should be partitioned into c clusters, where c is given as input. This can be done by clustering method. The target dimensionality r is an input to the algorithm, however, after clustering it can be modified based on the maximum rank of the clusters.

In the second step, all of the clusters are expanded to guarantee that each pair of neighboring clusters have enough shared points. The expanded patches are then embedded to an r-dimensional space. This can be done by PCA. It is also possible to use any other conventional dimensionality reduction method, such as Isomap or MVU to improve the embedding quality, however with higher computational cost.

In the next step, the neighboring patches are detected, and the connectivity graph $\mathcal{G}$ and its Laplacian matrix $\mathbf{L}$ will be formed. After that $\mathbf{Z}, \mathbf{F}, \mathbf{B}_{i,j}, \mathbf{K}_{i,j}$, and $\mathbf{K}$ are computed to setup the SDP optimization of (2.6). SDP can be solved by any standard solver, and $\mathbf{A}$—subsequently $\mathbf{R}$ by decomposing it—will be obtained. At the end, the unfolded manifold in an rc-dimensional space will be calculated, and

$\{C_1, .., C_c\} \leftarrow \mathit{Partition}(\mathbf{X}, c, r)$
$\forall i = 1..c: \quad \mathbf{X}_i = \mathit{Embed}(\mathit{Expand}(C_i), r)$
Form $\mathcal{E}$ and calculate $\mathbf{L}$ based on the neighborhoods
$\forall (i,j) \in \mathcal{E}: \quad \mathit{Find}(\mathbf{X}_{i,j})$
Compute $\mathbf{Z}, \mathbf{F}, \mathbf{B}_{i,j}, \mathbf{K}_{i,j}$ and $\mathbf{K}$
Solve SDP to obtain $\mathbf{A}$
$\mathbf{R} \leftarrow \mathit{Decompose}(\mathbf{A})$
$\mathbf{Y} \leftarrow \mathbf{RF}$
Return PCA($\mathbf{Y}$)

Algorithm 1: Isometric Patch Alignment (IPA)

[7]The maximum rank of the patches can be used as a rough estimation for the intrinsic dimensionality of the manifold.

by applying PCA the unnecessary dimensions will be removed to embed it in a low-dimensional space.

3. Experiments

The first experiment was conducted on a Swiss roll manifold. Although this manifold is a toy example, due to its highly non-linear global structure, many techniques are not able to unfold it. To make this manifold more challenging, by adding a rectangular hole, we turned it to a non-convex manifold. Many dimensionality reduction methods can not handle non-convex manifolds, and therefore, fail to preserve the shape of this rectangle. For example, MVU, Isomap, and tSNE are not able to provide a faithful low-dimensional representation, and distort the rectangular shape of the hole to a circular/ellipitcal shape (Figure 2-middle). Some other methods such as LTSA and LLE result in a rectangular shape, but they change the scales, thereby destroying the shape of manifold (Figure 2-bottom-left). In these comparisons, to have a better visualization of the pattern of data, 700 points have been sampled from a regular grid over the manifold.

Despite the observable qualitative difference between the embeddings in Figure 2, we conducted a quantitative study on this experiment. We used a well-known criterion, the *k-nearest neighbor intersection*, to study the quality of embedding for each method. In this criterion, for each data point, two sets of k-nearest neighbors are formed; one set contains the neighbors in the input space and the other set in the embedding space. Then, these sets are compared together [**2**]; the more points they have in common, the better the quality of the embedding is. Mathematically speaking, it can be computed by

$$\psi_{KNN} = 1 - \frac{1}{kn} \sum_{i=1}^{n} |\mathcal{N}_{\mathbf{x}_i} \cap \mathcal{N}_{\mathbf{y}_i}|, \tag{3.1}$$

where $\mathcal{N}_{\mathbf{x}_i}$ is the set of indices of the k-nearest neighbors of $\mathbf{x}_i$ in the input space, and in a similar way, $\mathcal{N}_{\mathbf{y}_i}$ is the k-nearest neighbor set of $\mathbf{y}_i$, where $\mathbf{y}_i$ is the low-dimensional representation of $\mathbf{x}_i$ in the unfolding space. To form an error measure,

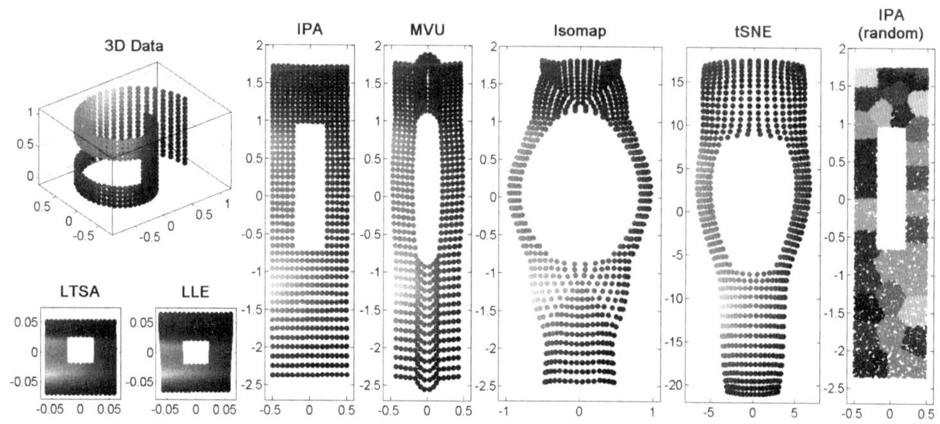

FIGURE 2. Different embeddings of a 3D Swiss roll with a rectangular hole on it.

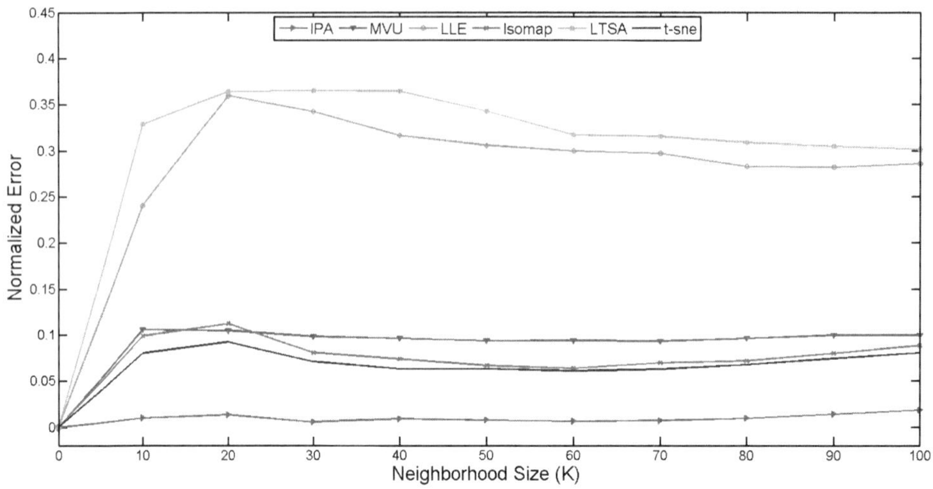

FIGURE 3. Evaluating the quality of embedding for different methods applied to the modified Swiss-roll. The KNN Intersection error is measured for different values of neighborhood size: $1 \le k \le 100$.

ψ_{KNN} is normalized between zero and one, where zero indicates that all of the k-nearest neighbor sets are the same, and so the neighborhood is completely preserved during the embedding.

We assessed the quality of embedding for all the compared methods in Figure 2 by the aforementioned measure. The result is depicted in Figure 3. As shown, the errors of IPA is significantly less than the errors of the other methods. In other words, the neighborhood of each point is better preserved by the proposed methods in embedding.

Based on this experiment, we can sort the quality of embedding in the following order: IPA (best), tSNE, Isomap, MVU, LLE, and LTSA (worst). Since LLE and LTSA change the scales, they have the worst results in this experiment. The errors of Isomap, MVU and tSNE are very similar; although they have better results compared to LLE and LTSA, the magnitude of error in these methods is apparently higher than that of IPA. It is important to note that for large neighborhood sizes preserving the neighborhood while the manifold is being unfolded, is not possible. This is the reason why for $k \ge 50$ the errors increase for IPA.

In the next experiment 20,000 points were sampled randomly from the same Swiss roll manifold. Handling this number of data points is usually difficult for most of non-linear methods. Even the simplest methods such as Isomap have to run a part of their optimization procedure on a $20{,}000 \times 20{,}000$ matrix, which may take a long time to converge on a conventional personal computer. We used k-means clustering to partition the points into 30 clusters, and embedded them in $r = 2$ dimensions, using PCA. The final embedding is shown in Figure 2-right, in which the hole still has a rectangular shape and the Swiss roll is perfectly unfolded. The overall running time of IPA (including the clustering) did not exceed one minute on one core of a 3.2 GHz CPU. In fact, none of our experiments with IPA took more than a minute.

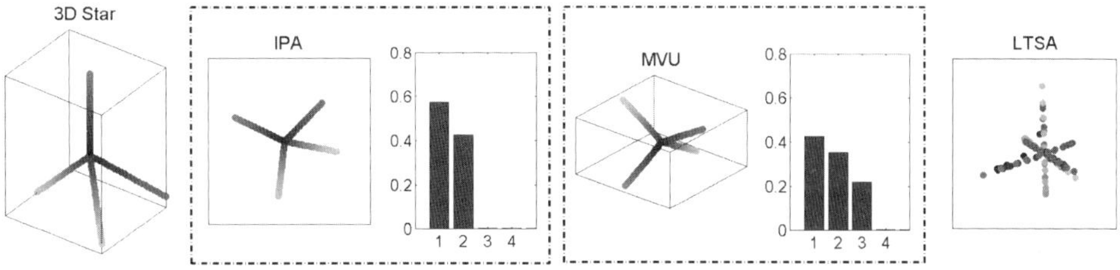

FIGURE 4. 3D Star and its embedding by IPA, MVU and LTSA, with their Eigenvalue spectra.

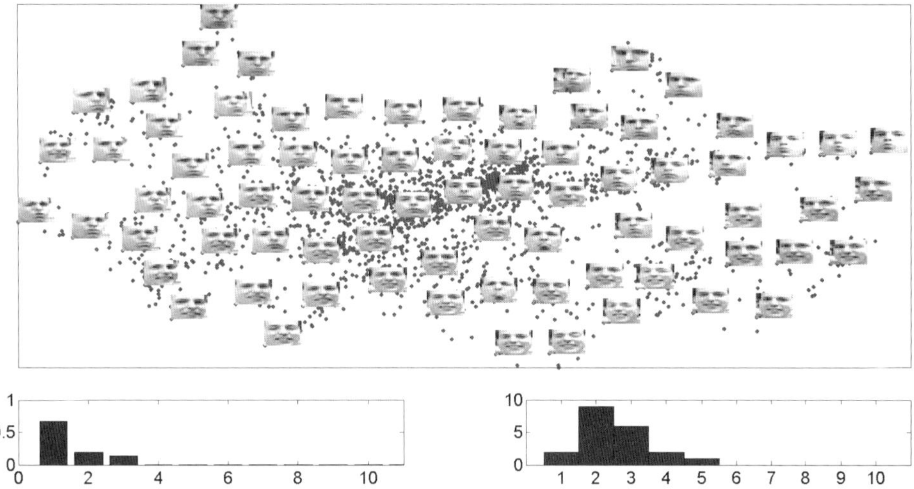

FIGURE 5. The embedding of the Frey images. The Eigenvalue spectrum of the embedding is shown at bottom-left, and the rank histogram of the patches is shown at bottom-right.

The second toy example is a star shape consisting of four line segments passing through a single point in a three-dimensional space. It has been shown that the methods which unfold by way of stretching (e.g., MVU), fail to embed this simple dataset into low dimensions [**13**]. Applying MVU to this set resulted in a three-dimensional star (Figure 4-middle), while IPA simply embedded it in the target dimensionality ($r = 2$). In addition, the result of embedding by LTSA is depicted in Figure 4-right, which shows that LTSA (and so LLE) was not able to handle this simple shape.

3.1. Real-world image sets. We studied the quality of embedding on two real-world datasets. Since real data usually has a complex structure, MPPCA is applied for clustering the data points into low-dimensional patches. The first dataset consists of 1965 face images in different moods. These images were partitioned into 20 patches of dimensionality $r = 2$. In the embedding shown in Figure 5, the orientation of the faces has been captured, while the images showing similar moods have been gathered together. The Eigenvalue spectrum of the embedding (Figure 5-bottom) indicates that the embedding is of rank three, and the two first dimensions capture about 90% of the variation.

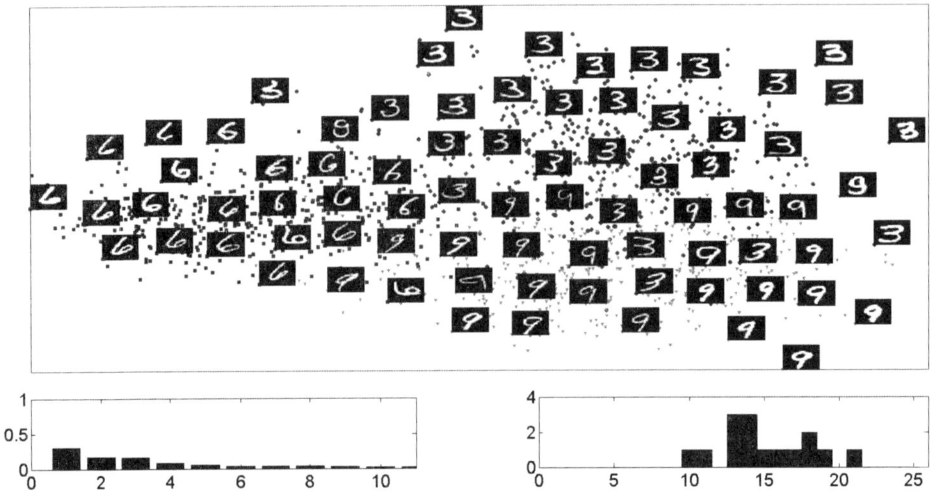

FIGURE 6. The embedding of three digits 3-6-9 of the MNIST dataset. The Eigenvalue spectrum of the embedding is shown at bottom-left, and the rank histogram of the patches is shown at bottom-right.

For the next experiment, a more challenging dataset was considered. This set is called MNIST and contains many handwritten samples of all digits. MNIST can be modeled by a union of manifolds rather than a single underlying manifold; it is possible to unfold each of them individually, but showing all of these disjoint embeddings together is meaningless. We selected a subset of the digits[8] (3, 6, and 9), and from each of them took 1000 samples. These points were partitioned into 15 patches, and based on their rank histogram (Figure 6-bottom) we set $r = 14$. The first two dimensions of the embedding were used for visualization in Figure 6 where the samples of each digit are shown with a single color. In this embedding, the variation of images have been considered, while the samples of different digits are clearly separated.

To study the quality of embedding by different methods, in this experiment, we computed the KNN Intersect measures for each method. It is important to note that the intrinsic dimensionality of these manifolds is more than 50, and consequently, the task of dimensionality reduction on this set is really challenging. The result is shown in Figure 7.

3.2. Protein structure determination. Computing the three-dimensional structure of biomacromolecules such as proteins is one of the most important and challenging problems in structural bioinformatics. In protein Nuclear Magnetic Resonance (NMR), the structure of a protein is calculated from a set of experimentally-determined short-range upper bounds between proximal hydrogen atoms and the domain knowledge about proteins, such as bond angles and bond lengths [**4**]. Since the size of protein molecules, i.e., number of atoms, is usually in the range of a few thousands, current SDP solvers cannot be directly applied.

[8]Since IPA preserves the local distances, we do not expect that all of the digit classes will be separated in the embedding; we chose 3 digits whose submanifolds are close to each other.

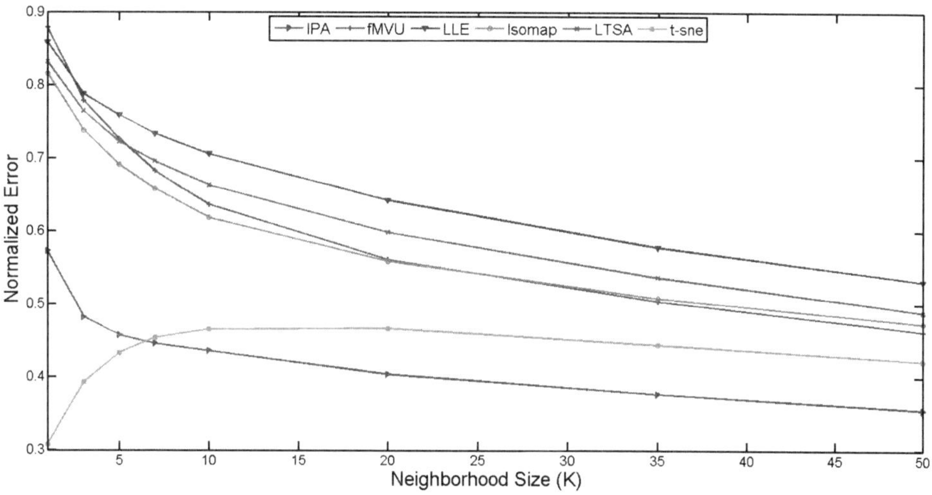

FIGURE 7. Measuring the quality of embedding for different methods applied to MNIST. This plot shows the k-nearest neighbor intersection error of different methods.

Here, we observe that IPA is naturally suitable for protein structure determination, because a protein molecule is virtually a set of interconnecting patches. For example, two consecutive peptide planes can be modeled as two patches that share the corresponding alpha carbon. Moreover, by using IPA we can significantly reduce the problem size. Imagine the planar indole functional group in the Tryptophan side chain that contains 16 atoms; this group can be represented by two columns of an $\mathbf{R}_i$. To solve the protein structure, we first find the patches in the protein molecule and then determine their common points.

We tested IPA on several protein datasets made from the protein structures deposited in the Protein Data Bank [**1**]. These datasets were made as follows: (i) downloaded structures were parsed and the ground-truth coordinates were recorded, (ii) all hydrogen atom pairs closer than 6 Å were enumerated and to make the test more realistic half of the pairs were randomly discarded, and (iv) upper bounds were formed by adding 10% multiplicative noise. Some of the solved structures together with the reference structures from PDB are shown in Figure 8. As can be seen in this figure, in each case, the two structures are very close, which shows that IPA has been able to mitigate the effect of the added noise to the upper bounds. It is important to note that many other methods, such as LTSA, LLE, and tSNE, are not suitable to be used in this application, and using MVU is usually very time consuming for the large molecules.

4. Conclusion

We proposed a novel dimensionality reduction method called IPA, which has a low computational cost. In IPA, first the data is clustered into low-dimensional overlapping patches, and then all of the patches are rearranged such that their border points are matched. We showed that the rearrangement can be computed by solving a small semi-definite program which is convex and so, has a unique solution. In addition, we proved that matching two patches will align them, and so

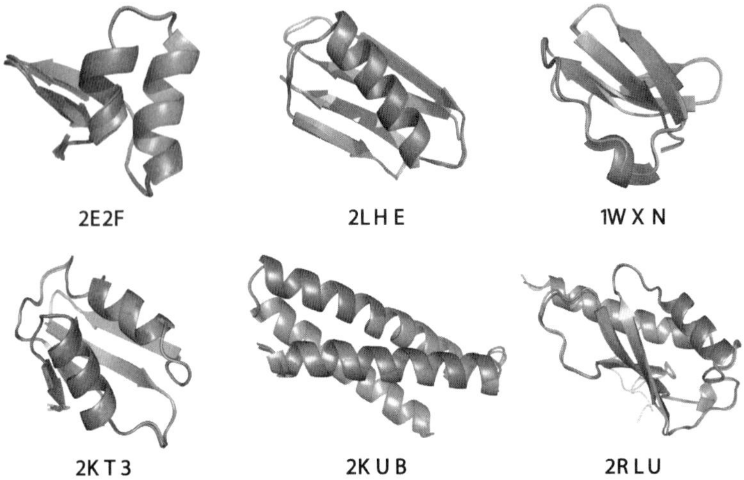

FIGURE 8. Superimposition of protein structures determined by IPA in orange and reference structures from PDB in green (better viewed in color).

rearranging the patches of a manifold will unfold it. Finally, by performing a wide range of experiments, we demonstrated the superiority of IPA to the conventional methods in different applications, including manifold learning, data visualization, and protein structure determination.

References

[1] F. C. Bernstein, T. F. Koetzle, G. J. B. Williams, jr. Meyer Edgar F., M. D. Brice, J. R. Rodgers, O. Kennard, T. Shimanouchi, and M. Tasumi, *The protein data bank: A computer-based archival file for macromolecular structures*, Journal of Molecular Biology **112** (1977), no. 3, 535–542, DOI 10.1016/0003-9861(78)90204-7.

[2] L. Chen and A. Buja, *Local multidimensional scaling for nonlinear dimension reduction, graph drawing, and proximity analysis*, J. Amer. Statist. Assoc. **104** (2009), no. 485, 209–219, DOI 10.1198/jasa.2009.0111. MR2504374 (2010d:62151)

[3] M. Fazel, H. Hindi, and S. P. Boyd, *A rank minimization heuristic with application to minimum order system approximation*, Proceedings of the 2001 American Control Conference (Arlington, VA, 2001), American Automatic Control Council, Evanston, IL, 2001, pp. 4734 –4739.

[4] P. Güntert, *Structure calculation of biological macromolecules from NMR data*, Quarterly Reviews of Biophysics **31** (1998), no. 2, 145–237.

[5] J. Ham, D. D. Lee, S. Mika, and B. Schölkopf, *A kernel view of the dimensionality reduction of manifolds*, Proceedings of the Twenty-First International Conference on Machine Learning, ACM, New York, NY, 2004, DOI 10.1145/1015330.1015417.

[6] G. E. Hinton, P. Dayan, and M. Revow, *Modeling the manifolds of images of handwritten digits*, IEEE Transactions on Neural Networks **8** (1997), 65–74, DOI 10.1109/72.554192.

[7] I. T. Jolliffe, *Principal component analysis*, Springer Series in Statistics, Springer-Verlag, New York, 1986. MR841268 (87m:62192)

[8] N. Kambhatla and T. K. Leen, *Dimension reduction by local principal component analysis*, Neural Computation **9** (1997), no. 7, 1493–1516, DOI 10.1162/neco.1997.9.7.1493.

[9] P. Khajehpour Tadavani and A. Ghodsi, *Low dimensional localized clustering* (*LDLC*), IEEE 12th International Conference on Data Mining (Brussels, 2012), IEEE Comput. Sci., Los Alamitos, CA, 2012, pp. 936–941, DOI 10.1109/ICDM.2012.134.

[10] L. Parsons, E. Haque, and H. Liu, *Subspace clustering for high dimensional data: a review*, ACM SIGKDD Explorations Newsletter **6** (2004), no. 1, 90–105, DOI 10.1145/1007730.1007731.
[11] S. T. Roweis and L. K. Saul, *Nonlinear dimensionality reduction by locally linear embedding*, Science **290** (2000), no. 5500, 2323–2326, DOI 10.1126/science.290.5500.2323.
[12] B. Schölkopf and A. J. Smola, *Learning with kernels*, Adaptive Computation and Machine Learning, MIT Press, Cambridge, MA, 2001.
[13] B. Shaw and T. Jebara, *Minimum volume embedding*, Proceedings of the Eleventh International Conference on Artificial Intelligence and Statistics (San Juan, PR, 2007), JMLR: Workshop and Conference Proceedings, vol. 2, 2007, pp. 460–467.
[14] B. Shaw and T. Jebara, *Structure preserving embedding*, Proceedings of the 26th International Conference on Machine Learning (Montréal, 2009), Omnipress, Madison, WI, 2009, pp. 937–944.
[15] J. B. Tenenbaum, V. de Silva, and J. C. Langford, *A global geometric framework for nonlinear dimensionality reduction*, Science **290** (2000), 2319–2323, DOI 10.1126/science.290.5500.2319.
[16] M. E. Tipping and C. M. Bishop, *Mixtures of probabilistic principal component analysers*, Neural Computation **11** (1999), no. 2, 443–482, DOI 10.1162/089976699300016728.
[17] D. Wang, C. Ding, and T. Li, *K-subspace clustering*, Machine Learning and Knowledge Discovery in Databases (Bled, 2009), Lecture Notes in Computer Science, vol. 5782, Springer, Berlin, 2009, pp. 506–521.
[18] K. Q. Weinberger, B. D. Packer, and L. K. Saul, *Nonlinear dimensionality reduction by semidefinite programming and kernel matrix factorization*, Proceeding of the Tenth International Workshop on Artificial Intelligence and Statistics (Barbados, 2005), 2005, pp. 381–388.
[19] K. Q. Weinberger and L. K. Saul, *Unsupervised learning of image manifolds by semidefinite programming*, Proceedings of the 2004 IEEE Computer Society Conference on Computer Vision and Pattern Recognition (Washington, DC, 2004), IEEE Comput. Soc., Los Alamitos, CA, 2004, pp. 988–995, DOI 10.1109/CVPR.2004.1315272.
[20] K. Q. Weinberger, F. Sha, Q. Zhu, and L. K. Saul, *Graph Laplacian regularization for large-scale semidefinite programming*, Advances in Neural Information Processing Systems, Vol. 19, MIT Press, Cambridge, MA, 2007, pp. 1489–1496.
[21] Z. Zhang and H. Zha, *Principal manifolds and nonlinear dimension reduction via local tangent space alignment*, SIAM Journal of Scientific Computing **26** (2002), no. 1, 313–338, DOI 10.1137/S1064827502419154.

School of Computer Science, University of Waterloo, 200 University Ave. W, Waterloo, ON N2L 3G1, Canada
E-mail address: pkhajehp@uwaterloo.ca

Department of Electrical and Computer Engineering, University of Toronto, 40 St. George St., Toronto, ON M5S 2E4, Canada
E-mail address: babak@psi.utoronto.ca

Department of Statistics and Actuarial Science, University of Waterloo, 200 University Ave. W, Waterloo, ON N2L 3G1, Canada
E-mail address: aghodsib@uwaterloo.ca

Published Titles in This Subseries

622 **S. Ejaz Ahmed, Editor,** Perspectives on Big Data Analysis, 2014

606 **Chantal David, Matilde Lalín, and Michelle Manes, Editors,** Women in Numbers 2, 2013

605 **Omid Amini, Matthew Baker, and Xander Faber, Editors,** Tropical and Non-Archimedean Geometry, 2013